Springer Series in

MATERIALS SCIENCE

The Springer Series in Materials Science covers the complete spectrum of materials physics, including fundamental principles, physical properties, materials theory and design. Recognizing the increasing importance of materials science in future device technologies, the book titles in this series reflect the state-of-the-art in understanding and controlling the structure and properties of all important classes of materials.

Please view available titles in *Springer Series in Materials Science* on series homepage http://www.springer.com/series/856

T. Kakeshita
T. Fukuda
A. Saxena
A. Planes
Editors

Disorder and Strain-Induced Complexity in Functional Materials

With 177 Figures

Springer

Editors

Professor Tomoyuki Kakeshita
Professor Takashi Fukuda

Osaka University, Graduate School of Engineering, Division of Materials and Manufacturing
Yamada-oka, Suita 2-1, 565-0871 Osaka, Japan
E-mail: kakeshita@mat.eng.osaka-u.ac.jp, fukuda@mat.eng.osaka-u.ac.jp

Dr. Avadh Saxena

Los Alamos National Laboratory
Theoretical Division, T-4, MS B262
Los Alamos, NM 87545, USA
E-mail: avadh@lanl.gov

Professor Antoni Planes

Universitat de Barcelona
Departament d'Estructura i Constituents
de la Matèria
Facultat de Física
Diagonal 647, 08028 Barcelona, Spain
E-mail: toni@ecm.ub.es

Springer Series in Materials Science ISSN 0933-033X
ISBN 978-3-642-20942-0 e-ISBN 978-3-642-20943-7
DOI 10.1007/978-3-642-20943-7
Springer Heidelberg Dordrecht London New York

Library of Congress Control Number: 2011936130

Printed on acid-free paper

Springer is part of Springer Science+Business Media (www.springer.com)

Preface

There is a paradigm shift in our understanding of the properties and behaviour of complex functional materials with multiple ordered phases and competing interactions. One novel aspect is that the underlying lattice provides an elastic template on which charge, spin, dipolar and other degrees of freedom couple to provide a number of emergent functionalities. The role of disorder in the presence of long-range dipolar and elastic forces is to lead to nanoscale inhomogeneity, which is responsible for the observed behaviour as well as frustration in the material – thus a strong sensitivity to external perturbations and possibly glassy response in certain regimes as well as anomalous avalanche phenomena.

This book brings together an emerging consensus on our understanding of the complex functional materials including ferroics, perovskites, multiferroics and magnetoelastics. The common theme is the existence of many competing ground states and frustration as a collusion of spin, charge, orbital and lattice degrees of freedom in the presence of disorder and (both dipolar and elastic) long-range forces. An important consequence of the complex unit cell and the competing interactions is that the emergent materials properties are very sensitive to external fields, thus rendering these materials with highly desirable, technologically important applications enabled by cross-response.

The idea for this book was born at the workshop *Jim Krumhansl Symposium: Complex Materials at the Cross-Roads* held at Osaka, Japan, during November 9–13, 2008. This workshop was a sequel to a previous workshop on *Interplay of Magnetism and Structure in Functional Materials* held at Benasque, Spain, during February 9–13, 2004. The Benasque workshop formed the basis of a book (*Magnetism and Structure in Functional Materials*, Springer, 2005), which was dedicated to Jim Krumhansl, a retired professor from Cornell University who passed away in May 2004. Much of the research reported in this as well as in the previous book was inspired by Prof. Krumhansl's overarching vision identifying common themes between solid-state physics, materials science and biology.

The topics covered in the present book are interdisciplinary in nature written by researchers from physics, materials science and engineering backgrounds.

Therefore, the book is addressed to both the experts and researchers getting into the field of functional materials with disorder and glassy behaviour including graduate students. It contributes to the fields of physics, materials science and nanotechnology. In general, the book represents a developing subject.

The carefully chosen 15 chapters written by internationally recognized experts in their respective fields cover general introduction to ferroics and multiferroics, principles of emergent complexity in materials science with a particular emphasis on magnetic shape memory alloys, glassy phenomena including strain glass and martensites, soft electronic matter, hysteresis and avalanches, high-resolution structural and magnetic visualization techniques, neutron scattering and shuffle-based transitions, defects in ferroelectrics and other ferroic materials, precursor phenomena, magnetostrucutral coupling and magnetocaloric properties, Heusler materials and magnetic martensites as well as first principles and mesoscopic modelling. Beyond illustrating some common threads (such as metastability, nonlinearity and disorder) between biological and materials functionality, the book concludes with a chapter that lays out clearly the future research directions.

Each chapter reviews the current state of the topic and provides sufficient background material for a graduate student or a new researcher to get started in this exciting field. At the same time, each chapter provides open questions for the experts to ponder and advance the field further.

Overall, the book provides an emergent paradigm shaped by the many advances made over the past decade in synthesis, characterization, modelling and fundamental understanding as well as technological applications of a variety of complex functional materials.

We gratefully acknowledge financial support from the Global COE Program "Center of Excellence (COE) for Advanced Structural and Functional Materials Design" at the University of Osaka (Suita campus), Japan. We thank Ms. Yuko Kuroda for her careful assistance in editing the book.

Osaka, Japan *Tomoyuki Kakeshita*
Osaka, Japan *Takashi Fukuda*
Los Alamos, USA *Avadh Saxena*
Barcelona, Spain *Antoni Planes*

Contents

1 Domain Boundary Engineering in Ferroic and Multiferroic Materials: A Simple Introduction 1
Ekhard K.H. Salje and Jason C. Lashley
1.1 Introduction 1
1.2 Multiferroic Domain Boundaries 2
1.3 Highly Conducting Interfaces 7
1.4 The Dynamics of Domain Movement and Ferroic Switching 9
1.5 Conclusions 15
References 16

2 Phase Diagrams of Conventional and Inverse Functional Magnetic Heusler Alloys: New Theoretical and Experimental Investigations 19
P. Entel, M.E. Gruner, A. Hucht, A. Dannenberg, M. Siewert, H.C. Herper, T. Kakeshita, T. Fukuda, V.V. Sokolovskiy, and V.D. Buchelnikov
2.1 Introduction and Computational 20
2.2 Crystal Structures of Half- and Full-Heusler alloys 21
2.3 Phase Diagrams of $Ni_2Mn_{1+X}Z_{1-X}$ (Z=Ga, In, Sn, Sb) Heusler alloys 25
2.4 Phase Diagrams of $Ni_{2+x}Mn_{1-x}Z$ (Z= Ga, In, Sn, Sb) Heusler alloys 30
2.5 Phase Diagrams of $Co_2Ni_{1-X}Z_{1+X}$ (Z= Ga, Zn) Heusler alloys 33
2.6 Conclusions and Future Aspects of Magnetic Heusler alloys 41
References 43

3 Ni–Mn–X Heusler Materials ... 49
Ryosuke Kainuma and Rie Y. Umetsu
3.1 Introduction ... 49
3.2 Atomic Ordering and Magnetic Properties in $Ni_2Mn(Ga_xAl_{1-x})$ Alloys ... 50
3.2.1 Atomic Ordering ... 51
3.2.2 Magnetic Properties ... 52
3.3 Magnetic Properties in Off-Stoichiometric $Ni_2Mn_{1+y}In_{1-y}$ Alloys ... 55
3.4 Martensitic Transformation and Magnetic Properties in NiMnIn Alloy ... 58
3.5 Concluding Remarks ... 62
References ... 63

4 Magnetic Interactions Governing the Inverse Magnetocaloric Effect in Martensitic Ni–Mn-Based Shape-memory Alloys ... 67
S. Aksoy, M. Acet, T. Krenke, E.F. Wassermann, M. Gruner, P. Entel, L. Mañosa, A. Planes, and P.P. Deen
4.1 Introduction ... 68
4.2 The Inverse Magnetocaloric Effect Around a Structural Transitions in a Ferromagnetic System ... 68
4.2.1 Conventional and Inverse Magnetocaloric Effects in $Ni_{50}Mn_{34}In_{16}$ ... 70
4.2.2 Magnetic Coupling in Ni–Mn-Based Martensitic Heusler Alloys ... 71
4.2.3 Magnetic Exchange Constants in Ni–Mn-Based Martensitic Heusler Alloys ... 74
4.3 Conclusion ... 75
References ... 76

5 Magnetic Field-Induced Strain in Ferromagnetic Shape Memory Alloys Fe-31.2Pd, Fe_3Pt, and Ni_2MnGa ... 79
Takashi Fukuda and Tomoyuki Kakeshita
5.1 Introduction ... 79
5.2 Martensitic Transformation in Fe-31.2Pd, Fe_3Pt, and Ni_2MnGa ... 81
5.3 Magnetic Field-Induced Strain in Fe-31.2Pd, Fe_3Pt, and Ni_2MnGa ... 82
5.4 Condition for Rearrangement of Martensite Variants by Magnetic Field ... 86
5.5 Origin of Martensitic Transformation in Fe_3Pt ... 90
5.6 Summary ... 93
References ... 93

6 **Soft Electronic Matter: Inhomogeneneous Phases in Strongly Correlated Condensed Matter** ... 95
Peter B. Littlewood
6.1 Introduction ... 95
6.2 A Microscopic View ... 96
6.3 Example 1: La_2NiO_4 ... 99
6.4 Example 2: Colossal Magnetoresistance in Manganites ... 100
6.4.1 The Basics: Double Exchange and Jahn–Teller ... 100
6.4.2 Competing and Cooperating Phases in Manganites ... 103
6.4.3 Ginzburg–Landau Theory for Manganites ... 105
6.5 Example 3: Superconductivity and Magnetism in $CeCoIn_5$ ... 108
6.6 Concluding Remarks ... 110
References ... 110

7 **Defects in Ferroelectrics** ... 113
Wenwu Cao
7.1 Introduction ... 113
7.2 Vacancies in Perovskite Ferroelectric Materials ... 115
7.3 Doping of Aliovalent Defects ... 117
7.4 Defects and Dielectric Properties ... 119
7.5 Grain Boundary and Positive Temperature Coefficient Resistor ... 122
7.6 Domain Walls as a Type of Mobile Defects ... 125
7.7 Size Effects and Surface in Ferroelectric Materials ... 129
7.8 Summary ... 131
References ... 132

8 **High-Resolution Visualization Techniques: Structural Aspects** ... 135
D. Schryvers and S. Van Aert
8.1 Earlier Results on Tweed Patterns in Ni–Al ... 136
8.2 Matrix Deformation and Depletion from Precipitation in Ni–Ti ... 137
8.3 Minimal Strain at Austenite – Martensite Interface ... 140
8.4 Internal Strain Control in Ni–Ti Micro-Wires ... 142
8.5 Strain Effects in Metallic Nano-beams ... 142
8.6 Future Prospects ... 144
References ... 148

9 **High-Resolution Visualizing Techniques: Magnetic Aspects** ... 151
Yasukazu Murakami
9.1 Introduction ... 151
9.2 Magnetic Imaging by TEM ... 152
9.2.1 Lorentz Microscopy ... 153
9.2.2 Electron Holography ... 154
9.2.3 Instrumentation for Magnetic Domain Observations ... 156
9.3 Study of Magnetic Microstructure in Colossal Magnetoresistive Manganite ... 157
9.3.1 Ferromagnetic Domain Nucleation and Growth ... 158

9.3.2 Determination of Magnetic Parameters of a Nanoscale Region 162
9.4 Magnetic Imaging of Ferromagnetic Shape-Memory Alloys 164
9.4.1 Impact of APBs on the Local Magnetization Distribution 165
9.4.2 Magnetic Pattern Formation Triggered by Premartensitic Lattice Anomaly 169
9.5 Concluding Remarks 172
References 173

10 Understanding Glassy Phenomena in Materials 177
David Sherrington
10.1 Introduction 177
10.2 Spin Glasses: A Brief Review 178
10.3 Martensites 181
10.4 Relaxors 188
10.5 Models, Simulations and Analysis 193
10.6 Conclusion 195
References 197

11 Strain Glass and Strain Glass Transition 201
Xiaobing Ren
11.1 Disorder–Order and Disorder–Glass Transition in Nature: Anticipation of a Strain Glass Transition and Strain Glass 201
11.2 Phase Diagram of Strain Glass: Crossover from LRO to Glass Due to Point Defects 204
11.3 Signatures of Strain Glass and Analogy with Other Glasses 207
11.4 Novel Properties of Strain Glass 211
11.5 Origin of Strain Glass and Theoretical Modeling/Simulations 214
11.6 Strain Glass as a Solution to Several Long-Standing Puzzles About Martensite 219
11.7 Summary 223
References 223

12 Precursor Nanoscale Textures in Ferroelastic Martensites 227
Pol Lloveras, Teresa Castán, Antoni Planes, and Avadh Saxena
12.1 Introduction 227
12.2 Structural Precursor Textures in Cubic Ferroelastics 230
12.2.1 Tweed Textures 230
12.2.2 Effect of Elastic Anisotropy on the Morphology of Structural Precursor Nanostructures 232
12.3 Phenomenological Modeling 235

12.4 Numerical Simulation Results 237
12.4.1 Effect of the Elastic Anisotropy on Structural Precursors: From Cross-Hatched to Mottled Morphology 238
12.4.2 Effect of the Disorder: Frozen Glass State 240
12.4.3 Thermomechanical Behaviour 242
12.5 Conclusions 243
References 244

13 Metastability, Hysteresis, Avalanches, and Acoustic Emission: Martensitic Transitions in Functional Materials 249
Martin-Luc Rosinberg and Eduard Vives
13.1 Introduction 249
13.2 What Can We Learn from Simple Models? 251
13.2.1 Relationship Between Hysteresis and the Distribution of Metastable States 252
13.2.2 Influence of the Driving Mechanism and the Effect of Long-Range Forces 256
13.3 What Can We Learn from Acoustic Emission Detection? 258
13.3.1 Pulse-Counting Technique 259
13.3.1.1 Transition Temperature 260
13.3.1.2 Athermal and Adiabatic Character of the Transition 260
13.3.1.3 Learning 261
13.3.1.4 Dependence on the Driving Mechanism 262
13.3.1.5 Correlation with Calorimetry 263
13.3.2 Statistical Analysis of Single Events 264
13.3.2.1 Exponent Universality Classes 265
13.3.2.2 Learning Process 266
13.3.2.3 Influence of the Driving Mechanism 268
13.3.3 Future Trends for the AE Technique in the Study of Structural Transitions 269
13.4 Concluding Remarks 269
References 270

14 Entropy–Driven Conformations Controlling DNA Functions 273
A.R. Bishop, K.Ø. Rasmussen, A. Usheva, and Boian S. Alexandrov
14.1 Introduction 274
14.2 Transcription Initiation, Transcriptional Start Sites, and DNA Breathing Dynamics 275
14.3 DNA Repair 283
14.4 Bioinformatics and DNA Breathing Dynamics 286
14.5 Conclusions 288
References 289

15 Conclusion and Outlook 293
Per-Anker Lindgård
15.1 Outlook 299
References 300

Index 303

Contributors

Mehmet Acet Physics Department, University of Duisburg-Essen, 47048 Duisburg, Germany, mehmet.acet@uni-due.de

S. Aksoy Faculty of Engineering & Natural Sciences, Sabanci University, 34956 Istanbul, Turkey, saksoy@sabanciuniv.edu

Boian S. Alexandrov Theoretical Division, Los Alamos National Laboratory, Los Alamos, New Mexico, USA, boian@lanl.gov

A. R. Bishop Theoretical Division, Los Alamos National Laboratory, Los Alamos, NM, USA, arb@lanl.gov

V.D. Buchelnikov Condensed Matter Physics Department, Chelyabinsk State University, 454021 Chelyabinsk, Russia, buche@csu.ru

Wenwu Cao Department of Mathematics, The Pennsylvania State University, University Park, PA 16802, USA, cao@math.psu.edu

Teresa Castán Facultat de Física, Departament d'Estructura i Constituents de la Matèria, Universitat de Barcelona, Diagonal 647, 08028 Barcelona, Catalonia, Spain

Institut de Nanociència i Nanotecnologia, Universitat de Barcelona, Catalonia, Spain, teresa@ecm.ub.es

A. Dannenberg Faculty of Physics & CeNIDE, University Duisburg-Essen, 47048 Duisburg, Germany, antje@thp.uni-duisburg.de

P.P. Deen European Spallation Source ESS AB P.O Box 176, SE-221 00 Lund, Sweden, pascale.deen@esss.se

Peter Entel Faculty of Physics & CeNIDE, University Duisburg-Essen, 47048 Duisburg, Germany, entel@thp.uni-duisburg.de

Takashi Fukuda Graduate School of Engineering, Osaka University, Suita, Osaka 565-0871, Japan, fukuda@mat.eng.osaka-u.ac.jp

M.E. Gruner Faculty of Physics & CeNIDE, University Duisburg-Essen, 47048 Duisburg, Germany, Markus.Gruner@uni-due.de

H.C. Herper Faculty of Physics & CeNIDE, University Duisburg-Essen, 47048 Duisburg, Germany, heike.herper@uni-due.de

A. Hucht Faculty of Physics & CeNIDE, University Duisburg-Essen, 47048 Duisburg, Germany, fred@thp.uni-duisburg.de

Ryosuke Kainuma Department of Materials Science, Tohoku University, Sendai, Japan, kainuma@material.tohoku.ac.jp

Tomoyuki Kakeshita Graduate School of Engineering, Osaka University, Suita, Osaka 565-0871, Japan, kakesita@mat.eng.osaka-u.ac.jp

T. Krenke Thyssen Krupp Electrical Steel GmbH, D-45881 Gelsenkirchen, Germany, thorsten.krenke@thyssenkrupp.com

Jason C. Lashley Los Alamos National Laboratories, Los Alamos, NM USA, j.lash@lanl.gov

Per-Anker Lindgård Materials Research Division, Risó, DTU, National Laboratory for Sustainable Energy, 4000-Roskilde, Denmark, hanne.frederiksen@mail.tele.dk

Peter B. Littlewood Cavendish Laboratory, Cambridge University, JJ Thomson Avenue, Cambridge CB3 0HE, UK, pbl21@cam.ac.uk

Pol Lloveras Facultat de Física, Departament d'Estructura i Constituents de la Matèria, Universitat de Barcelona, Diagonal 647, 08028 Barcelona Catalonia, Spain

Institut de Nanociència i Nanotecnologia, Universitat de Barcelona, Catalonia, Spain, pol@ecm.ub.es

L. Mañosa Departament d'Estructura i Constituents de la Matèria, Facultat de Física, Universitat de Barcelona Diagonal 647, 08028 Barcelona, Catalonia (Spain), lluis@ecm.ub.es

Yasukazu Murakami Institute of Multidisciplinary Research for Advanced Materials, Tohoku University, Sendai, Japan, murakami@tagen.tohoku.ac.jp

Antoni Planes Facultat de Física, Departament d'Estructura i Constituents de la Matèria, Universitat de Barcelona, Diagonal 647, 08028 Barcelona Catalonia, Spain

Institut de Nanociència i Nanotecnologia, Universitat de Barcelona, Catalonia, Spain, toni@ecm.ub.es,antoniplanes@ub.edu

Kim Ø. Rasmussen Theoretical Division, Los Alamos National Laboratory, Los Alamos, NM, USA, kor@lanl.gov

Xiaobing Ren Ferroic Physics Group, National Institute for Materials Science, Tsukuba, Japan, REN.Xiaobing@nims.go.jp

Martin-Luc Rosinberg Laboratoire de Physique Théorique de la Matière Condensée, Université Pierre et Marie Curie, 4 Place Jussieu, 75252 Paris, France, mlr@lptmc.jussieu.fr

Ekhard K.H. Salje University of Cambridge, Downing Street, Cambridge CB 2 3EQ, Cambridge UK, es10002@esc.cam.ac.uk

Avadh Saxena Theoretical Division, Los Alamos National Laboratory, Los Alamos, NM 87545, USA

Institut de Nanociència i Nanotecnologia, Universitat de Barcelona, Barcelona Spain, avadh@lanl.gov

Dominique Schryvers EMAT, University of Antwerp, Groenenborgerlaan 171, 2020 Antwerp, Belgium, nick.schryvers@ua.ac.be

David Sherrington Santa Fe Institute, 1399 Hyde Park Road, Santa Fe, NM 87501, USA

Rudolf Peierls Centre for Theoretical Physics, University of Oxford, 1 Keble Rd., Oxford OX1 3NP, UK, d.sherrington1@physics.ox.ac.uk

M. Siewert Faculty of Physics & CeNIDE, University Duisburg-Essen, 47048 Duisburg, Germany, mario@thp.uni-due.de

V.V. Sokolovskiy Condensed Matter Physics Department, Chelyabinsk State University, 454021 Chelyabinsk, Russia, vsokolovsky84@mail.ru

Rie Y. Umetsu Institute for Materials Research, Tohoku University, Sendai, Japan, rieume@imr.tohoku.ac.jp

Anny Usheva Beth Israel Deaconess Medical Center, Harvard Medical School, Boston, MA, USA, ausheva@bidmc.harvard.edu

Sandra Van Aert EMAT, University of Antwerp, Groenenborgerlaan 171, 2020 Antwerp, Belgium, sandra.vanaert@ua.ac.be

Eduard Vives Facultat de Física, Departament d'Estructura i Constituents de la Materia, Universitat de Barcelona, Martí i Franquès 1, 08028 Barcelona, Catalonia, Spain

Institut de Nanociència i Nanotecnologia (IN2UB), Universitat de Barcelona, Barcelona, Catalonia, Spain

Department of Physics, University of Warwick, Coventry CV4 7AL, UK, eduard@ecm.ub.es

E.F. Wassermann Physics Department, University of Duisburg-Essen, D-47048 Duisburg, Germany, eberhard.wassermann@uni-due.de

Chapter 1
Domain Boundary Engineering in Ferroic and Multiferroic Materials: A Simple Introduction

Ekhard K.H. Salje and Jason C. Lashley

Abstract Multiferroic behavior is commonly described as a bulk phenomenon where, at least, two of the three ferroic properties, ferromagnetism, ferroelectricity, and ferroelasticity, coincide. This notion is enlarged to contain as another "useful" property electrical conductivity. While bulk applications are potentially useful, we describe the recent development where the same properties are restricted to domain boundaries or interfaces, while the adjacent domains are not active elements themselves. This means that the information is restricted to thin, nearly two-dimensional slabs of some 2 nm thickness. The information density is, thus, extremely high, while conducting interfaces can serve as wires to connect the active elements. In this chapter, we discuss the underlying physical principles for the "engineering" of interfacial multiferroics.

1.1 Introduction

Multiferroicity combines at least two of the three ferroic properties of a material: ferroelasticity, ferroelectricity, or ferromagnetism. Its investigation has a long tradition with significant work on boracites in the 1960s [1] and a continuous stream of activities on ceramics with perovskite-like structures [2–8]. In addition, it was realized that "ferroelastics" and "martensites" describe the same materials properties that have simply different historic traditions for their naming (so that "ferroelastic" alloys are usually called "martensites" and have often, but not

E.K.H. Salje (✉)
University of Cambridge, Downing Street, Cambridge CB 2 3EQ, Cambridge, UK
e-mail: es10002@esc.cam.ac.uk

J.C. Lashley
Los Alamos National Laboratory, Los Alamos, NM, USA
e-mail: j.lash@lanl.gov

T. Kakeshita et al. (eds.), *Disorder and Strain-Induced Complexity in Functional Materials*, Springer Series in Materials Science 148,
DOI 10.1007/978-3-642-20943-7_1,

always, step-wise phase transitions [9], while ceramics and minerals, which have "martensitic" properties, are usually called "ferroelastics"). Magnetic martensites, for example, are multiferroics and follow the same physical mechanism as the very large number of ferromagnetic *cum* ferroelastic ceramics.

The motivation for the investigation of multiferroic materials is that better memory devices can be made from such compounds [10, 11]. For this purpose, ferroelasticity is of minor importance because reading/writing mechanisms will rely on magnetic or electric fields so that the key for the development of multiferroics is the combination of ferroelectric and ferromagnetic properties. The role of ferroelasticity is nevertheless often a key for the performance of such devices: coupling between different ferroic properties can be "strain-induced" where both properties couple strongly with some lattice distortion (via magnetostriction and electrostriction or piezoelectricity, etc.) and, thus, couple with each other. Strain-induced coupling occurs on an atomic scale [12, 13] or on a mesoscopic scale [14–16], whereby the latter allows further development of enhanced strains via microengineering resolution patterning and processes.

Current work on structural multiferroics was revived with work focussing mainly on $BiFeO_3$ [17–20]. Two further developments have occurred after 2000, which may lead to even more effective multiferroic device materials. Firstly, coupling with charge carriers is now subject to much research. Here, the combination between a magnetically or electrically written signal and its reading via high conductivity regions in a material has become an attractive proposition [21–25]. Such regions can be grain boundaries, interfaces, or twin boundaries or be part of exsolution patterns or amorphized/glassy clusters [26–28]. The second development leads this idea even further: why not to take such interfacial regions as active elements of the multiferroic properties themselves. This restricts the size of the active element to a few nm in thickness, while the crystal simply serves as matrix in which such heterostructures are located. It is the purpose of this introduction to highlight some of the developments that lead to the emerging field of "domain boundary engineering" which, potentially, brings the size of active elements, say in memory devices, from currently 0.1 μm to well below 50 nm and also allows a truly 3D arrangement of multiferroic elements.

1.2 Multiferroic Domain Boundaries

Domain boundaries, in particular twin boundaries, which are discussed now, show reduced chemical bonding with many of the structural constituents. With respect to elastic, magnetic, or electric susceptibilities, one expects domain boundaries to be "softer" than the bulk, although such softening can hardly be measured macroscopically because the volume proportion of interfaces is relatively small compared with the bulk. An exception is – in some measure – relaxor materials where the relaxor regions themselves have wall properties (for order–disorder systems) and show indeed strong finite size effects and soft susceptibilities [29–31].

Nevertheless, twin boundaries can reach several parts-per-million of the sample volume so that large signals from internal ferroic properties of the boundaries may even compete with the macroscopic properties of the bulk. The advantage of the localized wall properties is their high information density: they are contained within very thin sheets of twin walls and can be addressed spatially with very high resolution. This means that the storage density of information encrypted in twin walls is extremely high so that wall-related devices could theoretically outperform bulk devices by several orders of magnitude.

The second ingredient is the condition that structural gradients of twin walls extend over several interatomic distances. In Fig. 1.1, a high resolution electron microscopy image of a twin wall in $NdGaO_3$ is shown where the imaging condition was optimized for atoms inside the twin wall, while atoms outside the wall are slightly out of imaging condition by inclination of their lattice plane from the plane of diffraction. The "thickness" of the twin wall can now be estimated by simply counting the number of unit cells in the wall. The resulting wall thickness of ca. 2 nm compares well with results from diffraction experiments at low temperatures in interfaces and surfaces [32–34]. The wall thickness increases when the transition point is approached. Careful analysis of the diffuse diffraction of wall-related signals in $LaAlO_3$ showed that the wall thickness increases according to the predictions of Landau–Ginzburg theory for a second-order phase transition [8, 33]. In first-order martensitic transitions, the effect is smaller although the increase follows still the scaling of the correlation length of the phase transition, which leads to significant increases near the transformation point in compounds such as NiTi and NiTiFe [35, 36].

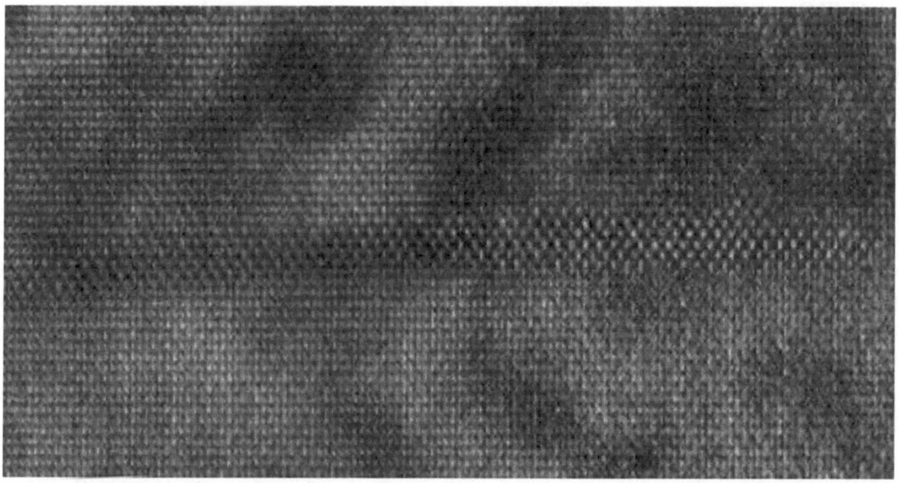

Fig. 1.1 Transmission electron microscopy image of NdGaO3 (Pbnm) near a {101} twin boundary in the middle of the figure. The unit cell is $a = 0.5426$ nm, $b = 0.5502$ nm, and $c = 0.7706$ nm for the orthorhombic cell and $a = b = 0.3864$ nm and $c = 0.3853$ nm for a perovskite-related cell. The thickness of the interface is 2w $\sim$ 6 unit layers or $\sim$2.3 nm (photograph courtesy G. Van Tendeloo, Antwerp)

The discovery of ferroic properties of interfaces is often related to computer simulations for materials design and the theoretical exploration of extreme physical properties in solids. Such research expands from inorganic materials to biological samples in life sciences. We first discuss an example in solid state physics where sizeable spontaneous polarization was predicted in {100} twin walls of $CaTiO_3$, a definitely nonpolar material [37, 38]. Theoretical simulations [39–41] of these walls show an extremely rich texture of the local polarization at and close to the walls. Local distortions include a strong antiferroelectric component, and local nonzero contributions perpendicular to the wall plane, which do not contribute to the macroscopic net dipole moment. Individual Ti displacements of 2 pm off the octahedron center give rise to a net polarization corresponding to a displacement of 0.6 pm in the direction of the bisector of the twin angle. The effect is intrinsically coupled with the appearance of twin boundaries in the matrix, which was already previously identified as locality of oxygen vacancies in $CaTiO_3$ [37, 41].

While indirect evidence for the polar behavior of twin walls has been reported before [42], as well as in antiphase boundaries, APBs [43], and grain boundaries [44], the results in $CaTiO_3$ are very instructive as it was the first clear indication of twin wall polarity and the underlying structural mechanism for the coupling between strain and dipole moments. $CaTiO_3$ is orthorhombic in its low-temperature form (space group Pnma) and is purely ferroelastic. No ferroelectric features have ever been recorded. The TiO_6 octahedron, on the other hand, is well known for its tendency to form polar groups where the Ti position is offcentered with respect to the geometrical center of the surrounding oxygen atoms. Such polar structures exist in compounds such as $BaTiO_3$, $PbTiO_3$, and others. The known competition with octahedral rotation [45] in the tetragonal and orthorhombic phases of $CaTiO_3$ suppresses the off-centering. It is, however, restored when the rotation angle vanishes or when the density of the material decreases. Both conditions are met inside the twin wall and it is thus not entirely unexpected that twin walls should show dipolar moments. What was unknown is the actual size of the polarization and the texture of the polarization field.

To investigate polar ordering in the ferroelastic walls of $CaTiO_3$, numerical simulations were performed based on an atomic-scale description of the walls in which atoms interact via empirically defined forces [37–42]. Periodic boundary conditions were used in three dimensions. Open boundary conditions in the direction perpendicular to the walls would imply surfaces, which would add unwanted complexity to the problem. Two twin walls are needed to conform to periodic boundary conditions. A supercell was built of 26 unit cells in the direction x perpendicular to the walls, six unit cells in the direction perpendicular to the plane of the twin angle (z), and ten unit cells in the bisector of the twin angle y (using the unit cell of the prototypic cubic structure). This gives a total of 7,800 atoms. Figure 1.2 shows the primary order parameter Q as a function of x, in the direction perpendicular to the wall. Q is a measure of the rotation around the y axis of the oxygen octahedra around each titanium atom, appropriately sign corrected. The dashed line indicates the fitted $Q \sim \tan h(x/w)$ functional form which is expected from Landau theory [8]. The wall width lies well within the experimental values

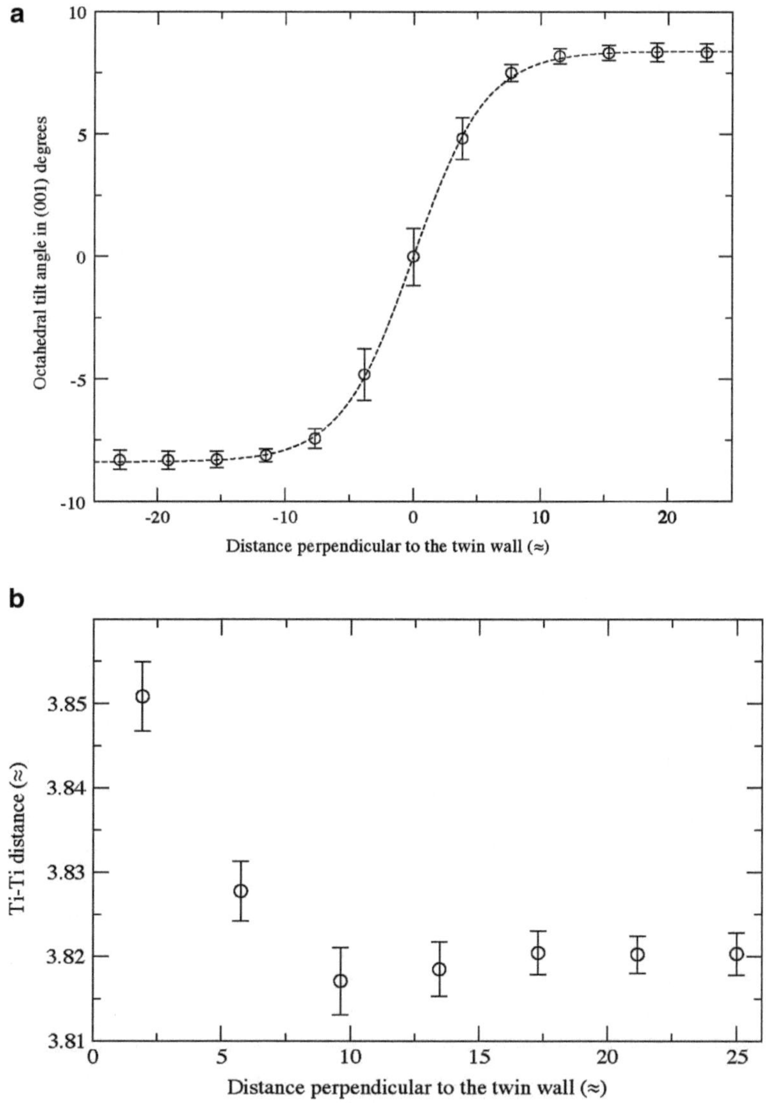

Fig. 1.2 (**a**), (**b**) Profile of the twin wall in $CaTiO_3$. The primary order parameter is the rotation of the TiO_6 octahedra. The twin boundary shows an inversion of the rotation angle (Fig. 1.1a) where the *dotted line* indicates the predictions from Landau–Ginzburg theory. The secondary order parameter is the widening of the unit cell, which is measured by the distance between two adjacent Ti positions. Figure 1.1b shows the increase of the Ti–Ti distance in the wall by ca. 1%, which is sufficient to induce off-centering of the Ti atom from the middle of the octahedra and also an increase of the mobility of defects

as determined previously [33]. The secondary order parameter of interest here is the off-centering of Ti from the center of charge of the corresponding oxygen octahedron. The largest displacements are of 2.0 pm, mostly along the z direction.

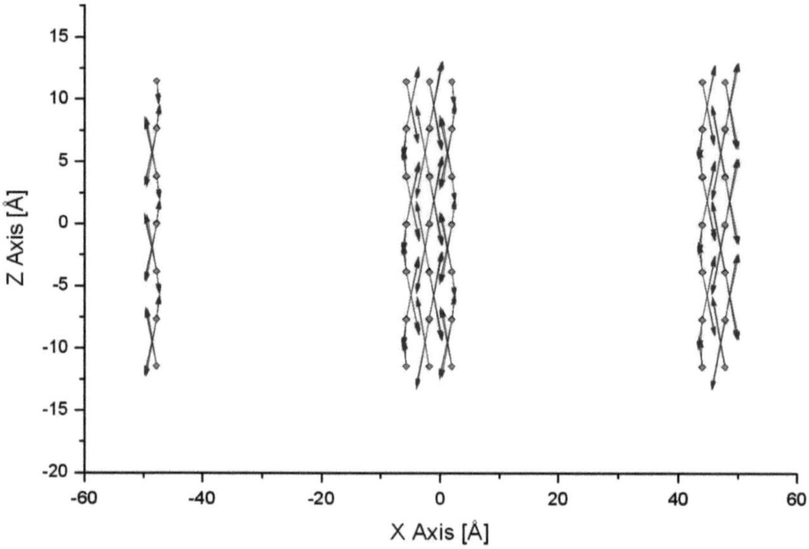

Fig. 1.3 Patterns of the off-centering of Ti from the midpoint of the oxygen octahedra in $CaTiO_3$. The graph shows the displacement patterns in the direction to the twin walls

Two types of off-centering are seen (Fig. 1.3) within the domain wall. The largest component is the one along the z direction antiferroelectric with alternating Ti positions shifted in opposite directions. In contrast, the off-centering along the y direction is ferroelectric and produces a net dipole moment for each domain wall equivalent to a net displacement of 0.6 pm per Ti atom (0.9 pm for the second force field). There is also a smaller antiferroelectric component along the direction perpendicular to the wall, x.

In addition to the appearance of polar properties of the walls, an increase of oxygen vacancies was also predicted. An oxygen vacancy gains ca. 1.1 eV when shifted from the bulk of the material into the twin wall [41]. While this effect is expected from the fact that twin walls in the geological context are known to be decorated by defects, we understand from these calculations that the geometrical requirement for the accommodation of defects may appear negligible, namely, ca.1% increase of a lattice spacing in $CaTiO_3$. Such small changes are typical for twin boundaries and other interfaces so that the observation that dopants are concentrated in interfaces is not unexpected. These localized dopants, on the other hand, can then be used systematically to modify the properties of the walls, e.g., their conductivity or polarity. Doping with magnetic ions may then lead to magnetic properties of the walls, while the same dopants would not necessarily enter the bulk.

The widening of the unit cell at the interface could also lead to a reduction of the local elastic response. This does not mean that the position of a twin wall can be shifted by external forces (which it can), but the compressibility of the wall itself is larger than the equivalent compressibility of the bulk. While such an effect has been seen [40], it appears that the effect is much smaller than could be expected by the simple reduction of the density in the wall. In fact, the reduction of the relevant elastic modulus as expected by the increase of the distance between nearest neighbous is partly compensated by the decrease of the distance of the next nearest neighbous [46] so that the relaxation of the structure compensates to a large extent the elastic softening due to the swelling of the interface.

1.3 Highly Conducting Interfaces

We have argued that twin walls can attract defects, which leads to the possibility to dope twin boundaries selectively, i.e. to introduce defects into the boundaries but not in the bulk. This possibility was first used to change the conductivity in WO_3 in 1998 [21] with the introduction of Na and oxygen vacancies in twin walls [47–49]. The chemical composition of the walls was very slightly modified (e.g. from WO_3 to $WO_{2.95}$), which induced a metal–insulator transition and, at low temperature, led to the appearance of superconductivity in twin walls. The fact that the dopants follow the trajectories of the twin walls means that nanopatterning of the superconducting structure is possible via the patterning of the twin boundaries and subsequent doping (Figs. 1.4 and 1.5).

Tungsten oxide, WO_3, and its substoichiometric derivatives, WO_{3-x}, are particularly well suited for this research because they display metal–insulator transitions, while they remain thermodynamically stable compounds. They display a multitude of structural phase transitions [50] mainly related to shape changes of the WO_6 octahedra and their rotations within an octahedral network. WO_3 easily releases oxygen and incorporates alkali ions and hydrogen. The facility with which oxygen is released under reducing conditions is less related to the chemical bonding of

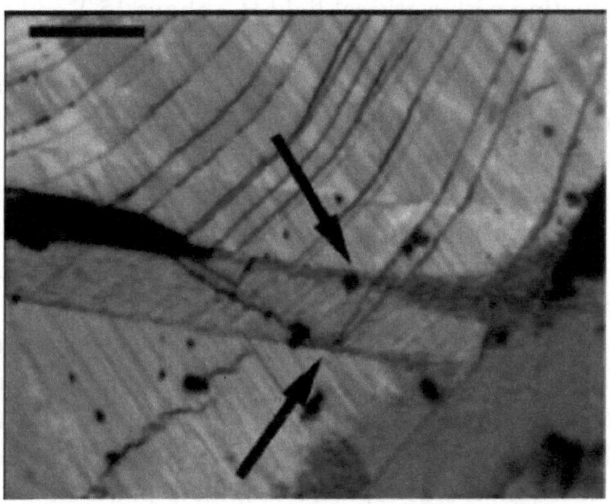

Fig. 1.4 superconducting twin walls (*arrows*) in WO_3 close to the crystal surface. The scale bar in the *top left corner* is 50 μm

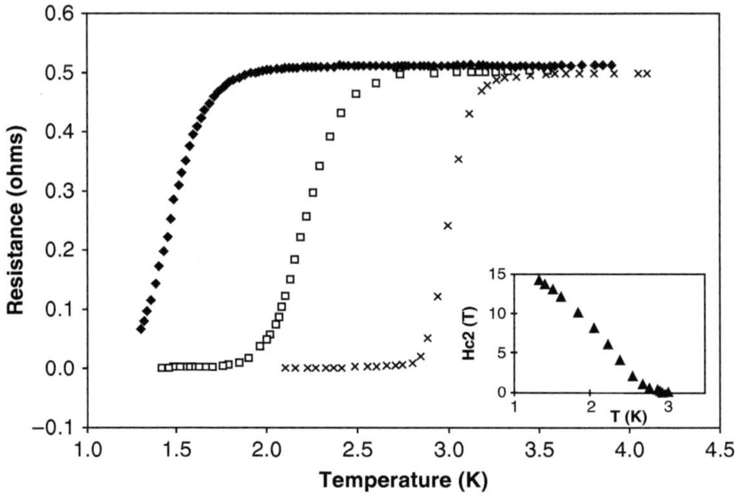

Fig. 1.5 Resistance of the superconducting twin wall in WO_3. The onset of superconductivity is 3 K; the critical field H_{c2} increases to 15 T at low temperatures

oxygen but rather to the low energy required to transfer the valence state of localized surplus electrons on the W^{6+} sites to W^{5+}. This tendency to form W^{5+} states near surfaces was directly confirmed by XPS/UPS experiments [51] and indirectly by STM imaging [52]. These W^{5+} states are not localized, however, and form bi-polarons in the low-temperature phase [53, 54]. WO_3 is a well-known electrochromic, solar cell, and catalytic material; it also displays the remarkable superconducting properties discussed before. Superconducting twin walls in WO_3 are chemically slightly reduced by inserting Na or removing O from the walls. The chemically modified walls (the changes are minor and analytically hard to detect) are then superconducting with a critical field H_{c2} above 15 T and a superconducting transition temperature T_C near 3 K. The surrounding matrix remains insulating so that this arrangement of superconducting twin boundaries with the formation of needle domains and domain junctions is potentially the key for engineering arrays of Josephson junctions and high sensitivity magnetic scanners. In addition, it has been suggested that surface layers, presumably similar to the interfacial structures in WO_3, may display superconductivity at temperatures up to $T_c = 91$ K (Na doping) and 120 K (H doping) [55]. These would constitute extreme values of T_c, which have not been reproduced independently, while the lower value in domain boundaries (3 K in [21]) has been directly observed by transport measurement and subsequently reproduced. In Fig. 1.6, we show the room temperature contrast as measured in AFM and PFM of a WO_3 surface. The highly conducting interfaces are clearly visible. The underlying bulk is piezoelectric, which ensures coupling with electric fields. In addition, it was reported that the ϵ-phase in WO_3 is ferroelectric [50] so that piezoelectric – ferroelectric – superconducting coupling becomes possible.

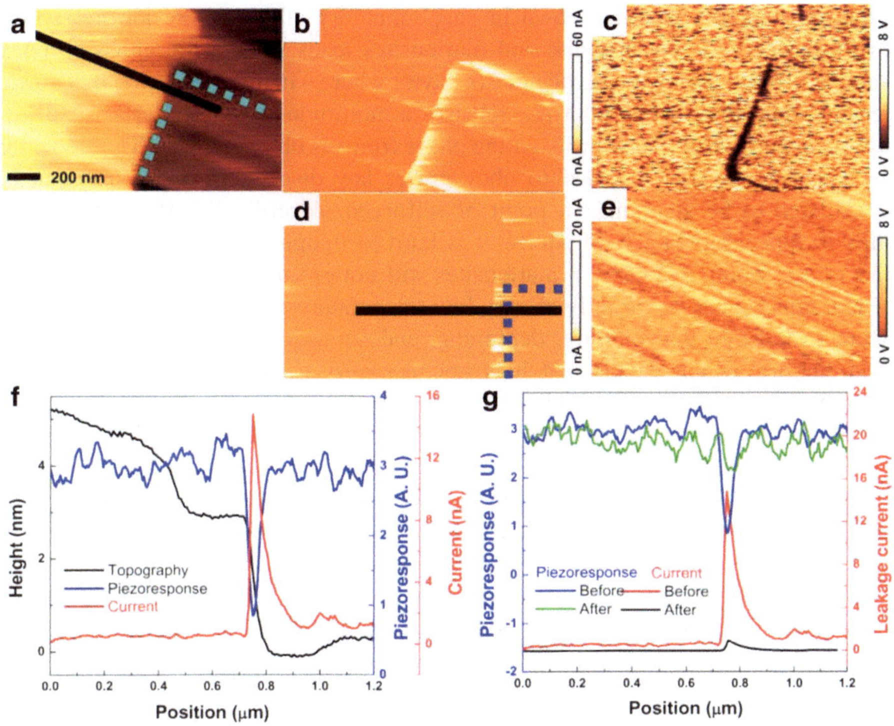

Fig. 1.6 (**a**) Topography, (**b**) current, and (**c**) PFM amplitude images of a freshly reduced WO_{3-x} single crystal. (**d**) Current and (**e**) PFM amplitude images of the same area of the WO_{3-x} single crystal as in (**a**) measured after several weeks. (**f**) Topography, current, and amplitude (piezoresponse) profiles acquired along the *black line* in (**a**), (**b**), and (**c**), respectively. (**g**) Current and amplitude (piezoresponse) profiles before and after an elapsed time of several weeks. The *dotted box* of (**a**) and (**d**) presents the steps to identify the location [56]

1.4 The Dynamics of Domain Movement and Ferroic Switching

If interfaces are taken to be the active elements of a material, the question arises whether such interfaces are stable under external forces or whether their location changes. This will decide their applicability: pinned interfaces will be used according to their internal dynamics, while mobile interfaces will change the size of the adjacent domains and, thus, operate similar classic ferroics where the size of the domains in the various orientations determines the response of a material with respect to external fields.

In multiferroics, the common view is that interfaces can move with external fields in a momentum-driven dynamics. The domains then propagate as classic front propagation [8] for large enough fields. For small field strength, this picture was shown to be wrong, however. Careful measurements under small thermal and elastic driving forces have revealed jerky front propagation and avalanche formation.

This phenomenon is well known in shape-memory alloys where the movement of interfaces between austenite and martensites leads to acoustic emission (AE) similar to Barkhausen-type avalanche behavior where – in the conventional picture – jerky propagation of one interface releases a multitude of other interfaces so that ultimately an avalanche of propagating phase fronts is observed [57]. Theoretically, such avalanches are expected to obey power law distributions [58] and can be considered to be at or near a point of selforganized criticality [59]. While this idea is appealing for its simplicity, it is hard to imagine that the randomness of the various pinning centers in martensites will not extend to extremely low values where pinning can occur only at very low temperatures. In fact, most experiments seem to indicate that pinning, depinning, and acoustic emission (AE) dynamics is a-thermal, which means that it is not thermally activated. A key experiment was recently performed [60] where the transition in a $Cu_{67.64}Zn_{16.71}Al_{15.65}$ shape-memory alloy was investigated calorimetrically, whereby the thermal driving force was minimized. The transition was scanned at rates of some mK/h so that each avalanche could be observed as an individual peak in the latent heat. The resulting DTA curve is shown in Fig. 1.7. It consists of two components: the jerks (Fig. 1.8) and a continuous background.

The entropy of the transition is not affected by the jerks and is the same on heating and cooling. Besides for the strongest avalanches, no memory effect was observed for the individual jerks. The statistical analysis of the jerks is the same as of AE spectra (Fig. 1.8) and follows a power law of the energy of the jerks: $P(E) \sim E^{-\varepsilon}$ with an exponent close to $-\varepsilon \sim -2$. This observation shows that the AE exponents are identical with or close to the energy exponents and not the size exponents (Fig. 1.9).

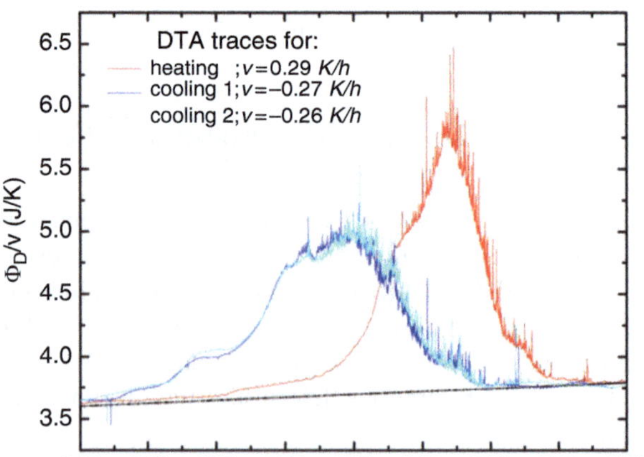

Fig. 1.7 DTA traces for cooling and heating experiments. The heating rate was 0.29 K/h, the cooling rates were 0.27 and 0.26 K/h. Note the coexistence of smooth front propagation and thermal spikes (jerks) even at very low thermal driving forces

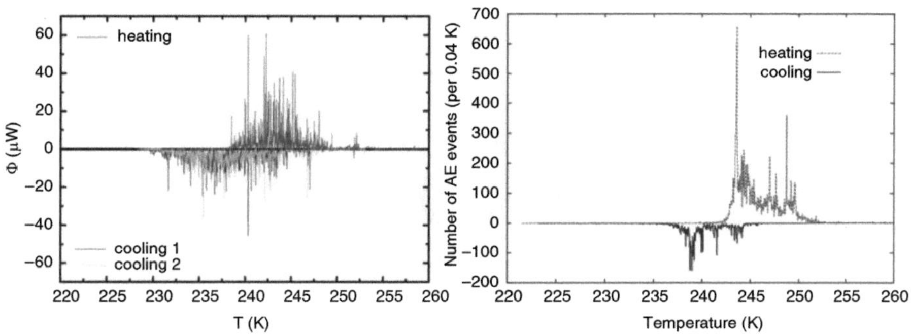

Fig. 1.8 Spikes in the calorimetric measurement after removing the smooth baseline (*left*) and acoustic emission (AE) signals (*right*) of the same sample. The sign of the peaks has been inverted for clarity between heating and cooling experiments [60]

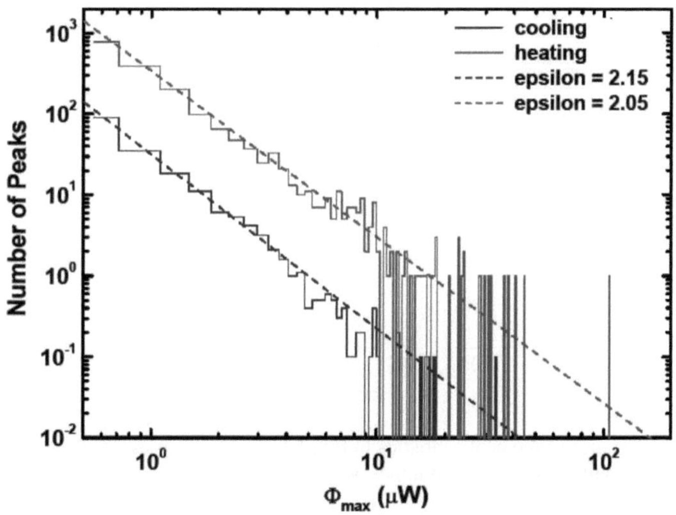

Fig. 1.9 Statistical analysis of the heating and cooling curves of the DTA traces in Fig. 1.3. Data corresponding to cooling experiments have been shifted one decade downward in order to clarify the picture [60]

A direct observation of the jerks in elastic measurements depends on the smallness of the applied forces. The transition in $Cu_{74.08}Al_{23.13}Be_{2.79}$ was investigated in a very careful dynamical mechanical analyzer (DMA) experiment. The frequency of the three-point-bending excitation was chosen as 0.1 Hz, the applied forces were extremely small (<50 mN max. amplitude), and the heating/cooling rate was <0.14 K/h. The mechanical loss is $\sim$ tan(δ) and shows spiky avalanche behavior (Fig. 1.10) similar to those in Fig. 1.8. Statistical analysis of the jerks leads again to a power law. The exponent (-1.3) is significantly smaller than the energy exponent of the calorimetric measurement, even though the uncertainty of the fitted exponent

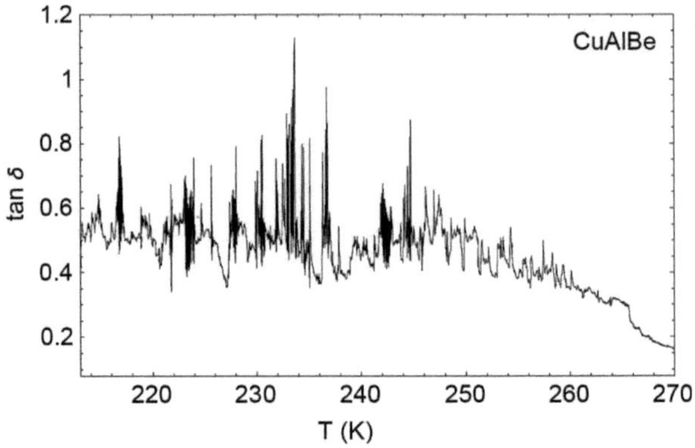

Fig. 1.10 Phase lag tan(δ) of a $Cu_{74.08}\,Al_{23.13}\,Be_{2.79}$ single crystal recorded in three-point-bending mode at 0.1 Hz and a heating rate of 0.15 K/h [61]

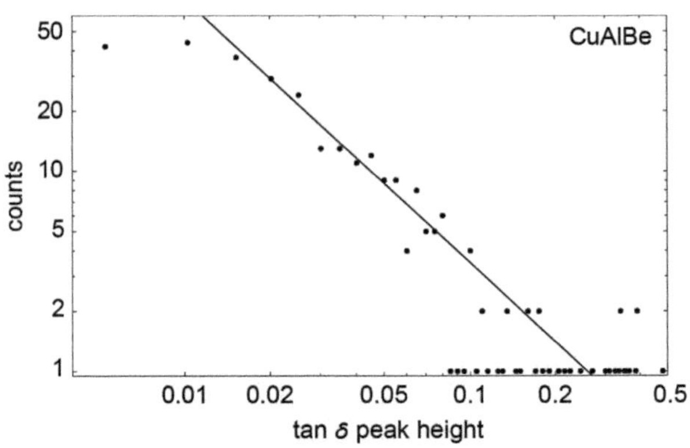

Fig. 1.11 Log–log plot of the peak statistics in Fig. 1.5. The *line* represents the best fit to a power law $P(E) \sim E^{-\varepsilon}$ with $\varepsilon = 1.3$ and an upper bound of 1.6 [61]

is very large. As an upper bound the exponent of -1.6 was estimated in [61]. As the elastic response is measured under oscillatory stress, one may expect that the exponent is related to field binning and related to the amplitude exponent $-\tau$, which was calculated to be in the order of -1.5 in MFT and near -1.3 in simulations [62] (Fig. 1.11).

The movement of interfaces between martensitic variants and ferroelastic twin walls depends theoretically on the dimension of the interface. Planar interfaces have the dimension 2 ($D = 2$), while the tip of a moving needle domain is a line in three dimensions and represents the case $D = 1$. Boundaries well beyond the depinning threshold move freely as solitary waves; their behavior has been well described in the literature [63].

In Fig. 1.12, the trajectory of a needle domain in $LaAlO_3$ is shown. The advancing or retracting needle domain is pinned by defects that are mostly located

Fig. 1.12 Shape of a needle domain inside the ferroelastic matrix. The advancing front ($D = 1$) is pinned by a small number of defects

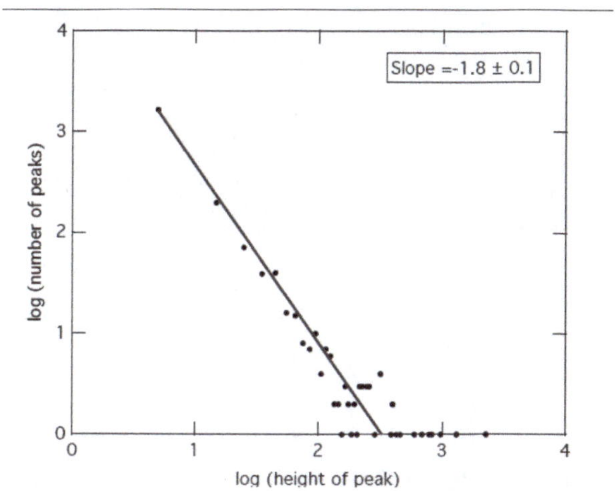

Fig. 1.13 Power law dependence of the energy emission of a single advancing needle domain ($D = 1$). The exponent is estimated to be ~ -1.8

at the advancing edge of the hull-shaped domain. Pinning is then described as the local fixation of a line in three-dimensional space, $D = 1, d = 3$. It is not trivial that pinning should occur at all in this scenario: the Larkin length of the edge in elastic systems is assumed to be large and very strong pinning centers are required in order to obtain the pinning of the advancing needle (Fig. 1.13).

In this context, the recent contribution of Proville [64] is relevant. He showed that in cases where the Larkin length is larger than the system size one can still expect avalanches with a finite depinning force. This observation calls into question the traditional way how the Larkin length is simulated in computer experiments: the elasticity of the interface is simply represented by interatomic springs between

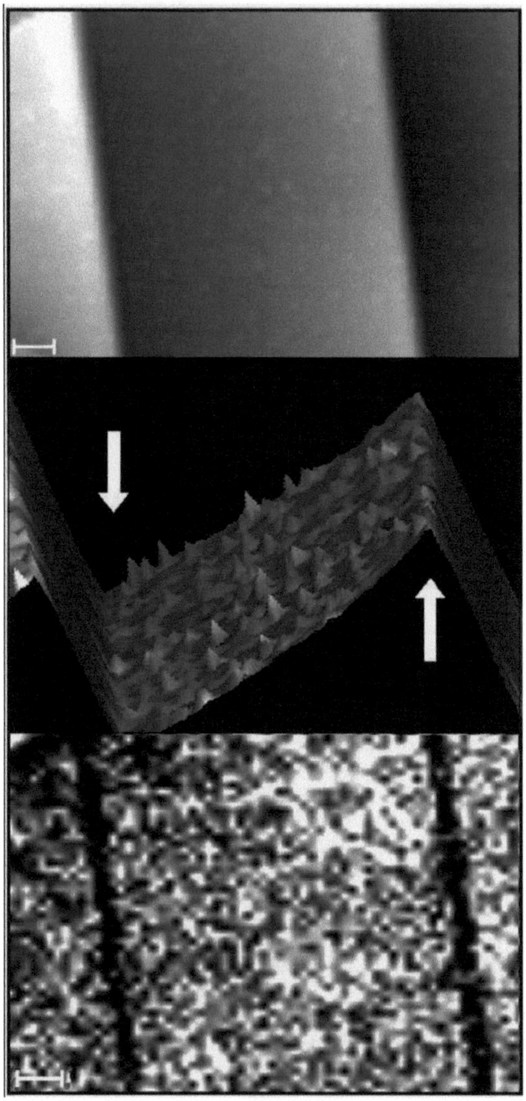

Fig. 1.14 Optical (*top*) and AFM images of twin boundaries in $Pb_3(PO_4)_2$ with high number of Ba defects (courtesy U. Bismayer, University of Hamburg) [33,34]. Scale bar is 6 μm. The pinning by Ba is very weak indicating that the Larkin length is very long

atoms in the wall. This is not the sole obvious mechanism in ferroelastic systems, however. It was shown [65] that the two major forces acting on the interface – in addition to the wall energy itself – are the "anisotropy energy" (which is minimal for walls with an orientation where the compatibility relation is satisfied). Any rotation of a wall segment requires significant energy $\sim \cos\varphi$ (in a local approximation), where φ is the angle between the stress-free equilibrium direction and the actual direction of the wall segment. The second important energy is the wall "bending energy," which resists any curvature of the wall and is particularly large in case of thick walls [65]. These energies will act against the roughening of walls and ensures that twin walls remain globally planar even under doping conditions, which would normally lead to rough walls in magnetic systems. Such walls would meander in order to take advantage of as many defects as feasible to obtain a maximum pinning force. The AFM image in Fig. 1.9 shows the case of Ba-doped $Pb_3(PO_4)_2$ where no significant pinning is achieved by the addition of the Ba atoms. The moving wall does not meander and remains smooth. It can wrap around defects whenever they encounter such defects or defect clusters. The characteristic length is then given by the geometry of the defect and the details of the elastic interactions and is much longer than the classic Larkin length. This leads to the question: how individual defects contribute to the wall pinning. It is no wonder, therefore, that a simple "elastic" model which leads to the formulation of the Larkin length cannot – in most realistic cases – correctly describe the pinning behavior in martensitic, ferroelastic, and other ferroic domain structures (Fig. 1.14).

1.5 Conclusions

Multiferroic materials are well established. For some systems, such as $BiFeO_3$, industrial applications seem to be within our reach during the next decade. The key is a firm understanding of the coupling between the various order parameters in the bulk. The theory for the various mechanisms is also well established even in cases where the exact atomic nature of the coupling mechanism remains unclear. The open question is what new developments we can expect in the next 5 years?

The first development is to consider conductivity and even superconductivity as an equally important feature as multiferroicity. The treatment of high carrier densities and pairing mechanisms can follow the same path as the treatment of coupling between elastic, magnetic, and electric degrees of freedom.

Disruptive technologies can be envisaged if we succeed to use interfaces as active elements in multiferroic and other functional materials. This idea is new and finds its expression in the terminology of "domain boundary engineering." Theoretically, coupling between various properties in domain boundaries is easier than in the bulk: the coupling need not be related to terms such as $Q_1^2 Q_2^2$ in the relevant Hamiltonian (where Q is understood as an order parameter, which is constant over a length scale of several interatomic distances, at least). Instead, we have many more coupling phenomena at our disposal, such as the all important gradient coupling. Strong coupling such as seen in flexo-elasticity of the type $Q_1 \nabla Q_2$ and the interference of the gradient term $(\nabla Q)^2$ for each order parameter opens the door for a multitude of novel effects that contain hereto unknown structural states on a length scale of the thickness of interfaces (i.e. some nm). The same terms apply when we consider conducting interfaces that can play the role of electric wiring in devices.

References

1. E. Ascher, H. Rieder, H. Schmid, H. Stossel, Some properties of ferromagnetoelectric nickel-iodine boracite Ni3B7O13. J. Appl. Phys. **37**, 140 (1966)
2. S. Kinge, M. Crego-Calama, D.N. Reinhoudt, Self-assembling nanoparticles at surfaces and interfaces. ChemPhysChem **1**, 20 (2008)
3. M. Fiebig, Revival of the magnetoelectric effect. J. Appl. Phys. D **38**, R123 (2005)
4. J. Wang, J.B. Neaton, H. Zheng, V. Nagarajan, S.B. Ogale, B. Liu, D. Viehland, V. Vaithyanathan, D.G. Schlom, U.V. Waghmare, N.A. Spaldin, K.M. Rabe, M. Wuttig, R. Ramesh, Epitaxial BiFeO3 multiferroic thin film heterostructures. Science **299**, 1719 (2003)
5. A. Lubk, S. Gemming, N.A. Spaldin, First-principles study of ferroelectric domain walls in multiferroic bismuth ferrite. Phys. Rev. B **80**, art. 104110, (2009)
6. N.A. Spaldin, M. Fiebig, The renaissance of magnetoelectric multiferroics. Science **309**, 391 (2005)
7. J.B. Neaton, C. Ederer, U.V. Waghmare, N.A. Spaldin, K.M. Rabe, Elastic behavior associated with phase transitions in incommensurate Ba2NaNb5O15. Phys. Rev. B **71**, art. 014113 (2005)
8. E.K.H. Salje, Phase Transitions in ferroelastic and co-elastic crystals (Cambridge University Press, Cambridge, UK, 1993)
9. J.C. Lashley, S.M. Shapiro, B.L. Winn, et al., Observation of a continuous phase transition in a shape-memory alloy. Phys. Rev. Lett. **101**, art. 135703 (2008)
10. W. Eerenstein, N.D. Mathur, J.F. Scott, Multiferroic and magnetoelectric materials. Nature **442**, 759 (2006)
11. E.K.H. Salje, Multiferroic domain boundaries as active memory devices: Trajectories towards domain boundary engineering. ChemPhysChem **11**, 940 (2010)
12. S. Marais, V. Heine. C. Nex, et al., Phenomena due to strain coupling in phase – transitions. Phys. Rev. Lett. **66**, 2480 (1991)
13. M.A. Carpenter, E.K.H. Salje, Elastic anomalies in minerals due to structural phase transitions. Eur. J. Mineral. **10**, 693 (1998)
14. S.H. Lim, M. Murakami, W.L. Sarney, et al., The effects of multiphase formation on strain relaxation and magnetization in multiferroic BiFeO3 thin films. Adv. Func. Mater. **17**, 2594 (2007)
15. R.D. James, M. Wuttig, Magnetostriction of martensite. Phil Mag. A **77**, 1273 (1998)
16. K. Mori, M. Wuttig, Magnetoelectric coupling in terfenol-D/polyvinylidenedifluoride composites, Appl. Phys. Lett. **81**, 100 (2002)
17. J. Wang, J.B. Neaton, H. Zheng, et al., Epitaxial BiFeO3 multiferroic thin film heterostructures. Science **299**, 1719 (2002)
18. J. Seidel, L.W. Martin, Q. He, et al., Conduction at domain walls in oxide multiferroics. Nat. Mater. **8**, 229 (2009)
19. N. Hur, S. Park, P.A. Sharma, et al., Electric polarization reversal and memory in a multiferroic material induced by magnetic fields. Nature 429, 392 (2004)
20. C. Ederer, N.A. Spaldin, Weak ferromagnetism and magnetoelectric coupling in bismuth ferrite, Phys. Rev. B **71**, art. 060401 (2005)
21. A. Aird, E.K.H. Salje, Sheet superconductivity in twin walls: experimental evidence of WO3-x. J. Phys.: Condens. Matter **10**, L377 (1998)
22. B. Nagaraj, T. Sawhney, S. Perusse, et al., (BaSr)TiO3 thin films with conducting perovskite electrodes for dynamic random access memory applications. Appl. Phys. Lett. **74**, 3194 (1999)
23. S. Ramesh, V.P. Kumar, P. Kistaiah, et al., Preparation, characterization and thermo electrical properties of co-doped Ce0.8- xSM0.2CaxO2 (-) (delta) materials. Solid State Ionics **181**, 86 (2010)
24. T. Shimada, S. Tomoda, T. Kitamura, Ab initio study of ferroelectric closure domains in ultrathin PbTiO3 films. Phys. Rev. B **81**, art 144116 (2010)
25. S.E. Barnes, S. Maekawa, Current-spin coupling for ferromagnetic domain walls in fine wires. Phys. Rev. Lett. **95**, art. 107204 (2005)

26. E.K.H. Salje, J. Chrosch, R.C. Ewing, Is "metamictization" of zircon a phase transition? Am. Mineral. **84**, 1107 (1999)
27. S. Rios, E.K.H. Salje, M. Zhang, et al., Amorphization in zircon: evidence for direct impact damage, J. Phys.: Condens. Matter **12**, 2401 (2000)
28. S. Sarkar, X.B. Ren, K. Otsuka, Evidence for strain glass in the ferroelastic-martensitic system Ti50-xNi50+x. Phys. Rev. Lett. 95, art. 205702 (2005)
29. B. Noheda, D.E. Cox, G. Shirane, et al., Phase diagram of the ferroelectric relaxor (1-x)PbMg1/3Nb2/3O3-xPbTiO(3), Phys. Rev. B **66**, art. 054104 (2002)
30. A. Levstik, Z. Kutnjak, C. Filipic, et al., Glassy freezing in relaxor ferroelectric lead magnesium niobate. Phys. Rev. B **57**, 11204 (1998)
31. G.A. Samara, The relaxational properties of compositionally disordered ABO(3) perovskites. J. Phys.: Condens. Matter **15**, R367 (2003)
32. J. Chrosch, E.K.H. Salje, Near-surface domain structures in uniaxially stressed SrTiO3. J. Phys.: Condens. Matter **85**, 722 (1999)
33. J. Chrosch, E.K.H. Salje, Temperature dependence of the domain wall width in LaAlO3. J. Appl. Phys. **85**, 722 (1999)
34. B. Wruck, E.K.H. Salje, M. Zhang, et al., On the thickness of ferroelastic twin walls in lead phosphate Pb3(PO4)2 – an X-ray diffraction study. Phase Transit. **48**, 135 (1994)
35. E.K.H. Salje, H. Zhang, A. Planes, et al., Martensitic transformation B2-R in Ni-Ti-Fe: experimental determination of the Landau potential and quantum saturation of the order parameter. J. Phys.: Condens. Matter **20**, art. 275216 (2008)
36. E.K.H. Salje, H. Zhang, D. Schryvers, et al., Quantitative Landau potentials for the martensitic transformation in Ni-Al. Appl. Phys. Lett. **90**, art. 221903 (2007)
37. M. Calleja, M.T. Dove, E.K.H. Salje, Trapping of oxygen vacancies on twin walls of CaTiO3: a computer simulation study. J. Phys.: Condens. Matter **15**, 2301 (2003)
38. W.T. Lee, E.K.H. Salje, L. Goncalves-Ferreira, et al., Intrinsic activation energy for twin-wall motion in the ferroelastic perovskite CaTiO3. Phys. Rev. B **73**, art. 214110 (2006)
39. L. Goncalves-Ferreira, S.A.T. Redfern, E. Artacho, et al., Ferrielectric twin walls in CaTiO3. Phys. Rev. Lett. **101**, art. 097602 (2008)
40. L. Goncalves-Ferreira, S.A.T. Redfern, E. Atacho, et al., The intrinsic elasticity of twin walls: Ferrielectric twin walls in ferroelastic CaTiO3. Appl. Phys. Lett. **94**, art. 081903 (2009)
41. L. Goncalves-Ferreira, S.A.T. Redfern, E. Artacho, et al., Trapping of oxygen vacancies in the twin walls of perovskite. Phys. Rev. B **81**, art. 024109 (2010)
42. P. Zubko, G. Catalan, P.R.L. Welche, A. Buckley, J.F. Scott, Strain-gradient-induced polarization in SrTiO3 single crystals. Phys. Rev. Lett. **99**, art.167601 (2007)
43. A.K. Tagantsev, E. Courtens, L. Arzel, Prediction of a low-temperature ferroelectric instability in antiphase domain boundaries of strontium titanate. Phys. Rev. B **64**, art.224107 (2001)
44. J. Petzelt, T. Ostapchuk, I. Gregora, I. Rychetsky, S. Hoffmann-Eifert, A.V. Pronin, Y. Yuzyuk, B.P. Gorshunov, S. Kamba, V. Bovtun, J. Pokorny, M. Savinov, V. Porokhonskyy, D. Rafaja, P. Vanek, A. Almeida, M.R. Chaves, A.A. Volkov, M. Dressel, R. Waser, Dielectric, infrared, and Raman response of undoped SrTiO3 ceramics: Evidence of polar grain boundaries. Phys. Rev. B **64**, art.184111 (2001)
45. W. Zhong, D. Vanderbilt, Competing structural instabilities in cubic perovskites. Phys. Rev. Lett.**74**, art. 2587 (1995)
46. E.K.H. Salje, A pre-martensitic elastic anomaly in nanomaterials: elasticity of surface and interface layers. J. phys.: Condens. Matter **20**, art. 485003 (2008)
47. A. Aird, E.K.H. Salje, Enhanced reactivity of domain walls in WO3 with sodium. Eur. Phys. J. B **15**, 205 (2000)
48. A. Aird, M.C. Domeneghetti, F. Mazzi, et al., Sheet superconductivity in WO3-x: crystal structure of the tetragonal matrix, J. Phys.: Condens. Matter **10**, L569 (1998)
49. R.A. de Groot, F.M. Mueller, P.G. van Engen, K.H.J. Buschow, New class of materials: Half-metallic ferromagnets. Phys. Rev. Lett. **50**, 2024 (1983)

50. E.K.H. Salje, S. Rehmann, F. Pobell, D. Morris, K.S. Knight, T. Herrmanndörfer, M.T Dove, Crystal structure and paramagnetic behaviour of epsilon-WO3-x. J. Phys.: Condens. Matter **9**, 6563 (1997)
51. E. Salje, A.F. Carley, M.W. Roberts, Effect of reduction and temperature on the electronic core levels of tungsten and molybdenum in WO3 and WxMo1-xO3 – photoelectron spectroscopy study. J. Solid State Chem. **29**, 237 (1979)
52. F.H. Jones, K. Rawlings, J.S. Foord, R.G. Egdell, J.B. Pethica, B.M.R. Wanklyn, S.C. Parker, P.M. Olive, An STM study of surface structures on WO3(001). Surf. Sci. **359**, 107 (1996)
53. O.F. Schirmer, E. Salje, Conduction bipolarons in low-temperature crystalline WO3-x. J. Phys.: Condens. Matter **13**, 1067 (1980)
54. O.F. Schirmer, E. Salje, W5+ polaron in crystalline low-temperature WO3 electron spin resonance and optical absorption. Solid State Comm. **33**, 333 (1980)
55. S. Reich, G. Leitus, R. Popovitz-Biro, A. Goldbourt, S. Vega, A possible 2D H (x) WO3 superconductor with a T (c) of 120 K. J. Supercond. Novel Magnetism **22**, 343 (2009) and reference given there
56. Y. Kim, M. Alexe, E.K.H. Salje, Nanoscale properties of thin twin walls and surface layers in piezoelectric WO3-x. Appl. Phys. Lett. **96**, art. 032904 (2010)
57. E. Vives, J. Ortin, L. Manosa, et al., Distribution of avalanches in martensitic transformations. Phys. Rev. Lett. **72**, 1694 (1994)
58. E. Bonnot, E. Vives, L. Mañosa, et al., Acoustic emission and energy dissipation during front propagation in a stress-driven martensitic transition. Phys. Rev. B **78**, art. 094104 (2008)
59. M.C. Kuntz, J.P. Sethna, Noise in disordered systems: The power spectrum and dynamic exponents in avalanche model. Phys. Rev. B **62**, art. 11699 (2000)
60. M.C. Gallardo, J. Manchado, F.J. Romero, et al., Avalanche criticality in the martensitic transition of Cu67.64Zn16.71Al15.65 shape-memory alloy: A calorimetric and acoustic emission study. Phys. Rev. B **81**, art.174102 (2010)
61. E.K.H. Salje, L. Koppensteiner, M. Reinecker, et al., Jerky elasticity: Avalanches and the martensitic transition in Cu74.08Al23.13Be2.79 shape-memory alloy. Appl. Phys. Lett. **95**, art. 231908 (2009)
62. A. Rosso, P. Le Doussal, K.J. Wiese, Avalanche-size distribution at the depinning transition: A numerical test of the theory. Phys. Rev. B **80**, art. 144204 (2009)
63. M.C. Kuntz, O. Perkovic, K.A. Dahmen, et al., Hysteresis, avalanches, and noise. Comput. Sci. Eng. **1**, 73 (1999)
64. L. Proville, Depinning of a discrete elastic string from a random array of weak pinning points with finite dimensions. J. Stat. Phys. **137**, 717 (2009)
65. E.K.H. Salje, Y. Ishibashi, Mesoscopic structures in ferroelastic crystals: Needle twins and right-angled domains. J. Phys.: Condens. Matter **8**, 8477 (1996)

Chapter 2
Phase Diagrams of Conventional and Inverse Functional Magnetic Heusler Alloys: New Theoretical and Experimental Investigations

P. Entel, M.E. Gruner, A. Hucht, A. Dannenberg, M. Siewert, H.C. Herper, T. Kakeshita, T. Fukuda, V.V. Sokolovskiy, and V.D. Buchelnikov

Abstract First-principles calculations allow to characterize the electronic and magnetic ground-state properties of the full-Heusler alloys of type X_2YZ. Functionality of the materials strongly depends on the type of elements and composition. A half-metallic state with 100% spin polarization at the Fermi level, which is an ideal spintronics material for tunneling devices, is, for instance, achieved for (X = Co, Y = Mn, and Z = Ge and Si). Replacing Co by Ni and Ge or Si by Ga yields the prototypical magnetic shape-memory compound Ni_2MnGa, which undergoes a (martensitic) tetragonal distortion at ca. 200 K, where the magnetic shape-memory features can be exploited by an external magnetic field and external stress in the martensitic state. Quite another functionality, the conventional or inverse magnetocaloric effect, is observed in the off-stoichiometric samples of (X = Ni, Y = Mn, and Z = Ga, In, Sn, and Sb), where the efficiency of the magnetocaloric effect depends on the size of the isothermal entropy change across the magnetostructural phase transition in an applied magnetic field. Here, we discuss how some of these material properties can be improved in order to obtain room temperature or higher operation temperatures needed for a technological breakthrough.

P. Entel (✉) · M.E. Gruner · A. Hucht · A. Dannenberg · M. Siewert · H.C. Herper
Faculty of Physics & CeNIDE, University Duisburg-Essen, 47048 Duisburg, Germany
e-mail: entel@thp.uni-duisburg.de; Markus.Gruner@uni-due.de; fred@thp.uni-duisburg.de; antje@thp.uni-duisburg.de; mario@thp.uni-due.de; heike.herper@uni-due.de

T. Kakeshita · T. Fukuda
Division of Materials and Manufacturing Science, Graduate School of Engineering, Osaka University, 2-1, Yamada-oka, Suita Osaka 565-0871, Japan
e-mail: kakeshita@mat.eng.osaka-u.ac.jp; fukuda@mat.eng.osaka-u.ac.jp

V.V. Sokolovskiy · V.D. Buchelnikov
Condensed Matter Physics Department, Chelyabinsk State University, 454021 Chelyabinsk, Russia
e-mail: vsokolovsky84@mail.ru; buche@csu.ru

T. Kakeshita et al. (eds.), *Disorder and Strain-Induced Complexity in Functional Materials*, Springer Series in Materials Science 148,
DOI 10.1007/978-3-642-20943-7_2,

2.1 Introduction and Computational

Over the last decade, a series of new complex functional materials have been synthesized; a recent discussion of their multiferroic, magnetoresistive, martensitic, and related magnetic shape-memory behavior as well as magnetocaloric and Invar properties can be found in [1]. Ferromagnetic full-Heusler alloys are intermetallic compounds with the generic formula X_2YZ and cubic $L2_1$ structure with X and Y being transition metal elements: X = Fe, Co, Ni, Pd, and Pt, usually, Y = Mn, and Z = Al, Ga, In, Si, Ge, Sn, and Sb. These fascinating materials may display half-metallic features [2, 3], which were originally predicted for the half-Heusler alloy NiMnSb with $C1_b$ structure by de Groot et al. [4]). Other alloys such as Ni_2MnGa show superelasticity and shape-memory behavior [5, 6], which are associated with large strain achieved in an external magnetic field first discovered by Ullakko et al. [7]. Further characteristics are connected with coexisting structural and magnetic phase transitions over a finite range of compositions [8] and associated conventional and inverse magnetocaloric effects [9, 10]. For a discussion of shape-memory and magnetocaloric effects, see [11]. The Heusler-based magnetic shape-memory alloys (MSMA) usually have too low Curie and martensite transformation temperatures, which so far prevent a technological breakthrough.

Since the discovery of the very large strains of up to 10% in Ni_2MnGa, an increasing activity can be found focusing, in particular, on MSMA to improve material properties such as operation temperatures and elastic behavior. New materials have been proposed [12], but so far no real breakthrough has been reported. With respect to the half-metallic ferromagnetic Heusler alloys used in tunneling magnetoresistance (TMR) devices, one meets a different kind of difficulty. Here, the problem arises from the rapid decrease of the TMR with increasing temperature [13, 14]. In this chapter, we discuss the similarity of electronic properties between ferromagnetic Heusler alloys, which, on the one hand, are of interest for spintronics devices and, on the other hand, for actuator devices with the goal to point out ways to circumvent the difficulties arising from the intrinsic properties of these materials. Co-based half-metallic ferromagnets, such as Co_2FeSi being particularly interesting for spintronics applications because of its very high Curie temperature of 1,100 K, unfortunately show low TMR values at room temperature due to interface states and nonquasiparticle many-particle effects. For a recent theoretical overview on half-metallic ferromagnets, see [15].

In this chapter, we discuss some basic properties of MSMA, which have been obtained by first-principles calculations and highlight the most important features of these materials on a microscopic level. Most of the calculations were done using the Vienna ab initio simulation package (VASP, version 4.6.28) [16] with PBE exchange correlation functional [17] and projector augmented wave method [18]. The eletronic density of states (DOS) calculations of Co_2MnGe involved [Co:$3d^94s^1$], [Mn:$3p^63d^64s^1$], [Ge:$3d^{10}4s^24p^2$], a cutoff energy of 360 eV, and 41^3k-points (Monkhorst pack), while for that of Ni_2MnGa involved [Ni:$3p^63d^94s^1$], [Ga:$3d^{10}4s^24p^1$], a cutoff energy of 460 eV, and 31^3k-points.

In all cases, ionic positions were relaxed. From the total energy calculations, one may infer information about structural and magnetic tendencies that can directly be related to particularities in the experimental phase diagrams such as the magnetostructural transition in case of Ni–Mn–Ga alloys. Large magnetic anisotropy energy, which is the basis or prerequisite of magnetic shape-memory behavior of Ni_2MnGa, was calculated using the full-potential local-orbital minimum-basis package (FPLO, version 5.00–19) [19] for a sufficiently large number of k-points (13.824 points in the full-Brillouin zone). In the ab initio calculations of magnetic exchange interaction constants, the SPR-KKR code was used together with the coherent potential approximation (CPA) for the nonstoichiometric cases [20, 21]. These calculations clearly reveal the competition of ferromagnetic and antiferromagnetic behavior in the Ni–Mn–(Ga, In, Sn, Sb) alloy series. Finally the phonon softening effects, which are present in most MSMA, were investigated by ab initio calculations of phonon dispersions using the direct method to obtain force constants from the displacements of the atoms in 3^3 supercells (108 atoms); see below.

2.2 Crystal Structures of Half- and Full-Heusler alloys

It is well known that the zinc-blende (B3), half-metal half-Heusler ($C1_b$), and half-metal full-Heusler ($L2_1$) crystal structures consist of four interpenetrating fcc sublattices: In the half-metal ferromagnet CrSb [22], two of the sublattices are unoccupied; see Fig. 2.1a (replacing Cr and Sb by C yields the diamond lattice). In the half-metal half-Heusler ferromagnet NiMnSb or CoMnSb [4], three of the four fcc sublattices are occupied and the second Ni- or Co-sublattice is empty (Fig. 2.1b), while in the half-metal full-Heusler Co_2MnGe [23] all four sublattices are occupied (Fig. 2.1c). The gap in the minority-spin channel of the half-metal full-Heusler Co_2MnGe closes if we replace Co by Ni which is the route toward materials

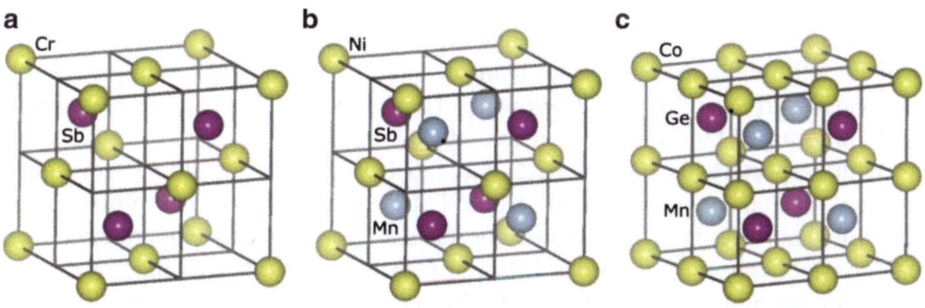

Fig. 2.1 The zinc-blende and Heusler crystal structures consist of four interpenetrating fcc lattices. (**a**) In the zinc-blende (XZ) structure such as CrSb, two sublattices are empty. (**b**) In the half-Heusler (XYZ) with $C1_b$ structure such as NiMnSb, one sublattice is unoccupied and Ni and Mn form a zinc-blende structure. (**c**) In the full-Heusler (X_2YZ) materials such as Co_2MnGe with $L2_1$ (bcc-like) structure, all sublattices are occupied

Table 2.1 Lattice constant, magnetic moment, and energy gap in the minority-spin channel of half-metal ferromagnets CrSb, NiMnSb, and Co_2MnGe as obtained from first-principles calculations [3, 4, 22, 23]. Where available, experimental data agree with the calculations. Note that the energy gap in the table is the total energy gap and differs from the spin-flip excitation energy E_{sf}, which is the energy of the valence band top, $-E_v$ (or conduction band bottom, E_c), of minority spin with respect to the Fermi energy E_F. The half-metallic gap is sometimes defined as the smaller one of E_v and E_c

System	a_0(Å)	$M(\mu_B/f.u.)$	E_g(eV)	References
CrSb	6.14	3.00	1.64	[3,22]
NiMnSb	5.92	3.96	0.46	[3,4]
Co_2MnGe	5.73	4.94	0.40	[3,23]

such as Ni_2Mn_{1+x}(Al, Ga, In Sn, Sb)$_{1-x}$, which exhibit the magnetic shape-memory effect (MSME) and/or magnetocaloric effect (MCE) (Table 2.1) [3, 11].

It is interesting to note that the origin of the gap in the full-Heusler compounds such as Co_2MnGe originates from the influence of two different crystal fields of octahedral and tetrahedral symmetry on the electronic states causing a splitting of the uneven e_u and t_{1u} antibonding d-electron states (of Co on the two Co-sublattices), which cannot hybridize with the even e_g- and t_{2g}-electron states of Mn [23–25]. Another feature of the half-metal Heusler compounds is the Slater–Pauling behavior of the magnetic moments [23]. First-principles calculations show that the total magnetic moment of the half-Heusler alloys is related to $M_{total} = Z_{total} - 18$, while it is $M_{total} = Z_{total} - 24$ for the full-Heusler alloys, where Z_{total} is the total number of valence electrons. This agrees with experimental observations. While all this is well documented in the literature [3, 23–26], we would like to point out its consequences for the MSMA using a rigid-band model in the argumentation.

Figure 2.2 shows the crossover from half-metallic Co_2MnGe in Fig. 2.2a to metallic Ni_2MnGa in Fig. 2.2b where the gap in the spin-down channel is closed because of one more valence electron and because of a rather complex change of the hybridization of the Ni-d and Mn-d states. For modulated 5M martensite, we obtain additional peaks in the DOS near E_F (see Fig. 2.2c), which renders the hybridization scenario even more complex. Although one should not emphasize too strongly the rigid-band picture, it shows that the martensitic behavior of Ni_2MnGa is connected with the splitting of the Ni-d states in the octahedral crystal field yielding an energy gap and a sharp Ni-e_g peak just below E_F. The splitting of the Ni-d states is similar to the splitting of the Co-d states in case of Co_2MnGe.

With respect to the use of half- and full-Heusler half-metallic ferromagnets in giant magnetoresistance (GMR) and TMR devices, two major obstacles are met. First, surface, interface, and defect states as well as finite-temperature effects can considerably decrease the 100% spin polarization at E_F and may render the band engineering of an all-Heusler GMR stack less effective; see Ambrose and Mryasov in [3]. Also, surface states are intrinsically related to different terminations and can hardly be avoided; see, for instance, the results of ab initio electronic structure calculations of $Co_2MnSi(001)/MgO$ [28]. The second obstacle is related to the observation that injection of spin-polarized electrons is not effective when

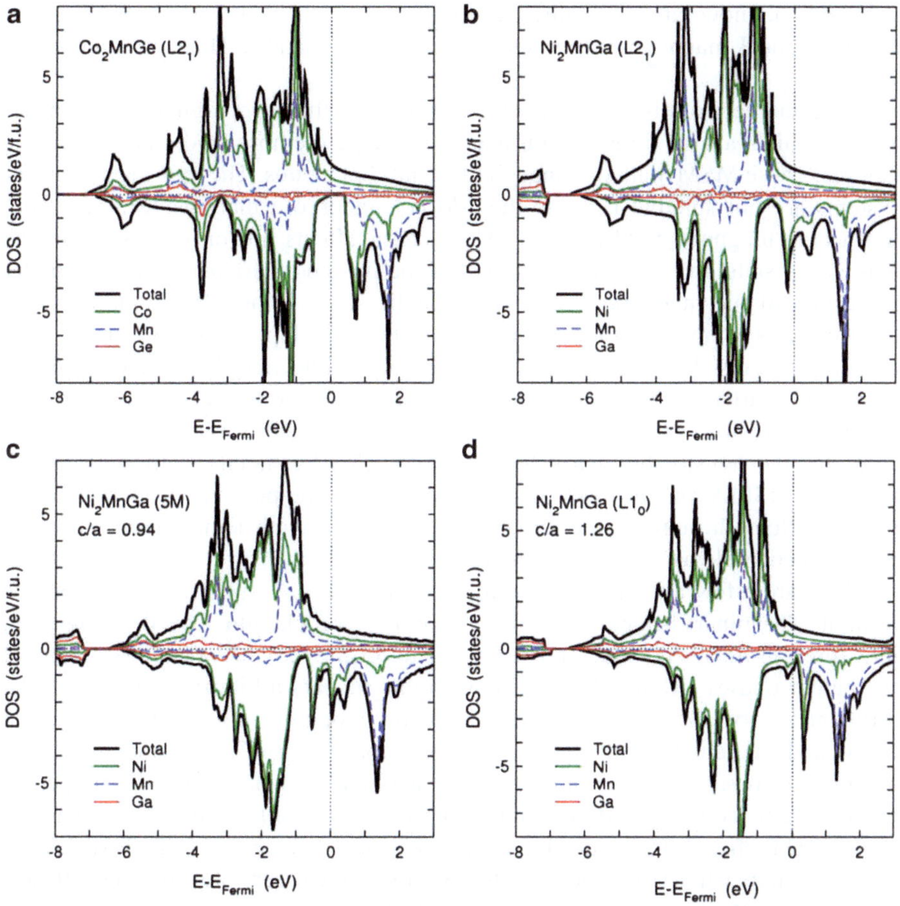

Fig. 2.2 Results of first-principles calculations of the electronic density of states (DOS) of (**a**) half-metallic full-Heusler Co_2MnGe, (**b**) MSMA Ni_2MnGa in austenite, and (**c**) 5 M and (**d**) $L1_0$ martensite states [25]. In a rigid-band picture, the DOS of Ni_2MnGa follows approximately from the DOS of Co_2MnGe if we fill in one more valence electron which places the Fermi energy E_F close to the high DOS peak of e_g symmetry. The system can lower its energy by a tetragonal distortion which shifts this peak to the energy range above E_F leading to the $L1_0$ structure. This explains the martensitic transformation in terms of the band Jahn–Teller effect [27]. For 7 M martensite ($c/a < 1$), the redistribution of states is discussed in [27] on the basis of spin-polarized neutron scattering consistently showing a depletion of the xy d-orbitals. Similar charge transfer effects are observed in our ab initio calculations of 5M martensite. Figure 2.2d shows that stability of the $L1_0$ structure is related to the tendency to reconstruct an energy gap in the minority-spin channel

using Ohmic contacts because of the so-called conductivity mismatch problem [29] meaning that TMR devices are left as the only tool to avoid this contact problem. On the other hand, a positive aspect is related to the observation that ferromagnetic Heusler alloys, which are half-metals or have at least a sufficiently high spin polarization, show large T_c values up to 10^3 K (for an overview of experimental and theoretical magnetic properties of various half-metallic Heusler compounds, see [30]).

Quite another promising feature is connected with the ferromagnetic full-Heusler alloys, which are of interest for magnetic shape-memory and magnetocaloric devices. These MSMA are no longer half-metals but metals. The half-metal Heusler systems are not suitable because they do not undergo a martensitic transformation (large magnetic field-induced (MFI) strain effects are only possible in the martensitic state due to the reorientation of the twinned martensitic variants under the action of the magnetic field). The metallic state of the MSMA implies that the Heusler systems do not any longer have the very high T_c values of the half-metallic systems. Hence, the major problem is to find new MSMA with higher Curie and martensite transformation temperatures. Some aspects of this problem are discussed in the following sections. Better functionality is achieved in a few cases for the nonstoichiometric ferromagnetic Heusler metals. Hence, in search for better MSMA, one has to systematically explore the physical properties of the Heusler alloys throughout the ternary phase diagram with temperature, pressure, and magnetic field as external parameters.

There are a few attempts to investigate shape-memory systems in a systematic fashion. The binary shape-memory systems may be classified by dividing them into three groups, A1, A2, and B2 type, according to the crystal structure of the parent phases. Although this classification appears arbitrary, it allows to put different binary alloy systems such as the thermoelastic martensites Ni–Ti and Ni–M and martensites with rubber-like behavior such as Cu–Au and In–Tl into certain categories. However, all schemes have their drawbacks. Elastic properties (pseudoelasticity, superelasticity, and rubber-like behavior [31]) may also not really serve as classification scheme since pseudoelasticity encompasses both superelasticity and rubber-like behavior; the latter occurs by the reversible movement of twin boundaries in the martensitic state; e.g., see [32]. Regarding ternary and quaternary MSMA, a possibility of characterization is to treat them as pseudobinary alloy systems such as pseudobinary Ni–Mn for the description of Ni–Mn–Z (Z = Ga, In, Sn, and Sb) and pseudobinary Fe–Ga for Fe–Co–Ga–Zn alloys. For example, the magnetoelastic properties of Fe–Ga solid solutions (with its GMR [33–36]) may also determine the magnetoelastic properties of the ternary or quaternary Fe–Ga-based systems.

Fe–Pd–Cu has recently been discussed by using combinatorial experimental methods [37]. The results show that, to some extent, one may treat Fe–Pd–Cu as a pseudobinary alloy system. A theoretical combinatorial study of full and inverse Heusler alloys was recently undertaken by Gilleßen et al. using first-principles calculations [38,39]. This investigation reveals that besides the full-Heusler systems X_2YZ the so-called inverse compounds (XY)YZ often have lower energy, which renders the investigation of the ternary Heusler systems more complex. Although concise rules for the formation of inverse Heusler structures are missing, one may say that the inverse structures are achieved by particular site preference of the

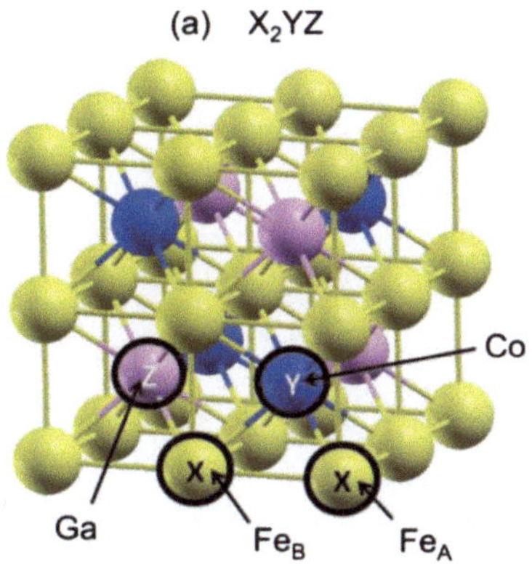

Fig. 2.3 (**a**) Conventional $L2_1$ ($Fm\bar{3}m$) Heusler structure of Fe_2CoGa where the iron atoms Fe_A and Fe_B (*yellow*) occupy the X-sites and Co (*blue*) and Ga (*violet*) occupy the Y- and Z-sites, respectively. (**b**) Inverse ($F\bar{4}3m$) Heusler structure of (FeCo)FeGa in which Co and Fe_B atoms have exchanged their atomic places (the corresponding iron atom is denoted now by Fe_C). The *arrow* indicates the interchange of X- and Y-atoms

elements that form the Heusler structure; for a discussion in case of Fe_2CoGa and (FeCo)FeGa, see [40]. The corresponding crystal structures are shown in Fig. 2.3.

In the following, we outline the main properties of Ni–Mn- and Co–Ni-based MSMA and discuss the martensitic tendencies in the ferromagnetic Heusler systems on the basis of a pseudobinary alloy description.

2.3 Phase Diagrams of $Ni_2Mn_{1+x}Z_{1-x}$ ($Z =$ Ga, In, Sn, Sb) Heusler alloys

The magnetic Heusler alloys Ni–Mn–Z have been extensively studied in recent years. Figure 2.4 shows the structural and magnetic phase diagram of $Ni_2Mn_{1+x}Z_{1-x}$ with Z = Ga, In, Sn, and Sb displaying the martensite start temperature denoted by $M_s(Z)$ as well as the Curie temperature $T_C^{A,M}(Z)$ for each alloy system as a function of the valence electron number per atom, e/a (constructed from the phase diagram of each alloy system [41]). The phase diagram shows that the martensite transformation temperature M_s, starting from the transformation of disordered fcc (A1) to chemically ordered $L1_0$ NiMn at high temperatures, scales linearly with the valence electron number per atom for all systems. Note that NiMn has a high Néel temperature well above 600 K; the magnetic moments of Mn are of the order of 4 μ_B and are perpendicular to the tetragonal axis and those of Ni are smaller than 0.6 μ_B (at 77 K). As M_s, also the Curie temperatures of $Ni_2Mn_{1+x}Z_{1-x}$ vary linearly with e/a in both austenite and martensite for all Z.

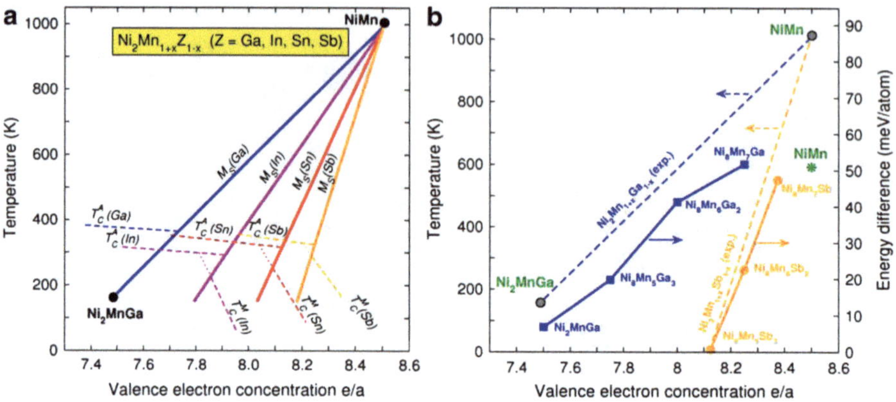

Fig. 2.4 (**a**) Martensite start temperature $M_s(Z)$ (*solid lines*); Curie temperatures $T_C^A(Z)$ and $T_C^M(Z)$ (*dashed lines*) of austenite and martensite phases of $Ni_2Mn_{1+x}Z_{1-x}$ with Z = Ga In, Sn, and Sb as a function of e/a. With increasing e/a each ternary alloy approaches the binary system NiMn. Among the stoichiometric compounds Ni_2MnGa ($e/a = 7.5$), Ni_2MnIn ($e/a = 7.5$), Ni_2MnSn ($e/a = 7.75$), and Ni_2MnSb ($e/a = 8$), only Ni_2MnGa transforms martensitically. The other materials transform at off-stoichiometric compositions. The decrease of T_C with increasing x is caused by the onset of antiferromagnetic correlations in the off-stoichiometric samples. $T_C^M(Z = \text{Ga})$ has not been measured. (**b**) Phase diagram obtained from energy differences, $E(c/a \le 1)$–$E(c/a > 1)$, of ab initio calculations (16-atom supercells and compositions as given in the figure) for the near-cubic structures ($c/a \le 1$) and the martensitic phases with $c/a > 1$. The temperature scale (*left*) is a translation of the energy scale (*right*). The slopes of the theoretical curves follow the experimental trends (*dashed curves* taken from the *left panel*) for both alloy systems. However, the estimated transformation temperature of NiMn from the CsCl (bcc) to tetragonal CuAu1 structure is lower than the experimental value

The increase of M_s with increasing e/a is related to the observation that already for small concentrations of x $Ni_2Mn_{1+x}Z_{1-x}$ develops antiferromagnetic correlations and that for $x = 1$ the system culminates in the antiferromagnet NiMn with its high martensite transformation temperature. However, the strict linearity of M_s is surprising and so far not explained. Also, with increasing x the ferromagnet becomes diluted which explains the decrease of T_C. This crossover from ferromagnetism to antiferromagnetism with increasing x is also visible when plotting the magnetic moments of $L2_1$ $Ni_2Mn_{1+x}Z_{1-x}$ alloys in a Slater–Pauling-like fashion (see Fig. 8 in [11]) showing the deviation from the Slater–Pauling curve for the off-stoichiometric systems when approaching the antiferromagnetic state of $L1_0$ NiMn.

We would like to emphasize that the martensite transformation from $L2_1$ to $L1_0$ structure in $Ni_2Mn_{1+x}Z_{1-x}$ and the crossover from ferromagnetic to antiferromagnetic order with increasing x are characteristic for all Ni–Mn-based Heusler alloys. The instability of the cubic phase with increasing concentration of Mn can again be related to the band Jahn–Teller effect [27], which leads to level splitting of the degenerate e_g states shifting E_F between the split levels. This stabilizes the martensite structure, which has nicely been demonstrated recently for $Ni_2Mn_{1+x}Sn_{1-x}$ [42, 43].

The increase of antiferromagnetic correlations with increasing x has a positive influence on the MCE since the resulting isothermal entropy change over the magnetostructural transition (near the crossing of M_s and T_C) in an external magnetic field and the resulting adiabatic temperature change may increase because of these correlations. On the other hand, the antiferromagnetic correlations have a negative influence on the MSME because they hinder to optimize both M_s and T_C simultaneously. There are a few successful attempts to increase M_s by adding a fourth element to the system, for example, by substituting 2% of In in $Ni_{50}Mn_{34}In_{16}$ ($e/a = 7.86$) by Ga, yielding isoelectronic $Ni_{50}Mn_{34}In_{14}Ga_2$ ($e/a = 7.86$), which pushes M_s toward the value of the corresponding $Ni_2Mn_{1+x}Ga_{1-x}$ composition, i.e., from 243 to 275 K in the present case [44]. However, this kind of tailoring of magnetic and structural transitions has not yet brought both transition temperatures to the desired operational regime of 100–200° above room temperature.

Since tetragonal distortions and atomic modulations of the structure help to enforce the antiferromagnetic correlations, a strong enough external magnetic field may shift M_s to lower temperatures as, e.g., in $Ni_{50}Mn_{34}In_{16}$ where a magnetic field of 5 T shifts M_s by 50 K [10]. This happens because in this case the austenitic phase shows less strong antiferromagnetic correlations and hence higher magnetization compared with the martensitic phase. However, in Ni–Mn–Ga, we find a positive shift of M_s. Note that the positive or negative change of M_s by an external magnetic field is important since it determines whether the MCE is of conventional or inverse nature, respectively. First-principles calculations are usually very successful in describing the magnetic properties of metallic systems. Figure 2.5 shows how the magnetic exchange parameters (calculated with the SPR-KKR code [20, 21]) change as a function of the distance between the atoms while passing from Ni_2MnGa to Ni_2 MnIn and to off-stoichiometric $Ni_{50}Mn_{34}In_{16}$ ($Ni_2Mn_{1.36}In_{0.64}$) and $Ni_{50}Mn_{32.5}Sb_{17.5}$ ($Ni_2Mn_{1.3}Sb_{0.7}$).

We have done extensive ab initio calculations of exchange parameters in order to follow the magnetic trends of the magnetic Heusler systems of type $Ni_2Mn_{1+x}Z_{1-x}$. We find that the antiferromagnetic correlations increase with increasing e/a ratio. Furthermore, the exchange parameters J_{ij} allow to determine the magnetic properties of the Heusler systems at finite temperatures with the help of the Heisenberg model and Monte Carlo simulations.

The Heisenberg model may be supplemented by terms that allow to simulate the austenite–martensite transformation in the spin model, for instance, by using the model of Cástan et al. [45]. The total Hamiltonian is then of the following form:

$$\mathcal{H} = \mathcal{H}_{\mathrm{m}} + \mathcal{H}_{\mathrm{lat}} + \mathcal{H}_{\mathrm{int}},$$

$$\mathcal{H}_{\mathrm{m}} = -\sum_{\langle ij \rangle} J_{\mathrm{m}}(i,j)\delta_{S_i,S_j} - g\mu_{\mathrm{B}} H_{\mathrm{ext}} \sum_{i} \delta_{S_i,S_g},$$

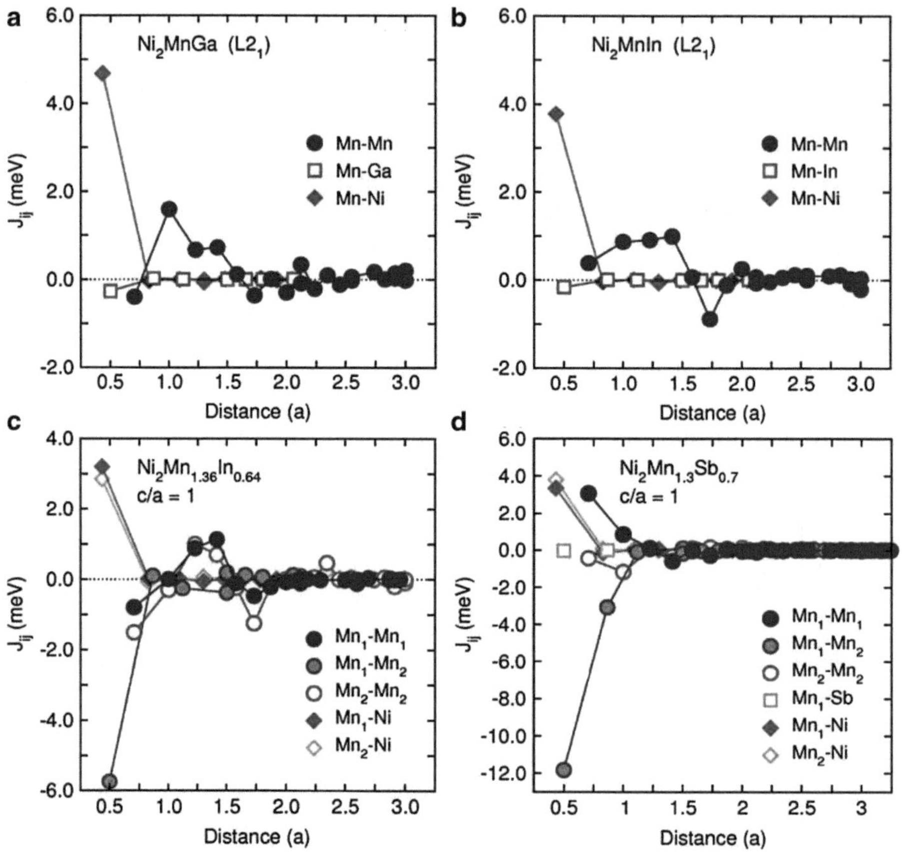

Fig. 2.5 Magnetic exchange interactions J_{ij} from ab initio calculations as a function of the distance between the atoms (in units of the lattice constant) in (**a**) cubic L2$_1$ phase of Ni_2MnGa ($e/a = 7.5$), (**b**) Ni_2MnIn ($e/a = 7.5$), (**c**) off-stoichiometric $Ni_{50}Mn_{34}In_{16}$ ($Ni_2Mn_{1.36}In_{0.64}$) with $e/a = 7.86$, and (**d**) $Ni_{50}Mn_{32.5}Sb_{17.5}$ ($Ni_2Mn_{1.3}Sb_{0.7}$) with $e/a = 8.15$ showing the development of antiferromagnetic correlations. These become stronger from Ga to In, Sn, and Sb. The index 1 refers to Mn atoms on the original Mn-sublattice, while 2 refers to Mn atoms on the In- and Sb-sublattices, respectively (Mn or Mn_1 is at the origin in each plot). Positive J_{ij} denote ferromagnetic coupling, while for $J_{ij} < 0$ the coupling is antiferromagnetic. With the onset of tetragonal distortion and transfromation to the martensite phase, the antiferromagnetic correlations further increase

$$\mathcal{H}_{\text{lat}} = -J \sum_{\langle ij \rangle} \sigma_i \sigma_j - K \sum_{\langle ij \rangle} \left(1 - \sigma_i^2\right) \left(1 - \sigma_j^2\right) - k_{\text{B}} T \ln(p) \sum_i \left(1 - \sigma_i^2\right)$$
$$- K_1 g \mu_{\text{B}} H_{\text{ext}} \sum_i \delta_{\sigma_i, \sigma_g} \sum_{\langle ij \rangle} \sigma_i \sigma_j ,$$

$$\mathcal{H}_{\text{int}} = 2U \sum_{\langle ij \rangle} \delta_{S_i, S_j} \left(\frac{1}{2} - \sigma_i^2\right) \left(\frac{1}{2} - \sigma_j^2\right) - \frac{1}{2} U \sum_{\langle ij \rangle} \delta_{S_i, S_j} .$$

The first term in $\mathcal{H}_m$ describes the interacting spins using a three-state and five-state Potts model for Ni and Mn, respectively, whereby the magnetic exchange constants are taken from first-principles calculations. The lattice Hamiltonian describes the austenite and martensite phases using the variables $\sigma = 0$ for the cubic structure and $\sigma = \pm 1$ for the tetragonally distorted structure, while $\mathcal{H}_{int}$ describes the coupling of the magnetic moments to the lattice. This allows to introduce two order parameters, ε (sum over the σ variables) and η (sum over the squared σ variables), where ε describes the degree of distortion and η can be used to describe the modulation of the atoms in the lattice planes. The external magnetic field is assumed to couple to one of the spin components labeled S_g ($S_g = 1$ for Ni and $S_g = 2$ for Mn) and to favor one of the martensite variants labeled σ_g ($\sigma_g = 1$); p is the degeneracy factor (i.e., number of martensite variants). For details, see [46,47].

Using the zero-temperature magnetic exchange interaction parameters as input, this Hamiltonian gives us the unique possibility to simulate finite-temperature magnetism and changes of magnetization with the onset of tetragonal distortion. In addition, we are able to simulate the influence of antiferromagnetic correlations and external magnetic field on magnetization and martensite transformation [45,46]. The corresponding results of Monte Carlo simulations of Ni–Mn–Z alloys agree well with experimental findings [46,47]. For example, the experimental magnetization curves of $Ni_2Mn_{1+x}Sb_{1-x}$ as a function of temperature and x shown in Fig. 2.6b can be described using the effective spin model, which allows to model the breakdown of magnetism in the martensite phase due to competing ferro- and antiferromagnetic interactions [46,48]. The simulations show that the decrease of magnetization with increasing x is an intrinsic effect and that it is due to the increasing importance of antiferromagnetic correlations in martensite with decreasing temperature.

Figure 2.7 shows the magnetic transition, structural transformation, and temperature variation of the order parameters m and ε of $Ni_{50}Mn_{34}In_{16}$ in zero magnetic

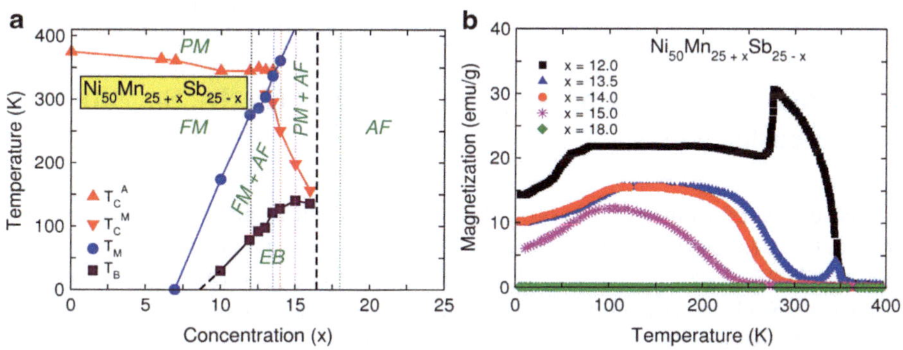

Fig. 2.6 (**a**) Experimental phase diagram and (**b**) variation in magnetization of $Ni_2Mn_{1+x}Sb_{1-x}$ for compositions along the *dotted lines* in (**a**). PM, FM, AF, and EB mark the paramagnetic, ferromagnetic, antiferromagnetic, and exchange bias regions, respectively. The sharp decrease of the magnetization when entering the martensitic phase is due to the onset of antiferromagnetic correlations [46]. Data adapted from [48]

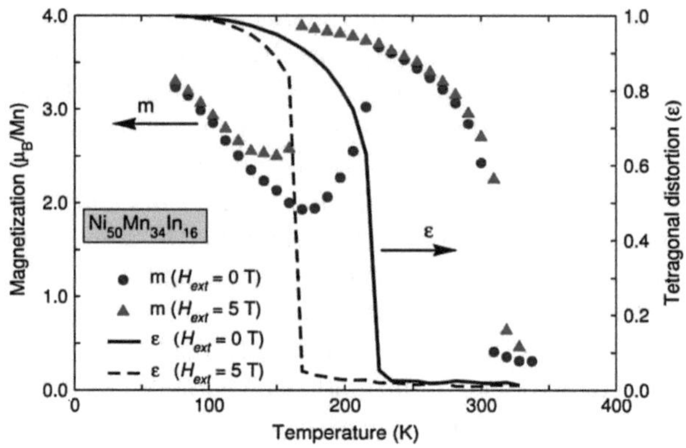

Fig. 2.7 Monte Carlo simulations of magnetization m and tetragonal distortion ε of $Ni_{50}Mn_{34}In_{16}$ ($Ni_2Mn_{1.36}In_{0.64}$, $e/a = 7.86$) in zero and 5 T magnetic field. The exchange parameters of martensite were determined for $c/a = 0.94$ by ab initio calculations. The simulations reproduce correctly the suppression of the martensite transformation temperature by 11 K/T as observed in experiment [10]

Table 2.2 Parameter values in meV (besides K_1 which is a dimensionless parameter) used in the Monte Carlo simulations of the reverse magnetic field effect in $Ni_{50}Mn_{34}In_{16}$ as shown in Fig. 2.7

c/a	$J^m_{Mn1-Mn1}$	$J^m_{Mn1-Mn2}$	$J^m_{Mn2-Mn2}$	$J^m_{Mn1-MnNi}$	$J^m_{Mn2-MnNi}$	J	K	U	K_1
1	−0.83	−5.74	−1.48	3.18	2.82	3.06	0.85	8.5	−5
0.94	0.258	−17.5	−0.82	4.59	3.02	3.06	0.85	12.3	−5

field and a magnetic field of 5 T obtained from the effective spin model by Monte Carlo simulations using the parameters listed in Table 2.2 (V.D. Buchelnikov, V.V. Sokolovskiy, P. Entel, unpublished simulation data). The simulations show the reverse MFI effect, i.e., the suppression of the martensite phase transformation by 50 K in a magnetic field of 5 T in agreement with experiment [10].

2.4 Phase Diagrams of $Ni_{2+x}Mn_{1-x}Z$ (Z = Ga, In, Sn, Sb) Heusler alloys

For Ni-excess concentrations in $Ni_{2+x}Mn_{1-x}Ga$, M_s and T_C approximately merge at $x = 0.18$ and stick together over an extended range of compositions up to $x = 0.27$, i.e., for $7.635 < e/a < 7.7025$ [8], which defines the region of magnetostructural transition. Such an extended range of magnetostructural transition is not observed for the alloy series $Ni_2Mn_{1+x}Z_{1-x}$ discussed above.

An interesting point is that for $Ni_{2+x}Mn_{1-x}Ga$, Ni_3Ga ($x = 1$) may be considered as the limiting binary system. The Ni − Ga alloy system is rather interesting in itself since it can exist in the B2 structure, Ni_3Ga_2 is hexagonal (B8), and Ni_3Ga possesses the $L1_2$ structure. Ni_3Al, Ni_3Ga, Ni_3In, and NiGa have been discussed in detail elsewhere [49]. While Ni_3Al is an itinerant ferromagnet, Ni_3Ga is considered to be an exchange enhanced paramagnet [50], disorder may induce antiferromagnetic correlation, and transition metal impurities such as Fe lead to giant magnetic moments of 60 μ_B per Fe atom. However, there is a mismatch between results of first-principles calculations and experiment regarding the evolution of magnetic moments in Ni_3Al, Ni_3Ga, and Ni_3In: While calculations report all three systems to have lower energy in the magnetic state, experiment shows only Ni_3Al to be magnetic with a small moment of 0.23 μ_B/cell [51]. For the other two binary systems Ni_3Sn and Ni_3Sb of the alloy series $Ni_{2+x}Mn_{1-x}Z$, one finds that Ni_3Sn is complex since it can exist in different forms with a $D0_3$ type of structure at high and a $D0_{19}$ type of structure at low temperatures and other structures such as $L1_2$, all lying close in energy [52]. The intermetallic compound Ni_3Sb with $D0_3$ structure has gained some interest because of the large diffusivity of Ni atoms [53]. The ternary phase diagram of Fe–Ni–Sb is discussed in [54].

So far, a systematic investigation of the $Ni_2Mn_{1+x}Z_{1-x}$ alloy series has not been undertaken for the $Ni_{2+x}Mn_{1-x}Z$ system. Therefore, we discuss here only the interesting aspects connected with the phase diagrams of $Ni_{2+x}Mn_{1-x}Ga$ [8, 55], which are shown in Figs. 2.8 and 2.9. Near-stoichiometric single crystals of Ni_2MnGa exhibit the most pronounced MFI strain effect of ~10% of all MSMA [5]. Merging of ferromagnetic transition and martensitic transformation around $x = 0.2$ can be described by strong magnetoelastic coupling. The latter range of compositions is as mentioned important for the MCE.

The phase diagrams in Figs. 2.8 and 2.9 are rather complex even near stoichiometry where the cubic $L2_1$ parent phase transforms to the premartensitic or

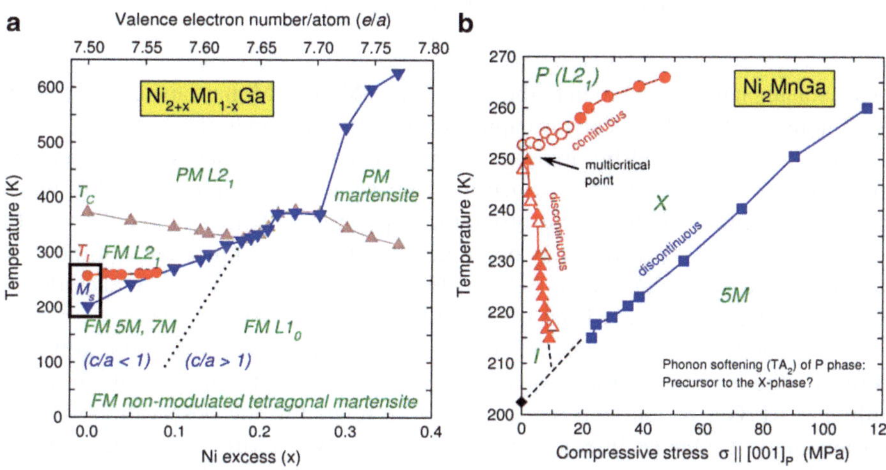

Fig. 2.8 (**a**) Ni-excess phase diagram of $Ni_{2+x}Mn_{1-x}Ga$. PM and FM denote paramagnetic and ferromagnetic phases in the cubic $L2_1$ and modulated 5M, 7M, and nonmodulated $L1_0$ structure, respectively. T_C and M_s are the Curie and martensite transformation temperatures [8, 55]. The *black box* encloses the premartensitic region. (**b**) Extending the *box* in the left phase diagram to the compressive stress axis, we obtain the phase diagram which displays the incommensurate X-phase. P and I denote the cubic parent and intermediate phases, respectively [56, 57]

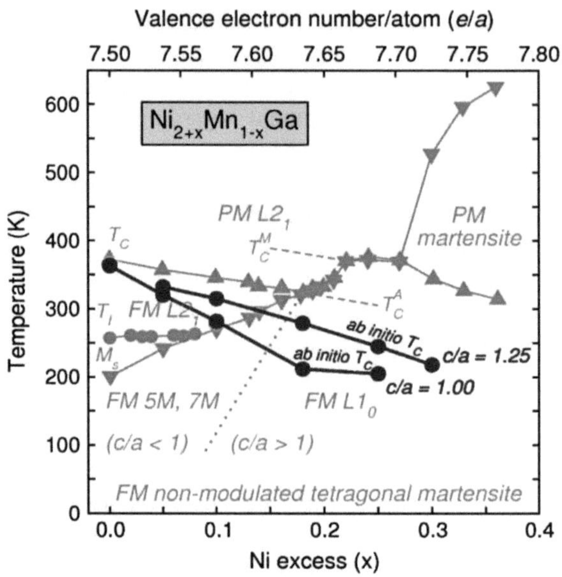

Fig. 2.9 The Ni-excess phase diagram of Fig. 2.8 with Curie temperatures of austenite ($c/a = 1$) and martensite ($c/a > 1$) phases as obtained from Monte Carlo simulations using the Heisenberg model and magnetic exchange parameters from ab initio calculations. The figure shows that the calculated Curie temperatures are too low because the KKR-CPA method [20] used in these calculations does not describe the influence of disorder effects accurately enough. For the stoichiometric compound, the calculated T_c agrees perfectly with the experimental value. Note also that the magnetic exchange in the Heusler alloys is indirect and mediated by the (s, p) conduction electrons so that the use of the Heisenberg model is rather crude

intermediate phase (I) and then to the modulated 5M and 7M martensites and finally to the nonmodulated $L1_0$ structure upon cooling. In addition to these phases, a new X-phase has been found by applying compressive stress, which seems to persist down to zero pressure [56, 57]. Since the I-phase disappears rather quickly when applying compressive stress, the pronounced phonon softening of the TA_2-[110] branch observed in Ni_2MnGa may be reinterpreted as being a precursor of the P–X transformation rather than of the P–I transformation. Further experimental evidence of this X-phase has recently been found by Karaca et al. [58].

As outlined before, it may be difficult to increase simultaneously both the Curie and the martensite transformations temperatures within the Ni–Mn–Z series with $Z =$ Al, Ga, In, Sn, and Sb. This may be related to the observation that the ferromagnetic exchange interaction, which primarily determines T_C, depends on the Ni–Mn interaction. Any dilution of the lattice leads to a decrease of T_C and simultaneously introduce, as shown above, antiferromagnetic Mn–Mn correlation, which further decreases T_C. On the other hand, M_s may be increased by alloying; however, this is limited by the magnetostructural transition (see Fig. 2.8). If one increases further the Ni-excess concentration, M_s can increase, but then the martensite transformation occurs in the paramagnetic and not in the ferromagnetic phase.

In conclusion, this means that in order to find better MSMA with considerably higher working temperature, one has to abandon NiMn-based Heusler alloys and address, for example, CoNi- or FeCo-based Heusler systems. The problem is then that the systems may no longer undergo a martensitic transformation. One remedy is to look for new systems with an e/a ratio, which pushes the systems away from half-metallicity, i.e., which shifts the Fermi level E_F away from the half-metallic gap region in the DOS, since E_F pinned to the pseudogap usually stabilizes the high-temperature cubic phase and prevents the system from undergoing a martensitic transformation. In the following, we briefly discuss the promising new system Co–Ni–(Ga, Zn) by outlining the differences to the Ni–Mn–Ga alloys.

2.5 Phase Diagrams of $Co_2Ni_{1-X}Z_{1+X}$ (Z = Ga, Zn) Heusler alloys

From the investigations in [38, 39], we know that when discussing new full-Heusler alloys X_2YZ, one has to consider the possibility that the inverse (XY)XZ Heusler systems may become important since for specific cases they have lower energies. This addresses a more general problem of the intermetallics pointing out that for each Heusler system a systematic investigation of the influence of atomic disorder on the physical properties is required. Here, we discuss a few of these aspects when considering the new CoNi-based Heusler systems.

Co–Ni–Al and Co–Ni–Ga MSM Heusler alloys and their corresponding phase diagrams have been investigated experimentally [59–61] and theoretically [62]. The ternary phase diagrams show that in a small stripe of the β-phase (disordered bcc-like B2) the systems indeed undergo a martensitic transformation to the tetragonal $L1_0$ phase. This small stripe is halfway in the middle of the ternary phase diagram, parallel to the binary Co–Ni axis on the Ni-rich side. Stoichiometric $L2_1$ Co_2NiGa ($e/a = 7.75$) lies just at the border of the martensitic phase. This system is also of interest because recent neutron scattering experiments on $Co_{48}Ni_{22}Ga_{30}$ ($Co_{1.92}N_{0.88}Ga_{1.2}$, $e/a = 7.42$) show that there is no pronounced softening of the TA_2-[110] phonon branch as in Ni_2MnGa [62].

The calculations have been carried out for the $L2_1$ and $L1_0$ structures of Co_2NiGa. In order to mimic the trends associated with nonstoichiometry, additional calculations have been done using a supercell with 16 atoms for $Co_8Ni_3Ga_5$, which corresponds to $Co_{50}Ni_{18.75}Ga_{31.25}$ ($e/a = 7.3125$). In all accompanying phonon calculations, we have employed the VASP [16] to obtain the forces and the dynamical matrices [62].

We first discuss the difference of the electronic structure of $L2_1$ and $L1_0$ Co_2NiGa compared with Ni_2MnGa. Figure 2.10 shows the resulting DOS of Co_2NiGa.

The DOS of cubic Co_2NiGa shows that E_F does not fall into a region which is close to the pseudogap region near E_F of Ni_2MnGa (see Fig. 2.2b). It also does not fall into a region of very high DOS in the spin-down channel. The stability

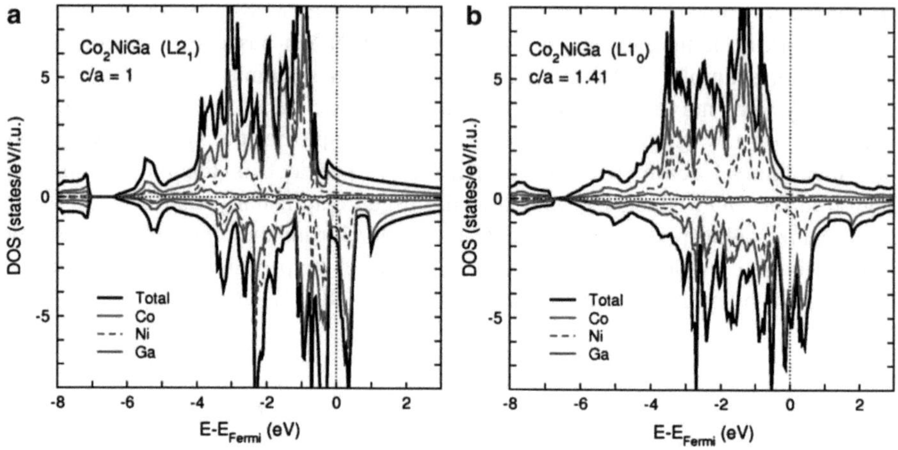

Fig. 2.10 Total and element-resolved DOS of (**a**) $L2_1$ and (**b**) $L1_0Co_2NiGa$ ($e/a = 7.75$). The Fermi energy E_F is now shifted to higher energies compared with Ni_2MnGa in Fig. 2.2b, d

of the martensite structure seems now to be connected with the position of E_F in a region of low DOS in the majority-spin channel and a low DOS of Ni-d states in the minority-spin channel (see Fig. 2.10b). Experiments showed that both Curie and martensite temperatures of CoNi-based MSMA may be slightly higher than room temperature. Monte Carlo simulations with ab initio exchange constants yield $T_c = 377$ K for Co_2NiGa to be compared with $T_c = 365$ K for Ni_2MnGa. Antiferromagnetic correlations seem to be absent in Co_2NiGa. The martensitic tendency is strong for Co_2NiGa judging from the energy variation of $E(c/a)$, which is shown in Fig. 2.11 in comparison to some other well-known and new Heusler systems.

The first-principles investigations of the Co–Ni–Ga systems have revealed further interesting and new features such as the absence of phonon softening in agreement with results of neutron scattering experiments on the near-stoichiometric $Co_{48}Ni_{22}Ga_{30}$ alloy with B2 structure and strong atomic disorder [62]. The calculated Fermi surface of the spin-down electrons of cubic $L2_1$ Co_2NiGa shows a completely different topology compared with the Fermi surface of Ni_2MnGa [62] (the Fermi surface of Ni_2MnGa has been discussed in [63]; see also Fig. 2.12). Here, we briefly summarize the results.

Figure 2.12 shows the individual Fermi surface sheets as well as the complete topology of the surfaces for spin-up and spin-down electrons of stoichiometric Ni_2MnGa.

Although the martensitic transformation in Ni_2MnGa occurs at 200 K, the magnetization is still close to the ground-state magnetization, which means that the strong nesting of the spin-down Fermi surface is only weakly reduced with increasing temperature. In addition, it has been argued that the weaker nesting of the majority-spin electrons should increase with decreasing magnetization [64].

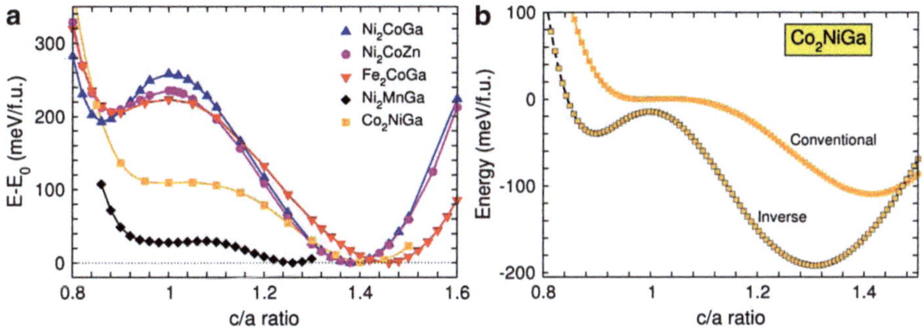

Fig. 2.11 (**a**) Variation of the total energy as a function of the tetragonal distortion c/a relative to the energy of the $L1_0$ structure for each (conventional) Heusler compound. (**b**) In some cases, the inverse Heusler structure becomes important because it has lower energy compared to the regular or conventional one. This is shown here for the case of Co_2NiGa (reference energy is the cubic state at $c/a = 1$ of the conventional Heusler structure). The $F\underline{4}3\,m$ inverse structure type of other Heusler compounds may [(FeCu)FeGa] or may not [(FeCo)FeGa] show tendencies for tetragonal distortions. This depends on the gain in binding energy due to nearest neighbor bond strengthening effects versus the cost of elastic energy when undergoing a martensitic transformation (see also [39]). For each displayed curve, the martensite transformation temperature can be estimated from the energy difference between the two energy minima

Although this enhancement of nesting of the majority-spin electrons has not been confirmed by our susceptibility calculations, it is obvious that stoichiometric Ni_2MnGa is an outstanding compound reflecting nesting behavior in both spin channels. All other alloys of the series $Ni_2Mn_{1+x}Z_{1-x}$ with $Z =$ In, Sn, and Sb undergo a martensite transformation at off-stoichiometric composition and do not show such a strong nesting behavior. Therefore, one has to search for another explanation different from nesting features why the other alloys undergo a structural transformation at off-stoichiometry. It may be that in spite of disorder the electron–phonon coupling in these alloys is very strong acting as a driving force for the structural transformation. The role of antiferromagnetic correlations in the structural transformation requires further investigations.

Comparing the Fermi surfaces of Co_2NiGa ($e/a = 7.75$) with those of Ni_2MnGa ($e/a = 7.5$), we find no Fermi surface nesting behavior for the spin-down electrons in Co_2NiGa (not shown here), but quite remarkably, the Fermi surfaces of spin-up electrons are nearly identical for both systems. Hence, there is weak nesting in case of Co_2NiGa associated with the spin-up electrons [62]. But this seems not to be sufficiently strong to drive the structural transformation or to cause pronounced softening of the transverse shear mode [62]. Co_2NiGa can also exist in the inverse Heusler structure (CoNi)NiGa which has a local energy minimum at $c/a = 0.86$, which is lower in energy compared with the $L2_1$ structure. However, since the phonon measurements were done on $Co_{48}Ni_{22}Ga_{30}$ with cubic bcc-like B2 structure (completely disordered) with the observation that for this particular alloy system no phonon softening is observed, we have concentrated the investigations on close in composition lying $L2_1$ Co_2NiGa and not on the inverse structure.

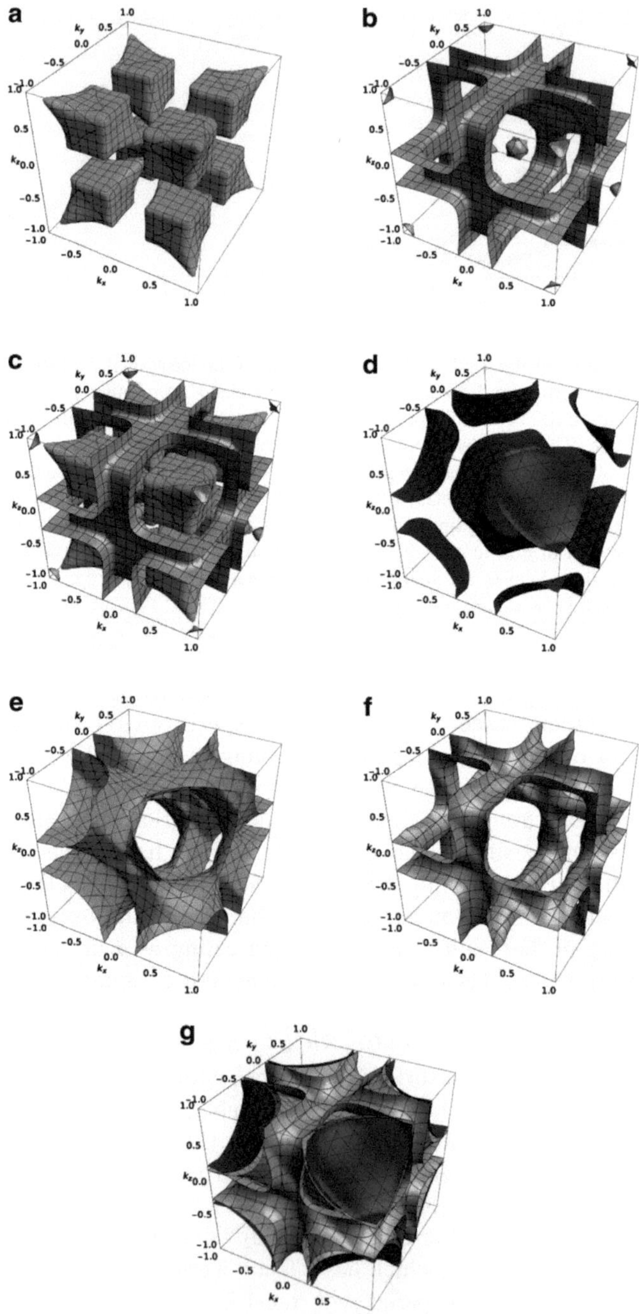

Fig. 2.12 Fermi surfaces of spin-up and spin-down electrons of Ni_2MnGa. (**a**)–(**b**) Fermi surface sheets of spin-down electrons. (**c**) Complete Fermi surface topology of spin-down electrons. (**d**)–(**f**) The three Fermi surface sheets of spin-up electrons. (**g**) Complete Fermi surface topology of spin-up electrons. There are contributions from different energy bands. Fermi surface nesting is most pronounced in (**b**) for the [110] direction, while it is less pronounced for the spin-up electrons in (**f**)

The absence of any acoustic phonon softening in Co_2NiGa may also be connected with the strong disorder in the bcc-like structure, which may hinder softening tendencies usually connected with the small elastic shear constant $C' = (C_{11}-C_{12})/2$. This, however, is rather intriguing since usually the existence of a martensitic instability is connected to softening of elastic constants, in particular to softening of C′ as well as to a large anisotropy constant $A = C_{44}/C'$ ($A = 1$ for an elastically isotropic material). C' may be considered as the resistance against a shearing stress across the {110} plane in [110] direction in a cubic crystal and can be related to the TA_2 mode softening. Elastic constants softening in martensite in transition-metal systems is discussed, for instance, in [65].

For Ni_2MnGa the room temperature elastic constants are listed in Table 2.3, which also contains the longitudinal constant $C_L = (C_{11} + C_{12} + 2C_{44})/2$ as well as the bulk modulus $B = (2C_{11} + C_{12})/3$. Room temperature is well above the premartensite phase transition at 265 K and is below the magnetic phase transition at 376 K of a perfect stoichiometric single crystal. The measurements [66, 67] confirm the softening when approaching the premartenistic phase transformation; C' softens by about 60% at the intermediate phase transformation [67]. The experimental values compare well with results of ab initio calculations of the elastic constants in austenite [68–70]. For completeness, some calculated values for martensite [68, 69] have been added to Table 2.3. The elastic constants measured by ultrasound compare well with the values computed from the slopes of corresponding phonon curves [71, 72]. The discrepancy in C' between different experiments results from the extreme sensitivity of C' on composition. Strong magnetoelastic coupling in Ni–Mn–Ga (and Ni–Mn–Z) can be observed when measuring the C_{ij} in an external magnetic field: They all increase with increasing magnetic field until saturation of the magnetic moments is reached.

Another interesting aspect is connected with C' since the martensite start temperature M_s does not only scale with e/a, but also seems to scale with C': M_s decreases linearly with increasing C' for $Ni_2Mn_{1+x}Ga_{1-x}$ [73].

For Co–Ni–Ga, systematic investigations of the elastic properties are not available. From the change of the DOS associated with the $L2_1$ to $L1_0$ transformation in

Table 2.3 Comparison of experimental elastic constants at 300 K of near-stoichiometric Ni_2MnGa in units of 10^{12} dyn/cm^2 (10^2GPa) with ab initio results

	Austenite					Martensite	
	Experiment		Ab initio			Ab initio	
C_{ij}	[66]	[67]	[68]	[69]	[70]	[68]	[69]
C_{11}	1.52	1.36		1.63	1.73		
C_{12}	1.43	0.92		1.51	1.41		
C_{44}	1.03	1.02		1.11	0.99		
C'	0.045	0.22	0.04	0.061	0.159	0.12 ($c/a = 0.94$) 0.30 ($c/a = 1.26$)	0.89 ($c/a = 1.26$)
C_L	2.50	2.22		2.68	2.566		
B	1.49	1.21		1.59	1.519		

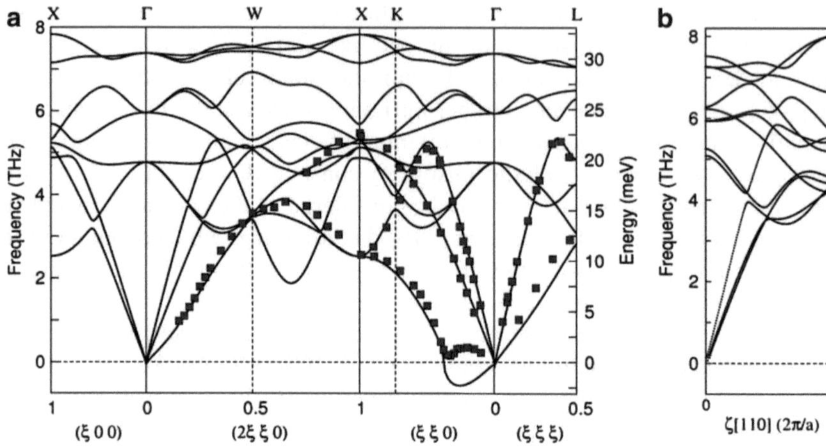

Fig. 2.13 (**a**) Ab initio phonon dispersion curves of $L2_1$ Ni_2MnGa [63] (M. Siewert, unpublished data) compared with the experimental results of Zheludev et al. [74] (*filled squares*). (**b**) Calculated [110] dispersions of $c/a = 1.25$ martensite for which the TA_2 branch is stable [71]

Fig. 2.10, we are tempted to attribute the stabilization of martensite to the position of E_F in the weak pseudogap of Co-d states in the majority-spin channel and to the more pronounced pseudogap of Ni-d states in the minority-spin channel. In addition, the electron–phonon coupling may be strong in these materials and may finally drive the formation of martensite.

We would like to add a few remarks regarding the importance of phonon measurements of MSMA. Phonons of MSMA (the dispersions of prototype Ni_2MnGa are shown in Fig. 2.13) may be important in connection with the self-accommodation of nanotwins down to the atomic scale in the modulated martensites [75] since any twin movement initiated by an external magnetic field may be supported by the large displacements of atoms associated with the softening of tetragonal shear modes.

The phonon self-energy, which is the source of the renormalization of the phonon frequencies, depends on the electron–phonon coupling strength and DOS of spin-polarized electron–hole excitations. In the second order of a fully renormalized electron–phonon matrix element (calculated from first principles using Kohn–Sham quasiparticles and eigenfunctions), its corresponding contribution to the dynamical matrix is given by [76, 77]

$$\mathcal{D}^{(2)}_{\nu\mu}(\mathbf{q}) = -\sum_{\mathbf{k}nn'} g^{\nu *}_{\mathbf{k}n,\,\mathbf{k+q}n'}\, g^{\mu}_{\mathbf{k+q}n',\,\mathbf{k}n}\, \frac{f_{\mathbf{k}n} - f_{\mathbf{k+q}n'}}{\varepsilon_{\mathbf{k}n} - \varepsilon_{\mathbf{k+q}n'}},$$

where f is the Fermi distribution function, $\varepsilon_{\mathbf{k}n}$ is the quasiparticle energy with momentum $\mathbf{k}$ and band index n, g is the electron–phonon coupling strength, and ν, μ are the Cartesian indices.

The second factor on the right-hand side of this relation is the Lindhard function, which can display large peaks in reciprocal space in case of Fermi surface nesting leading to a dip in the phonon dispersion (Kohn anomaly) and enhancing its linewidth, and finally, may lead to a structural instability. Then, the band Jahn–Teller effect associated with degenerated Ni-d states and the Kohn anomaly can be considered as the main driving force for the martensite transformation in case of Ni_2MnGa. The electron–phonon coupling function does usually not depend strongly on the phonon wavevector, but may lead to softening effects away from the nesting vector and may also lead to lattice instabilities. This shows that the calculation of the electron–phonon coupling strength and $D^{(2)}$ and their analysis are of primary importance for a complete understanding of the magnetic shape-memory Heusler alloys.

Another aspect of phonon renormalization is connected with the increase of phonon–phonon scattering with an increasing temperature, which increases the entropy arising from the lattice vibrations that may stabilize the cubic bcc-like $L2_1$ structure above the martensite transformation temperature [78]. The calculations in [78] show that lattice vibrations and magnestism are both important when discussing the sequence of structural transformations in Ni_2MnGa.

In the ab initio calculations in [78] the finite-temperature formalism relies on the assumption that one may approximately write the free energy of electrons, phonons, and magnons in an additive manner. At each temperature, the phonon dispersions are calculated in the quasiharmonic approximation for a series of volumes. The stationary condition of the free energy with respect to the volume V allows then to calculate from first principles the lattice constant a_0, thermal expansion coefficient α, specific heat C_V, phonon frequency $\omega_{\mathbf{q}\lambda}$, and the anharmonicity $\gamma_{\mathbf{q}\lambda}$ of the lattice vibrations from the mode Grüneisen parameters by using

$$\mathcal{F}(V,T) = \mathcal{F}_{\rm el}(V,T) + \mathcal{F}_{\rm ph}(V,T) + \mathcal{F}(V,T),$$

$$\mathcal{F}_{\rm ph}(V,T) = \frac{1}{\beta}\sum_{\mathbf{q}\lambda} \ln\{2\sinh(\beta\hbar\omega_{\mathbf{q}\lambda}(V)/2)\},$$

$$\frac{\partial\mathcal{F}(V,T)}{\partial V} = 0 \Rightarrow a_0(T), \alpha(T), C_V(T), \omega_{\mathbf{q}\lambda}(T),$$

$$\gamma_{\mathbf{q}\lambda} = -\frac{V}{\omega_{\mathbf{q}\lambda}(V)}\frac{\partial\omega_{\mathbf{q}\lambda}(V)}{\partial V}.$$

The electronic part of the free energy, $\mathcal{F}_{\rm el}$, can be taken from the electronic structure calculations at volume V using Fermi statistics for the occupation of the one-particle states. $\mathcal{F}_{\rm m}$ is identical in form to $\mathcal{F}_{\rm ph}$ if we replace the phonon energies by the magnon energies. The latter may approximately be obtained from fixed spin moment calculation at fixed magnetization m and volume V and Fourier transform of the magnetic exchange coupling constants J_{ij} (or by solving the Bethe–Salpeter equation for the transverse spin–spin correlation function). So far, this method has successfully been applied to a series of nonmagnetic transition

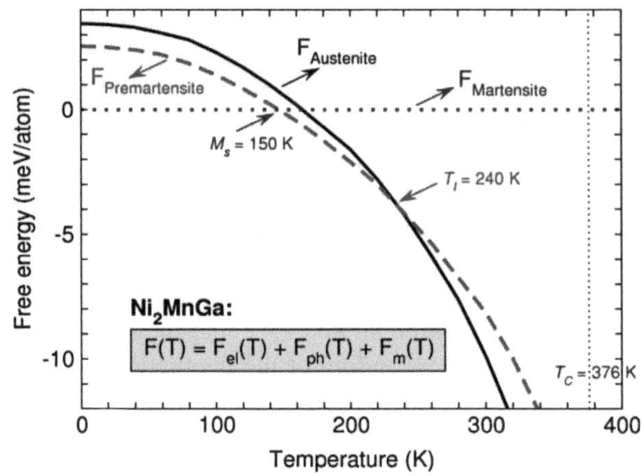

Fig. 2.14 Calculated free energies of Ni_2MnGa relative to martensite including electronic, vibrational, and magnetic contributions [78]. Premartensite is stable below $T_I = 240$ K, while martensite is stable below $M_s = 150$ K to be compared with the experimental values $T_I = 260$ K and $M_s = 200$ K of Fig. 2.8. Modulated phases were not considered in the calculations, which may in part explain the too low M_s (see [78])

metals [79, 80]. Although magnetic materials are still a challenge, application of this method to Ni_2MnGa with a rather crude approximation for $\mathcal{F}_m$ yields the correct order regarding the stability range of austenite, premartensite, and martensite phases. The corresponding phase diagram is shown in Fig. 2.14.

Concerning the calculation of the phonon part of the free energy, a slightly different procedure was proposed in [81, 82] where the phonon Hamiltonian with anharmonic terms is cast into a mean-field Hamiltonian of harmonic or quasiharmonic nature. *Intrinsic* temperature dependence enters the mean-field Hamiltonian due to the expectation value of the square of operators of the canonical coordinates of the underlying harmonic Hamiltonian:

$$\mathcal{H}^{\mathrm{MF}} = \frac{1}{2}\sum_{\mathrm{k,s}} \left(\mathcal{P}_{\mathrm{k,s}}^2 + \bar{\omega}_{\mathrm{k,s}}^2 \mathcal{D}_{\mathrm{k,s}}^2\right),$$

$$\bar{\omega}_{\mathrm{k,s}}^2 = \omega_{\mathrm{k,s}}^2 \left(1 + \frac{1}{2}\sum_{\mathrm{k_1,k_2}}\sum_{s_1,s_2} \mathcal{A}(\mathrm{k,k_1,k_2,s,s_1,s_2}) \frac{\mathcal{D}_{\mathrm{k1},s_2}\mathcal{D}_{\mathrm{k2},s_2}}{\mathcal{D}_{\mathrm{k,s}}\omega_{\mathrm{k,s}}^2} + \dots\right),$$

$$\mathcal{D}_{\mathrm{ks}} \approx \pm\sqrt{\left\langle \widehat{\mathcal{D}}_{\mathrm{ks}}^{\dagger}\widehat{\mathcal{D}}_{\mathrm{ks}}\right\rangle} = \pm\sqrt{\frac{\hbar}{\omega_{\mathrm{ks}}}\left[\frac{1}{2} + \bar{n}\left(\frac{\hbar\omega_{\mathrm{ks}}}{k_B T}\right)\right]}.$$

Here, A is the third-order anharmonic contribution and n is the Bose distribution function. Using as initial guess the harmonic frequencies from ab initio calculations, a self-consistent ab initio lattice dynamical (SCAILD) procedure can be used to calculate the renormalized frequencies, free energy, entropy, and thermal expansion [81, 82]. This method is now applied to Heusler alloys and is under current investigation in our group. Electronic as well as magnetic contributions to the free energy from magnons and magnon–magnon interactions must be added as before.

As mentioned, the first method outlined above has been applied to Ni_2MnGa yielding renormalized stable dispersion relations at finite temperatures [78]. However, the calculations have not yet allowed to settle the question whether the phonon softening observed for Ni–Mn–Ga must be considered as a precursor of intermediate martensite, the X-phase, or a precursor of martensite itself. In this context, one should remark that for appropriate off-stoichiometric compositions, the premartensitic phase in Ni–Mn–Ga can be suppressed, but the phonon softening of the austenite phase may still prevail as before. The phonon softening shown in Fig. 2.13 for the stoichiometric sample [71, 74] (M. Siewert, unpublished data) does not change too much when considering off-stoichiometry or disorder: The measurements of the phonon spectrum of $Ni_{49}Mn_{32}Ga_{19}$ ($Ni_{1.96}Mn_{1.28}Ga_{0.76}$) with $e/a = 7.71$ by Ener et al. yield very similar results compared to the stoichiometric case (S. Ener, J. Neuhaus, W. Petry, Unpublished data). When extending the phonon calculations to finite temperatures, the imaginary frequencies near the Γ-point become real due to the influence of phonon–phonon interactions that enhance the internal pressure and the entropy, although weak softening is still present at room temperature (M. Siewert, unpublished data).

The stabilization of the TA_2-[110] phonon dispersion of 5M Ni_2MnGa at finite T is a rather subtle problem since here, in addition to the phonon–phonon interaction, magnetization and the modulation of the atomic positions might be crucial for the stabilization of the modulated structure; for example, see the discussion in [63, 83].

2.6 Conclusions and Future Aspects of Magnetic Heusler alloys

Ni–Mn–Z Heusler alloys with Z = Ga, In, Sn, and Sb have a limited range of applicability because of the still too low operation temperatures when used in actual devices. New MSMA based on the full-Heusler crystal structure meet the difficulty that they should show comparable MFI effects that originate from large magneto-crystalline anisotropy and high mobility of twin boundaries. Theoretical tools to discuss twin boundary motion can be found in the review by Gruner et al. [83].

Finally, a critical remark concerns the magneto-crystalline anisotropy, which one may expect for the new Heusler alloys not containing manganese. Let us briefly consider the Ni–Mn–Ga alloys for which the magnetic anisotropy energy has been measured [84] and calculated [85, 86]. The anisotropy energy is small or vanishes in the $L2_1$ phase of Ni–Mn–Ga and increases to large values as a function of increasing tetragonal distortion in both experiment and theory. Ab initio results obtained with the full-potential code (FPLO [19]) are shown in Fig. 2.15; for technical details of

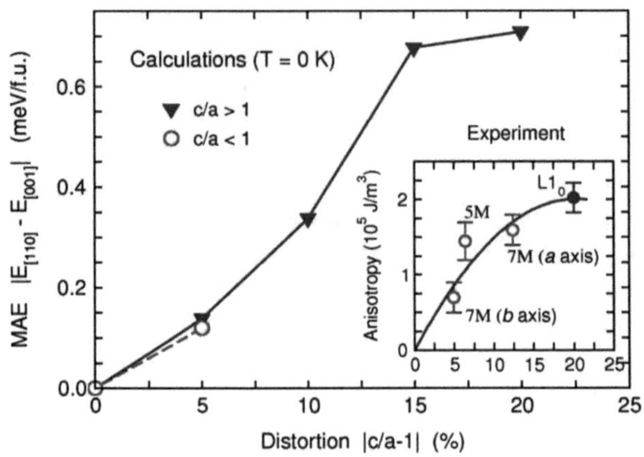

Fig. 2.15 Magneto-crystalline anisotropy energy per formula unit as a function of tetragonal distortion in Ni_2MnGa [86]. The *inset* shows the experimental results of Sozinov et al. [84]

Table 2.4 Results of ab initio calculations of a few promising new Heusler systems with conventional crystal structure. For each system, we have listed the e/a ratio, the lattice constant used in the calculations, the Curie temperature obtained from Monte Carlo simulation using ab initio exchange parameters, the energy difference between $L2_1$ and $L1_0$ phases, $\Delta E_{c/a} = E(L2_1) - E(L1_0)$, for Ni_2MnGa and Co_2NiGa, respectively, and between local minima at $c/a < 1$ and $c/a > 1$ for the other systems, $\Delta E_{c/a} = E(c/a < 1) - E(L1_0)$, as well as the mixing energy for the $L1_0$ $(c/a > 1)$ phases. The mixing energy is the energy difference between the energy of the alloy system and its corresponding constituents at their respective equilibrium volumes. Negative mixing energy points toward the formation of a stable alloy. The energies in the table are given in meV/atom

System	*e/a*	a_0(Å)	T_c(K)	$\Delta E_{c/a}$(meV/a)	$E_{mix}(L1_0)$(meV/a)
Ni_2MnGa	7.5	5.807	365	6	−1257
Co_2NiGa	7.75	5.695	377	28	−569
Fe_2CoGa	7.0	5.774	770	52	−463
$Fe_2CoGa_{0.75}Zn_{0.25}$	6.9375	5.779	802	50	−333
$Fe_2CoZn_{0.75}Ga_{0.25}$	6.8125	5.780	896	45	−71
Fe_2CoZn	6.75	5.782	925	44	51

calculation, we refer to [85, 86]. The [001] quantization axis is the easy axis for $c/a < 1$; it becomes the hard axis for $c/a > 1$. With increasing c/a ratio, the Mn contribution to the magnetic anisotropy energy starts to dominate. So, it remains an open question whether the new manganese-free magnetic Heusler alloys may develop comparable anisotropy energy with the onset of tetragonal distortion. This is under present study.

In Table 2.4, we have listed computational data of Ni_2MnGa, Co_2NiGa, and a few more Heusler systems of which the conventional and inverse type of structures (see [39]) must still be investigated in more detail (the lattice structure of the inverse type of Heusler compounds may be inferred from Fig. 2.1c by distributing the atoms on the lattice according to (XY)XZ; cf. Fig. 2.3).

Preliminary ab initio and Monte Carlo calculations of the new Heusler systems look promising [87] and the results listed in the table show that apart from Fe_2CoZn all other compounds have negative mixing energies, i.e., are stable against decomposition. The Curie temperatures of the Zn-containing Heusler alloys can be quite high and from the energy differences listed in the last but one column one may also expect elevated martensite transformation temperatures compared with $M\text{s}{\sim}200\,\text{K}$ of Ni_2MnGa. Since some of the stoichiometric systems such as Fe_2CoZn show signs of spinodal decomposition, future promising systems may be *nonstoichiometric* and *Ni-free* MSMA where spinodal tendencies can be suppressed by disorder effects or by adding a quarternary element. This emphasizes the important role of atomic disorder effects on the structural and magnetic properties of the Heusler systems, which is left as an outstanding theoretical problem. Preliminary results of phonon calculations for such complex alloy systems reveal that softening effects may exist comparable to Ni_2MnGa, although the softening of shear modes is not a necessary prerequisite for martenistic behavior as the case of Co_2NiGa proves [62].

Further pseudobinary systems such as $Fe_2Co_{1-x}Ga_{1+x}$ or $Fe_2Co_{1-x}Fe_xGa$ may be conceived; the latter for $x = 1$ leads to $Fe_{75}Ga_{25}$, which is close in composition to Galfenol which has a huge magnetostrictive coefficient; see, for example, [88] and references therein. As mentioned before, $Fe_{75}Ga_{25}$ possesses a rather complicated structural phase diagram with $D0_3$, B2-like, and $L1_2$ phases. The energetic scenario of very close in energy lying structures is rather intriguing since one may search for an appropriate third element which can stabilize a bcc-like Heusler alloy. This would bridge the gap between ternary intermetallics which exhibit the MSME and those binary intermetallics which show a huge magnetostrictive effect. First-principles calculations underline that $Fe_2Co_{1-x}(Ga, Zn)_{1+x}$ (with $e/a = 7$ for Fe_2CoGa) may also turn out to be a new, very promising MSMA system (cf. Table 2.4).

Acknowledgments P. Entel, M.E. Gruner A. Dannenberg, and M. Siewert acknowledge financial support by the DFG Priority Programme 1239 on *Magnetic Shape Memory Alloys*. Stimulating discussions with Prof. M. Acet, L. Mañosa and A. Planes were very helpful.

References

1. A. Planes, L. Mañosa, A. Saxena (eds.), *Magnetism and Structure in Functional Materials*, Springer Series in Materials Science **79** (Springer, Berlin, 2005)
2. J. Kübler, *Theory of Itinerant Electron Magnetism*, International Series of Monographs on Physics **106** (Clarendon Press, Oxford, 2000)
3. I. Galanakis, P.H. Dederichs (eds.), *Half-metallic Alloys – Fundamentals and Applications*, *Lecture Notes in Physics* **676** (Springer, Berlin, 2005)
4. R.A. de Groot, F.M. Mueller, P.G. van Engen, K.H.J. Buschow, New class of materials: Half-metallic ferromagnets. Phys. Rev. Lett. **50**, 2024 (1983)
5. A. Sozinov, A.A. Likhachev, N. Lanska, K. Ullakko, Giant magnetic-field- induced strain in NiMnGa seven-layered martensitic phase. Appl. Phys. Lett. **80**, 1746 (2002)

6. M.A. Marioni, R.C. O'Handley, S.M. Allen, S.R. Hall, D.I. Paul, M.L. Richard, J. Feuchtwanger, B.W. Peterson, J.M. Chambers, R. Techapiesancharoenkij, The ferromagnetic shape-memory effect in Ni–Mn–Ga. J. Magn. Magn. Mater. **35**, 290–291 (2005)
7. K. Ullakko, J.K. Huang, C. Kantner, R.C. O'Handley, V.V. Kokorin, Large magnetic-field-induced strains in Ni_2MnGa single crystals. Appl. Phys. Lett. **69** 1966 (1996)
8. V.V. Khovaylo, V.D. Buchelnikov, R. Kainuma, V.V. Koledov, M. Ohtsuka, V.G. Shavrov, T. Takagi, S.V. Taskaev, A.N. Vasiliev, Phase transitions in $Ni_{2+x}Mn_{1-x}Ga$ with a high Ni excess. Phys. Rev. B **72**, 224408 (2005)
9. M. Pasquale, C.P. Sasso, L.H. Lewis, L. Giudici, T. Lograsso, D. Schlagel, Magnetostructural transition and magnetocaloric effect in $Ni_{55}Mn_{20}Ga_{25}$ single crystals. Phys. Rev. B **72**, 094435 (2005)
10. T. Krenke, E. Duman, M. Acet, E.F. Wassermann, X. Moya, L. Mañosa, A. Planes, E. Suard, B. Ouladdiaf, Magnetic superelasticity and inverse magnetocaloric effect in Ni-Mn-In. Phys. Rev. B **75**, 104414 (2007)
11. A. Planes, L. Mañosa, M. Acet, Magnetocaloric effect and its relation to shape- memory properties in ferromagnetic Heusler alloys. J. Phys.: Condens. Matter **21**, 233201 (2009)
12. E. Karaca, I. Karaman, B. Basaran, Y. Ren, Y.I. Chumlyakov, H.J. Maier, Magnetic field-induced phase transformation in NiMnCoIn magnetic shape-memory alloys – A new actuation mechanism with large work output. Adv. Funct. Mater. **19**, 983 (2009)
13. K. Inomata, S. Okamura, A. Miyazaki, M. Kikuchi, N. Tezuka, M. Wojcik, E. Jedryka, Structural and magnetic properties and tunnel magnetoresistance for Co_2(Cr,Fe)Al and Co_2FeSi full-Heusler alloys. J. Phys. D: Appl. Phys. **39**, 816 (2006)
14. G.H. Fecher, C. Felser, Substituting the main group element in cobalt-iron based Heusler alloys: $Co_2FeAl_{1-x}Si_x$. J. Phys. D: Appl. Phys. **40**, 1582 (2007)
15. M.I. Katsnelson, V.Yu. Irkhin, L. Chioncel, A.I. Lichtenstein, R.A. de Groot, Half-metallic ferromagnets: From band structure to many-body effects. Rev. Mod. Phys. **80**, 315 (2008)
16. G. Kresse, D. Joubert, From ultrasoft pseudopotentials to the projector augmented- wave method. Phys. Rev. B **59**, 1758 (1999)
17. J.P. Perdew, K. Burke, M. Ernzerhof, Generalized gradient approximation made simple. Phys. Rev. Lett. **77**, 3865 (1996)
18. P.E. Blöchl, Projector augmented-wave method. Phys. Rev. B **50**, 17953 (1994)
19. K. Koepernik, H. Eschrig, Full-potential nonorthogonal local-orbital minimum- basis band-structure scheme. Phys. Rev. B **59**, 1743 (1999); http://www.fplo.de
20. H. Ebert, Fully relativistic band structure calculations for magnetic solids – Formalism and application, in: Electronic Structure and Properties of Solids, Lecture Notes in Physics, Vol. 535, ed. by H. Dreyssé (Springer, Berlin, 1999), p. 191
21. H. Ebert, Magneto-optical effects in transition metal systems, Rep. Prog. Phys. **59**, 1665 (1996)
22. B.G. Liu, Robust half-metallic ferromagnetism in zinc-blende CrS. Phys. Rev. B **67**, 172411 (2003)
23. I. Galanakis, P.H. Dederichs, N. Papanikolaou, Slater-Pauling behavior and origin of the half-metallicity of the full-Heusler alloys. Phys. Rev. B **66**, 174429 (2002)
24. I. Galanakis, P.H. Dederichs, N. Papanikolaou, Origin and properties of the gap in the half-ferromagnetic Heusler alloys. Phys. Rev. B **66**, 134428 (2002)
25. P. Entel, M.E. Gruner, A. Dannenberg, M. Siewert, S.K. Nayak, H.C. Herper, V.D. Buchelnikov, Fundamental aspects of magnetic shape memory alloys: Insights From *ab initio* and Monte Carlo studies. Mater. Sci. Forum **635**, 3 (2010)
26. S. Picozzi, A. Continenza, A.J. Freeman, Co_2MnX (X=Si, Ge, Sn) Heusler compounds: An ab initio study of their structural, electronic, and magnetic properties at zero and elevated pressure. Phys. Rev. B **66**, 094421 (2002)
27. P.J. Brown, A.Y. Bargawi, J. Crangle, K.U. Neumann, K.R.A. Ziebeck, Direct observation of a band Jahn-Teller effect in the martensitic phase transition of Ni_2MnGa. J. Phys.: Condens. Matter **11**, 4715 (1999)

28. B. Hülsen, M. Scheffler, P. Kratzer, Structural stability and magnetic and electronic properties of $Co_2MnSi(001)/MgO$ heterostructures: A density-functional theory study. Phys. Rev. Lett. **103**, 046802 (2009)
29. G. Schmidt, D. Ferrand, L.W. Molenkamp, A.T. Filip, B.J. van Wees, Fundamental obstacle for electrical spin injection from a ferromagnetic metal into a diffusive semiconductor. Phys. Rev. B **62**, R4790 (2000)
30. J. Thoene, S. Chadov, G. Fecher, C. Felser, J. Kübler, Exchange energies, Curie temperatures and magnons in Heusler compounds. J. Phys. D: Appl. Phys. **42,** 084013 (2009)
31. X. Ren, K. Otsuka, Origin of rubber-like behavior in metal alloys. Nature **389**, 579 (1997)
32. K. Otsuka, C.M. Wayman (eds.), *Shape Memory Materials* (Cambridge University Press, Cambridge 1998)
33. M. Wuttig, L. Dai, J. Cullen, Elasticity and magnetoelasticity of Fe-Ga solid solutions. Appl. Phys. Lett. **80**, 1135 (2002)
34. Q. Xing, T.A. Lograsso, Magnetic domains in magnetostrictive Fe-Ga alloys. Appl. Phys. Lett. **93**, 182501 (2008)
35. A.G. Khachaturyan, D. Viehland, Structurally heterogeneous model of extrinsic magnetostriction for Fe-Ga and similar magnetic alloys: Part I. Decomposition and confined displacive transfromation. Metal. Mater. Trans A **38**, 2308 (2007)
36. A.G. Khachaturyan, D. Viehland, Structurally heterogeneous model of extrinsic magnetostriction for Fe-Ga and similar magnetic alloys: Part II. Giant magnetostriction and elastic softening. Metal. Mater. Trans A **38**, 2317 (2007)
37. S. Hamann, M.E. Gruner, S. Irsen, J. Buschbeck, C. Bechtold, I. Kock, S.G. Mayr, A. Savan, S. Thienhaus, E. Quandt, S. Fähler, P. Entel, A. Ludwig, The ferromagnetic shape memory system Fe-Pd-Cu. Acta Mater. **58**, 5949–5961 (2010)
38. M. Gilleßen, R. Dronskowski, A combinatorial study of full Heusler alloys by first- principles computational methods. J. Comp. Chem. **30**, 1290 (2008)
39. M. Gilleßen, R. Dronskowski, A combinatorial study of inverse Heusler alloys by first-principles computational methods. J. Comp. Chem. **31**, 612 (2008)
40. N.K. Jaggi, K.R.P.M. Rao, A.K. Grover, L.C. Gupta, R. Vijayaraghavan, Le Dang Khoi: Mossbauer and NMR study of the site preference and local environment effects in Co_2Fe Ga & FeCoGa. Hyperfine Int. **4**, 402 (1978)
41. S. Aksoy, M. Acet, E.F. Wassermann, T. Krenke, X. Moya, L. Mañosa, A. Planes, P. Deen, Structural properties and magnetic transitions in martensitic Ni-Mn-Sb alloys. Phil. Mag. **89**, 2093 (2009)
42. M. Ye, A. Kimura, Y. Miura, M. Shirai, Y.T. Cui, K. Shimada, H. Nanatame, M. Taniguchi, S. Ueda, K. Kobayashi, R. Kainuma, T. Shishido, K. Fukushima, T. Kanomata, Role of electronic structure in the martensitic phase transition of $Ni_2Mn_{1+x}Sn_{1-x}$ studied by hard-x-ray photoelectron spectroscopy and *ab initio* calculation. Phys. Rev. Lett. **104**, 176401 (2010)
43. A. Planes, Controlling the martensitic transition in Heusler shape-memory materials. Physics **3**, 36 (2010)
44. S. Aksoy, T. Krenke, M. Acet, E.F. Wassermann, X. Moya, L. Mañosa, A. Planes, Tailoring magnetic and magnetocaloric properties of martensitic transitions in ferromagnetic Heusler alloys. Appl. Phys. Lett. **91**, 241916 (2007)
45. T. Castán, E. Vives, P.A. Lindgård, Modeling premartensitic effects in Ni_2MnGa: A mean-field and Monte Carlo simulation study. Phys. Rev. B **60**, 7071 (1999)
46. V.D. Buchelnikov, P. Entel, S.V. Taskaev, V.V. Sokolovskiy, A. Hucht, M. Ogura, H. Akai, M.E. Gruner, S.K. Nayak, Monte Carlo study of the influence of antiferromagnetic exchange interactions on the phase transitions of ferromagnetic Ni-Mn-X alloys (X=In, Sn, Sb). Phys. Rev. B **78**, 184427 (2008)
47. V.D. Buchelnikov, V.V. Sokolovskiy, H.C. Herper, H. Ebert, M.E. Gruner, S.V. Taskaev, V.V. Khovaylo, A. Hucht, A. Dannenberg, M. Ogura, H. Akai, M. Acet, P. Entel, A first-principles and Monte Carlo study of magnetostructural transition and magnetocaloric properties of $Ni_{2+x}Mn_{1-x}Ga$. Phys. Rev. B **81** (2010) 19pp

48. M. Khan, I. Dubenko, S. Stadler, N. Ali, Magnetostructural phase transitions in Ni_{50} $Mn_{25+x}Sb_{25-x}$ Heusler alloys. J. Phys.: Condens. Matter **20**, 235204 (2008)
49. L.S. Hsu, Y.K. Wang, G.Y. Guo, Experimental and theoretical study of the electronic structures of N_3Al, Ni_3Ga, Ni_3In, and NiGa. J. Appl. Phys. **92**, 1419 (2002)
50. R.P. Smith, Magnetic properties of Ni_3Al and Ni_3Ga: Emergent states and the possible importance of a tricritical point. J. Phys.: Condens. Matter **21**, 095601 (2009)
51. G.Y. Guo, Y.K. Wang, L.S. Hsu, First-principles and experimental studies of the electronic structures and magnetism in Ni_3Al, Ni_3Ga and Ni_3In. J. Magn. Magn. Mater. **239**, 91 (2002)
52. G. Ghosh, First-principles calculations of phase stability and cohesive properties of Ni-Sn intermetallics. Metal. Mater. Trans. A **40**, 4 (2009)
53. O.G. Randl, G. Vogl, W. Petry, Phonons – A diffusion motor in intermetallics? Physica B **219 & 220**, 499 (1996)
54. K.W. Richter, H. Ipser, An experimental investigation of the Fe-Ni-Sb ternary phase diagram. J. Phase Equil. **18**, 235 (1997)
55. P. Entel, V.D. Buchelnikov, V.V. Khovailo, A.T. Zayak, W.A. Adeagbo, M.E. Gruner, H.C. Herper, E.F. Wassermann, Modelling the phase diagram of magnetic shape memory alloys. J. Phys. D: Appl. Phys. **39**, 865 (2006)
56. H. Kushida, K. Hata, T. Fukuda, T. Terai, T. Kakeshita, Equilibrium phase diagram of Ni_2MnGa under [001] compressive stress. Scripta Mater. **60**, 96 (2009)
57. T. Kakeshita, T. Terai, M.Y. Yamamoto, T. Fukuda, Rearrangement of crystallographic domains driven by magnetic field in Fe_3Pt and CoO and new phase appearance in Ni_2MnGa. EDP Sciences, Proc. ESOMAT 2009, 04008 (2009)
58. H.E. Karaca, I. Karaman, B. Basaran, D.C. Lagoudas, Y.I. Chumlyakov, H.J. Maier, On the stress-assisted magnetic-field-induced phase transformation in Ni_2MnGa ferromagnetic shape memory alloys. Acta Mater. **55**, 4253 (2007)
59. M. Wuttig, J.L. Li, C. Craciunescu, A new ferromagnetic shape memory alloy system. Scripta Mater. **44**, 2393 (2001)
60. P.J. Brown, K. Ishida, R. Kainuma, T. Kanomata, K.U. Neumann, K. Oikawa, B. Ouladdiaf, K.R.A. Ziebeck, Crystal structures and phase transitions in ferromagnetic shape memory alloys based on Co-Ni-Al and Co-Ni-Ga. J. Phys.: Condens. Matter **17**, 1301 (2005)
61. R. Ducher, R. Kainuma, K. Ishida, Phase equilibria in the Ni-Co-Ga alloy system. J. Alloys Comp. **466**, 208 (2008)
62. M. Siewert, A. Dannenberg, M.E. Gruner, A. Hucht, S.M. Shapiro, G. Xu, D.L. Schlagel, T. Lograsso, P. Entel, Electronic structure and lattice dynamics of magnetic shape memory alloy Co_2NiGa. Phys. Rev. B **82**, 064420 (2010), 11pp
63. P. Entel, V.D. Buchelnikov, M.E. Gruner, A. Hucht, V.V. Khovailo, S.K. Nayak, A.T. Zayak, Shape memory alloys: A summary of recent achievements. Mater. Sci. Forum **583**, 21 (2008)
64. Y. Lee, J.Y. Rhee, B.N. Harmon, Generalized susceptibility of the magnetic shape-memory alloy Ni_2MnGa. Phys. Rev. B **66**, 054424 (2002)
65. E.F. Wassermann, Invar: Moment-volume instabilities in transition metals and alloys, in *Ferromagnetic Materials*, vol. 5, ed by K.H.J. Buschow, E.P. Wohlfarth (Elsevier, Amsterdam, 1990)
66. J. Worgull, E. Petti, J. Trivisonno, Behavior of the elastic properties near an intermediate phase transition in Ni_2MnGa. Phys. Rev. B **54**, 15695 (1996)
67. L. Mañosa, A. Gonzàles-Comas, E. Obradó, A. Planes, V.A. Chernenko, V.V. Kokorin, E. Cesari, Anomalies related to the TA_2-phonon-mode condensation in the Heusler Ni_2MnGa alloy. Phys. Rev. B **55**, 11068 (1997)
68. A. Ayuela, J. Enkovaara, R.M. Nieminen, Ab initio study of tetragonal variants in Ni_2MnGa alloy. J. Phys.: Condens. Matter **14**, 5325 (2002)
69. S.Ö. Kart, M. Uludoğan, I. Karaman, T. Çağin, DFT studies on structure, mechanics and phase behavior of magnetic shape memory alloys: Ni_2MnGa. Phys. Stat. Solidi A **205**, 1026 (2008)
70. Q.M. Hu, C.M. Li, R. Yang, S.E. Kulkova, D.I. Bazhanov, B. Johansson, Site occupancy, magnetic moments, and elastic constants of off-stoichiometric Ni_2MnGa from first-principles calculations. Phys. Rev. B **79**, 144112 (2009)

71. A.T. Zayak, P. Entel, J. Enkovaara, A. Ayuela, R.M. Nieminen, First-principles investigation of phonon softenings and lattice instabilities in the shape-memory system Ni_2MnGa. Phys. Rev. B **68**, 132402 (2003)
72. A.T. Zayak, P. Entel, K.M. Rabe, W.A. Adeagbo, M. Acet, Anomalous vibrational effects in nonmagnetic and magnetic Heusler alloys. Phys. Rev. B **72**, 054113 (2005)
73. Q.M. Hu, C.M. Li, S.E. Kulkova, R. Yang, B. Johansson, L. Vitos, Magnetoelastic effects in $Ni_2Mn_{1+x}Ga_{1-x}$ alloys from first-principles calculations. Phys. Rev. B **81**, 064108 (2010)
74. A. Zheludev, S.M. Shapiro, P. Wochner, L.E. Tanner, Precursor effects and premartensitic transformation in Ni_2MnGa. Phys. Rev. B **54**, 15045 (1996)
75. S. Kaufmann, U.K. Rößler, O. Heczko, M. Wuttig, J. Buschbeck, L. Schultz, S. Fähler, Adaptive modulations of martensite. Phys. Rev. Lett. **104**, 145702 (2010)
76. C.M. Varma, W. Weber, Phonon dispersion in transition metals. Phys. Rev. Lett. **39**, 1094 (1977)
77. C.M. Varma, W. Weber, Phonon dispersion in transition metals. Phys. Rev. B **19**, 6142 (1979)
78. M. Uijttewaal, T. Hickel, J. Neugebauer, M.E. Gruner, P. Entel, Understanding the phase transitions of the Ni_2MnGa magnetic shape memory system from first- principles. Phys. Rev. Lett. **102**, 035702 (2009)
79. S. Narasimham, S. Gironcoli, *Ab initio* calculation of the thermal properties of Cu: Performance of the LDA and GGA. Phys. Rev. B **65**, 064302 (2002)
80. B. Grabowski, T. Hickel, J. Neugebauer, *Ab initio* study of the thermodynamic properties of nonmagnetic elementary fcc metals: Exchange-correlation-related error bars and chemical trends. Phys. Rev. B **76**, 024309 (2007)
81. P. Souvatzis, O. Eriksson, *Ab initio* calculations of the phonon spectra and the thermal expansion coefficients of the 4d metals. Phys. Rev. B **77**, 024110 (2008)
82. P. Souvatzis, O. Eriksson, M.I. Katsnelson, S.P. Rudin, The self-consistent *ab initio* lattice dynamical method. Mater. Sci. **44**, 888 (2009)
83. M.E. Gruner, P. Entel, Simulating functional magnetic materials on supercomputers. J. Phys.: Condens. Matter **21**, 293201 (2009)
84. A. Sozinov, A.A. Likhachev, K. Ullakko, Crystal structures and magnetic anisotropy properties of Ni-Mn-Ga Hortensia phases with giant magnetic-field- induced strain. IEEE Trans. Magn. **38**, 2814 (2002)
85. J. Enkovaara, A. Ayuela, L. Nordström, R.M. Nieminen, Magnetic anisotropy in Ni_2MnGa. Phys. Rev. B **65**, 13422 (2002)
86. M.E. Gruner, P. Entel, I. Opahle, M. Richter, *Ab initio* investigation of twin boundary motion in the magnetic shape memory Heusler alloy Ni_2MnGa. J. Mater. Sci. **43**, 3825 (2008)
87. A. Dannenberg, M.E. Gruner, M. Wuttig, P. Entel, Characterization of new ferromagnetic Fe-Co-Zn-Ga alloys by *ab initio* investigations, ESOMAT 2009, 04004 (2009), EDP Sciences (2009)
88. Y.N. Zhang, J.X. Cao, R.Q. Wu, Rigid band model for prediction of magnetostriction of iron/gallium alloys. Appl. Phys. Lett. **96**, 062508 (2010)

Chapter 3
Ni–Mn–X Heusler Materials

Ryosuke Kainuma and Rie Y. Umetsu

Abstract In this chapter, the order–disorder phase transformation from the $B2$ to the $L2_1$-type phase and magnetic properties of $Ni_2Mn(Ga_xAl_{1-x})$ and $Ni_2Mn_{1+y}In_{1-y}$ alloys are introduced, and the influence of martensitic transformation on the magnetic properties in NiMnIn-based alloy is also discussed. The magnetic properties of these alloys are significantly sensitive to the degree of long-range order, the alloy composition, and the crystal structure, and the magnetism varies among paramagnetic, ferromagnetic, and antiferromagnetic properties, in dependence on these factors. On the whole, the magnetic properties are mainly governed by the magnetic moments of Mn atoms and can be understood by taking into account the sign and the strength of the magnetic exchange interactions in Mn–Mn and Ni–Mn pairs.

3.1 Introduction

Since Ullakko et al. [1] reported magnetic-field-induced strain (MFIS) in a Heusler-type Ni_2MnGa ferromagnetic shape memory alloy (FSMA) [2], FSMAs have attracted considerable interest as a new type of material applicable to actuators and sensors which can be controlled by external magnetic fields. The magnetic shape memory effect is caused by the magnetic-field-induced rearrangement of the martensite variants, this behavior being due to the high magnetocrystalline

R. Kainuma (✉)
Department of Materials Science, Tohoku University, Sendai, Japan
e-mail: kainuma@material.tohoku.ac.jp

R.Y. Umetsu
Institute for Materials Research, Tohoku University, Sendai, Japan
e-mail: rieume@imr.tohoku.ac.jp

T. Kakeshita et al. (eds.), *Disorder and Strain-Induced Complexity in Functional Materials*, Springer Series in Materials Science 148,
DOI 10.1007/978-3-642-20943-7_3,

anisotropy energy [3]. So far, a large MFIS of up to about 10% has been confirmed in off-stoichiometric Ni_2MnGa single crystals [4].

Furthermore, a new type of MFIS was reported in Co-doped off-stoichiometric NiMnIn and NiMnSn Heusler-type alloys in 2006 [5,6]. In the NiMnIn and NiMnSn alloys, the spontaneous magnetization of the martensite phase is much smaller than that of the austenite phase, and the martensitic transformation temperatures drastically decrease by the application of a magnetic field. Magnetic-field-induced reverse martensitic transformation (MFIRT), which is a kind of metamagnetic phase transition, is detected at temperatures just below the martensitic reverse transformation starting temperature. The new type of MFIS, called metamagnetic shape memory (MMSM) effect, has been confirmed using the MFIRT in predeformed specimens of $Ni_{45}Co_5Mn_{36.7}In_{13.3}$ [5] and $Ni_{43}Co_7Mn_{39}Sn_{11}$ [6] martensite alloys. Furthermore, many interesting phenomena such as the inverse magnetocaloric effect [7,8], the giant magnetoresistance effect [9], the giant magnetothermal conductivity [10], the exchange bias effect [11], etc., are derived from this unique transformation. Details on the basic physical properties for the NiMnX Heusler alloys, including the NiMnGa alloys, have recently been reviewed by Planes et al. [12].

In this chapter, it is shown that the magnetic properties of the austenite phase in the NiMn-based alloys drastically change depending on the degree of chemical long-range order and on the deviation from the stoichiometric composition. Furthermore, the drastic change of the magnetic property in martensitic transformation is also presented in relation to the tetragonal distortion yielded by the transformation.

3.2 Atomic Ordering and Magnetic Properties in $Ni_2Mn(Ga_xAl_{1-x})$ Alloys [13]

Figure 3.1 shows the atomic configurations of $B2$-type NiAl (space group $Pm3m$) and Heusler $L2_1$-type (space group $Fm3m$) Ni_2MnAl phases. In the Heusler $L2_1$-type structure of Ni_2MnAl alloy, Ni atoms occupy equivalent $4a$ and $4c$ (in Wyckhoff notation) positions, whereas $4b$ and $4d$ positions are occupied by Mn and Al atoms, respectively. The Heusler phase is one of the highly ordered structures on the basis of the $B2$ structure, in which the $4b$ and $4d$ positions are randomly occupied by Mn and Al atoms, and $A2$(disordered bcc) $\rightarrow B2$ and/or

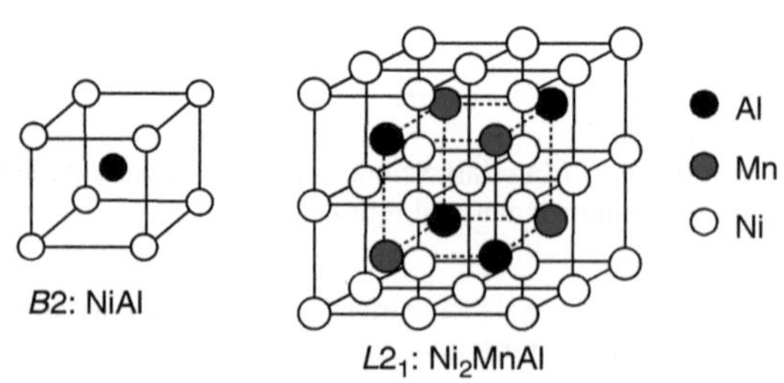

Fig. 3.1 Atomic configurations of $B2$-type NiAl and Heusler $L2_1$-type Ni_2MnAl phases

$B2 \rightarrow L2_1$ order–disorder phase transformations sometimes appear in the detectable temperature region [14, 15]. It is known that in the Ni_2MnAl alloy the $B2 \rightarrow L2_1$ order–disorder phase transformation temperature, $T_t^{B2/L21}$, is located at about 775 K [16] and that the $B2$ phase alloy obtained by quenching from temperatures higher than the $T_t^{B2/L21}$ possesses a conical antiferromagnetic structure with a Néel temperature, T_N, of about 300 K [17]. However, atomic ordering from the $B2$ to the $L2_1$ structure is induced by ageing at 673 K and its magnetic property changes from antiferromagntic to ferromagnetic [16]. Theoretical calculations have suggested that the magnetic moment and the Curie temperature, T_C, of the $L2_1$ phase in Ni_2MnAl alloy are comparable to those in Ni_2MnGa alloy [18–20]. However, the magnetic properties of the $L2_1$ phase in Ni_2MnAl alloy are in disagreement with the theoretical results, in contrast to those in the Ni_2MnGa alloy [2]. Actually, although off-stoichiometric Ni_2MnAl alloys show martensitic transformations to long-period stacking order structures, similar to those in Ni–Mn–Ga alloys [21], a $Ni_{53}Mn_{25}Al_{22}$ alloy exhibits only slight MFIS of about -100 ppm [22], which is only about 1/1,000 of that in Ni–Mn–Ga alloys. In the present section, the phase stability and the magnetic properties of $Ni_2Mn(Ga_xAl_{1-x})$ $(0 \leq x \leq 1)$ alloys are introduced in relation to the degree of long-range order of the $L2_1$ phase [13].

3.2.1 Atomic Ordering

According to the Bragg–Williams–Gorsky (BWG) approximation [23, 24], under the assumption that the X atom always occupies its own site, the $T_t^{B2/L2_1}$ in the $X_2Y_{1+y}Z_{1-y}$ (i.e., $XY(B2)$–$X_2YZ(L2_1)$–$XZ(B2)$) pseudobinary system is simply given by

$$T_t^{B2/L2_1} = \frac{3W_{YZ}^{X(2)}}{2k_B} \cdot (1 - y^2), \tag{3.1}$$

where $W_{YZ}^{X(2)}$ is the interchange energy between Y(4b) and Z(4d) atoms in the second nearest neighbor (NN) surrounded by X(4a and 4c) atoms in the first NN and k_B is the Boltzmann constant. Here, the $W_{YZ}^{X(2)}$ is defined as $W_{YZ}^{X(2)} \equiv \varepsilon_{YY}^{X(2)} + \varepsilon_{ZZ}^{X(2)} - 2\varepsilon_{YZ}^{X(2)}$ using the atomic bonding energy $\varepsilon_{ij}^{X(2)}$ between i and j atoms in the second NN. For the stoichiometric alloy with $y = 0$,

$$T_t^{B2/L2_1} = \frac{3W_{YZ}^{X(2)}}{2k_B}. \tag{3.2}$$

Equation (3.2) means that $T_t^{B2/L2_1}$ is simply proportional to $W_{YZ}^{X(2)}$, independent of the other pairwise interactions, and that $W_{YZ}^{X(2)}$ can easily be estimated from $T_t^{B2/L2_1}$.

Figure 3.2 shows the $T_t^{B2/L2_1}$ for $Ni_2Mn(Ga_xAl_{1-x})$ alloys [13] including $T_t^{B2/L2_1}$ for Ni_2MnAl [16] and Ni_2MnGa [25]. $T_t^{B2/L2_1}$ (1,071 K) of the Ni_2MnGa

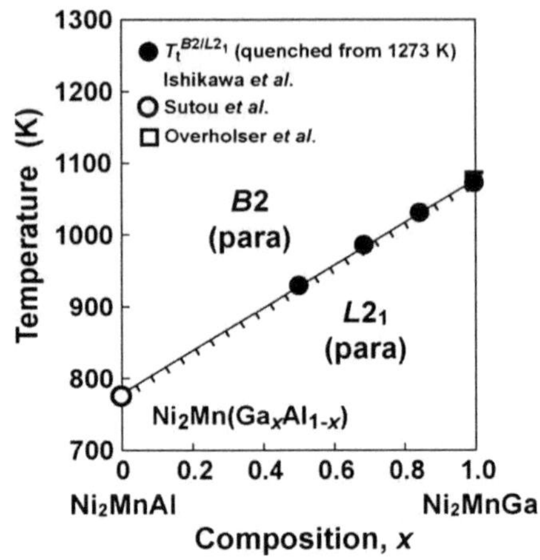

Fig. 3.2 Concentration dependence of $T_t^{B2/L2_1}$ for $Ni_2Mn(Ga_xAl_{1-x})$ alloys [13, 16, 25]

alloy is about 300 K higher than that (775 K) of the Ni_2MnAl alloy, and the $W_{MnGa}^{Ni(2)}$ and $W_{MnAl}^{Ni(2)}$ are evaluated by (3.2) as being $714k_B$ and $517k_B$, respectively. It is apparent that the $T_t^{B2/L21}$ linearly increases with increasing x. This suggests that the $L2_1$ phase in $Ni_2Mn(Ga_xAl_{1-x})$ alloys is monotonically stabilized by the substitution of Ga and that the effective interchange energy $W_{Mn(Al,Ga)}^{Ni(2)}$ between the Mn and (Ga, Al) site atoms in the substituted alloys is simply given by the weighted mean values between $W_{MnGa}^{Ni(2)}$ and $W_{MnAl}^{Ni(2)}$.

3.2.2 *Magnetic Properties*

Figure 3.3a and b show the thermomagnetization curves measured with a superconducting quantum interference device (SQUID) magnetometer under a magnetic field of 0.05 T for the $Ni_2Mn(Ga_xAl_{1-x})$ alloys annealed at a temperature adequate for each alloy before quenching in ice water. Here, a temperature which is about 20 K higher than the $T_t^{B2/L2_1}$ was selected as the annealing temperature for each alloy. It is seen that while the magnetic properties of the $x = 0.00$ and 0.50 alloys are antiferromagnetic as shown in the Fig. 3.3b, the $x = 0.68$ and 1.00 alloys exhibit a ferromagnetic behavior. It was confirmed by transmission electron microscopic (TEM) observation that for $x = 0.68$ and 1.00, the ordering from the $B2$ to $L2_1$ phase cannot be suppressed by quenching due to their high $T_t^{B2/L2_1}$, while it can be perfectly suppressed for $x = 0.00$ and 0.50. Figure 3.3c shows the thermomagnetization heating curves for the alloys annealed at 673 K for 1 day in order to heighten the degree of $L2_1$ long-range order. It is seen that annealing at 673 K induces a ferromagnetic feature in all the alloys, although the T_C of $x = 0.00$ is lower than that of the other alloys. Kinks observed at 100–200 K in the thermomagnetization curve for $x = 0.84$ and 1.00 correspond to the martensitic transformation. In the as-annealed specimens, dips associated with intermediate reverse transformation are also observed.

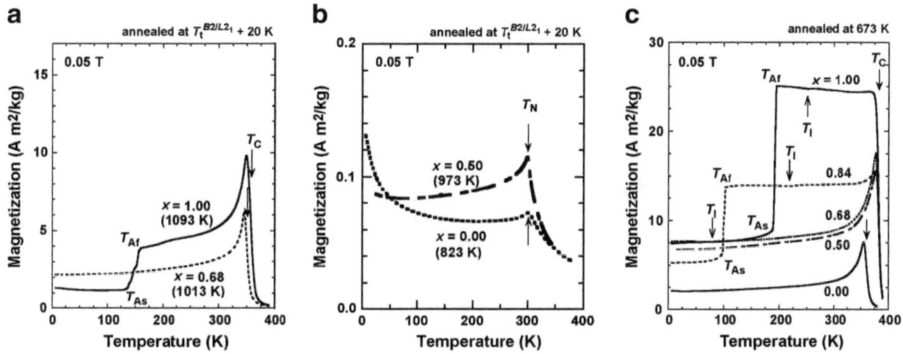

Fig. 3.3 Thermomagnetization curves measured in a magnetic field of 0.05 T for the $Ni_2Mn(Ga_xAl_{1-x})$ alloys (**a**) and (**b**) quenched from $T_t^{B2/L2_1} + 20$ K, and (**c**) further annealed at 673 K for 1 day. Here, T_C, T_{As} and T_{Af}, and T_I are the Curie temperature, the martensitic reverse transformation starting temperatures and finishing temperatures, and the intermediate phase transformation temperature, respectively [13]

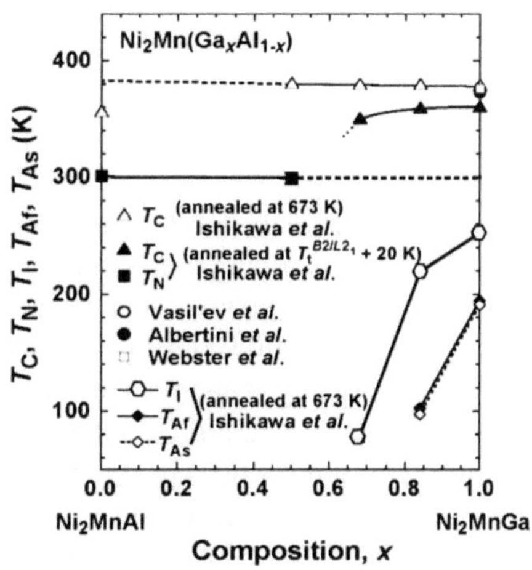

Fig. 3.4 Concentration dependence of the magnetic transition temperatures extracted from the data of Fig. 3.3 for $Ni_2Mn(Ga_xAl_{1-x})$ alloys [2, 13, 26, 27]

The transformation temperatures determined from Fig. 3.3 are plotted in Fig. 3.4 together with some other reported experimental data [2, 13, 26, 27]. It is interesting to note that the T_C for the alloys of $x \geq 0.5$ annealed at 673 K and the T_N for the as-quenched alloys with the $B2$ structure of $x = 0.0$ and 0.5 are almost constant at about 380 K and about 300 K, respectively. All the martensitic transformation temperatures monotonically decrease with decreasing Ga composition. It is apparent that only in the $x = 0.50$ alloy, the ordering condition can fully be varied from $B2$ to $L2_1$.

Figure 3.5 shows the magnetic transition temperatures (a) and the saturation magnetization M_s at 4.2 K (b) as a function of annealing temperature for the $Ni_2Mn(Ga_{0.5}Al_{0.5})$ specimens annealed at various temperatures for 1 day after quenching from 973 K, the values obtained from the specimen two-step-annealed in sequence of 873 K → 673 K after quenching from 973 K being used only for 673 K.

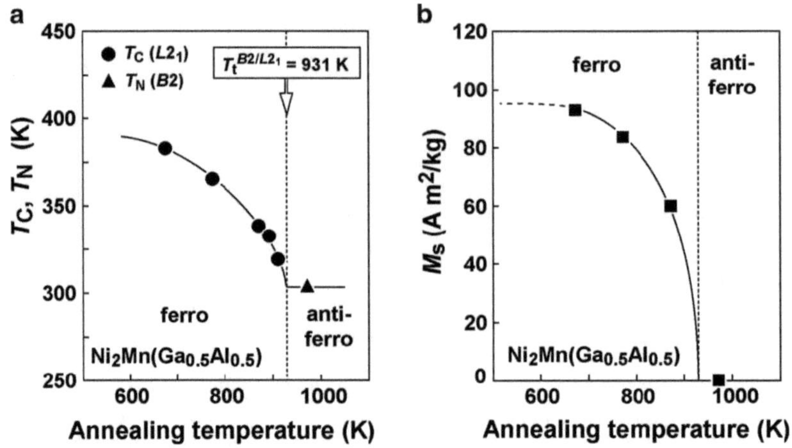

Fig. 3.5 (**a**) Magnetic transition temperatures of T_C or T_N as a function of the annealing temperature for the $Ni_2Mn(Ga_{0.5}Al_{0.5})$ alloys. (**b**) Saturation magnetization M_s measured at 4.2 K for the $Ni_2Mn(Ga_{0.5}Al_{0.5})$ alloys as a function of the annealing temperature [13]

Both the T_C and M_s parabolically decrease with increasing annealing temperature up to the ordering temperature $T_t^{B2/L2_1} = 931$ K. From these figures, it can be concluded that both the M_s and T_C are significantly sensitive to the annealing temperature. According to neutron diffraction investigations for Ni_2MnGa [28], the degree of long-range order between the Mn and Ga sites, which differs between the $B2$ and $L2_1$ structures, drastically changes in the temperature range from 773 K to 1,053 K ($= T_t^{B2/L2_1}$) and reaches an almost perfect degree of long-range order due to annealing at temperatures below $T = (3/4)T_t^{B2/L2_1}$. If this condition is available to the present $Ni_2Mn(Ga_{0.5}Al_{0.5})$ alloy with $T_t^{B2/L2_1}$ at 931 K, the maximum temperature to effectively obtain a high degree of long-range order is estimated to be about 700 K (= 931 K × 0.75). From this fact, it can be concluded that the drastic changes of the T_C and M_s, which depend on the annealing temperature in the range of 673–930 K shown in Fig. 3.5, result from the change of the degree of long-range order between the Mn and (Ga,Al) sites.

In the present $Ni_2Mn(Ga_xAl_{1-x})$ alloys, the antiferromagnetic property is observed only in the as-quenched specimens of $x = 0.00$ and 0.50 and the ferromagnetic feature showing the high T_C of about 380 K cannot be obtained only in the as-annealed specimen of $x = 0.00$ as shown in Fig. 3.4. These results can be explained by the difference in the $T_t^{B2/L2_1}$ and the degree of long-range order for the specimens, respectively. Since $B2$ to $L2_1$ order–disorder phase transformation is of the second order, ordering reaction is usually difficult to suppress by conventional quenching. However, when both the ordering temperature and the diffusivity of the alloy are sufficiently low, suppression of the ordering reaction is possible. In the

$Ni_2Mn(Ga_xAl_{1-x})$ alloys, the critical temperature in suppression of the $B2 \rightarrow L2_1$ ordering reaction by quenching is considered to be located between 931 and 989 K, corresponding to the $T_t^{B2/L2_1}$ for $x = 0.50$ and 0.68, respectively. On the other hand, the reason why the $T_t^{B2/L2_1}$ in the as-annealed Ni_2MnAl is lower than those in other alloys as shown in Fig. 3.4 is explained as resulting from the low degree of order. As mentioned above, an almost perfect degree of order is achieved by annealing at temperatures below $(3/4)T_t^{B2/L2_1}$. The $T_t^{B2/L2_1}$ in the Ni_2MnAl alloy is about 775 K, and an actually fully ordered condition is expected to be achieved by annealing at temperatures below 581 K. The annealing at 673 K performed in the present case is too high to obtain a high degree of order. This means that in the Ni_2MnAl alloy it is possible to increase both T_C and M_s by using annealing at lower temperatures, although the diffusivity at lower temperatures is very low. By linear extrapolation from the T_C determined in the present $Ni_2Mn(Ga_xAl_{1-x})$ alloys, the T_C for the fully ordered $L2_1$-type structure in the Ni_2MnAl alloy is evaluated to be about 380 K, which is slightly higher than that of the Ni_2MnGa alloy. On the other hand, the $B2$-type Ni(Mn,Ga) phase seems to be antiferromagnetic with a T_N of about 300 K.

The reason why the atomic ordering in the $Ni_2Mn(Ga_xAl_{1-x})$ alloys strongly affects the magnetic properties is explained as being due to difference in configuration of Mn atoms between the $B2$ and $L2_1$ structures. In many Mn-based Heusler alloys, it is known that the magnetic moment of the Mn atoms at the regular Mn$(4b)$ sites is ferromagnetically coupled with that at the same Mn$(4b)$ sites in the third NN [29]. In the perfectly ordered stoichiometric $L2_1$-type $Ni_2Mn(Ga, Al)$, all the Mn atoms are located only at the regular Mn$(4b)$ sites and no Mn$(4b)$–Mn$(4d)$ pair in the second NNs appears. However, with decreasing degree of long-range order, the number of Mn$(4b)$–Mn$(4d)$ in the second NN increases, and instead, that of the Mn$(4b)$–Mn$(4b)$ pairs in the third NN decreases. If the magnetic moment of the Mn atoms at the Mn$(4b)$ sites is antiferromagnetically coupled with that at the Ga or Al$(4d)$ sites, the decrease of the degree of long-range order might result in the decrease of the T_C and the M_s.

3.3 Magnetic Properties in Off-Stoichiometric $Ni_2Mn_{1+y}In_{1-y}$ Alloys [30]

Most martensitic transformations in the NiMn-based alloys have been reported at off-stoichiometric compositions typically described as $X_2Y_{1+y}Z_{1-y}$ [31]. In this section, the order–disorder phase transformation and the basic magnetic properties for the $Ni_2Mn_{1+y}In_{1-y}$ alloys are introduced

As mentioned above, according to the BWG approximation, the $T_t^{B2/L2_1}$ in the $XY(B2)$–$X_2YZ(L2_1) - XZ(B2)$ pseudobinary system is given by (3.1). This equation means that the $T_t^{B2/L2_1}$ is described by a parabolic curve as a function of y with the maximum at the stoichiometric composition of $L2_1$ phase (i.e., at $y = 0$)

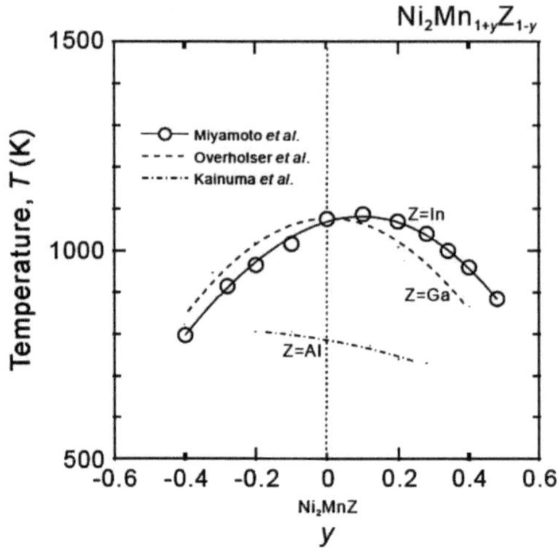

Fig. 3.6 Critical temperature of the $B2/L2_1$ order–disorder phase transformation for the $Ni_2Mn_{1+y}In_{1-y}$ [30], $Ni_2Mn_{1+y}Al_{1-y}$ [21] and $Ni_2Mn_{1+y}Ga_{1-y}$ [25] alloys

if $W_{YZ}^{X(2)}$ is constant. The values of $T_t^{B2/L2_1}$ experimentally determined for the $Ni_2Mn_{1+y}In_{1-y}$ specimens as well as those for the $Ni_2Mn(Ga_xAl_{1-x})$ specimens are plotted in Fig. 3.6, together with those for the NiMnGa [25] and NiMnAl [21] alloys. The values of $T_t^{B2/L2_1}$ are located in the temperature region from 800 to 1,100 K as are those in the NiMnGa alloys, and the maximum temperature appears at 22.5In ($y = 0.1$), which deviates from the stoichiometric composition of Ni_2MnIn ($y = 0.0$). The fact that the $T_t^{B2/L2_1}$ deviates from a parabolic curve suggests that one or both of the two assumptions giving the parabolic relation, i.e., the perfect occupancy of the X atoms on the X-site atoms and the independency of the $W_{YZ}^{X(2)}$ on composition and temperature, may not be valid in the present case. It is interesting to note that the $T_t^{B2/L2_1}$ in the NiMn–NiAl section deviates from the theoretical parabolic curve as well, but the maximum point of the $T_t^{B2/L2_1}$ is located at an Al composition higher than 25Al ($y = 0.0$), as shown in Fig. 3.6 [30]. In the case of the NiMn–NiGa section, the $T_t^{B2/L2_1}$ almost coincides with the theoretical parabolic curve [25]. It is not clear why such a difference on the maximum point of $T_t^{B2/L2_1}$ appears in these three systems.

As discussed in the previous section, it is known that the final annealing temperature is important to obtain the $L2_1$ phase with a high degree of order and that a fully high degree of order can be obtained by annealing at temperatures below $(3/4)T_t^{B2/L2_1}$. In the present study, the specimens were solution-treated at 1,173 K for 24 h followed by quenching into water, and then the 32In ($y = -0.28$) and 35In ($y = -0.4$) specimens with a low $T_t^{B2/L2_1}$ were annealed at 573 K for 3 days and the other ones were annealed at 673 K for 1 day, which meet the condition of below $(3/4)T_t^{B2/L2_1}$. Figure 3.7a and b show the concentration dependence of the T_C and the spontaneous magnetization per formula unit μ_m (μ_B/f.u.) at 4.2 K for the NiMnIn alloys, respectively. The T_C exhibits a very abnormal behavior, i.e., while

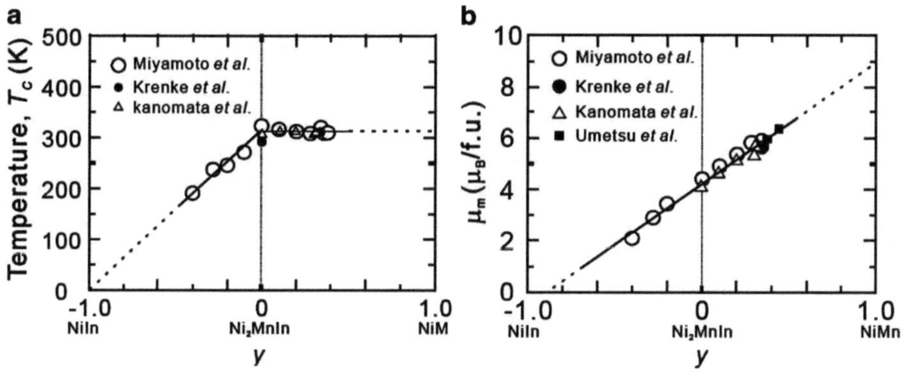

Fig. 3.7 (**a**) Curie temperature, T_C, and (**b**) spontaneous magnetization per formula unit, μ_m, determined at 4.2 K for the $Ni_2Mn_{1+y}In_{1-y}$ alloys with the $L2_1$ structure [30, 32–34]

being almost constant in the Mn-rich region of $y > 0$ as reported in some previous papers [32, 33], the T_C increases linearly with increasing y in the In-rich region of $y < 0$. On the other hand, the μ_m almost linearly increases with increasing y as shown in Fig. 3.7b [30, 32–34]. With regard to the concentration dependence of the μ_m, Kanomata et al. [33] have suggested that the magnetic moment of the Mn atoms at the regular Mn sites is ferromagnetically coupled with that of the Mn atoms not only at the regular Mn sites in the third NN but also at the In sites in the second NN. They have proposed a simple relation given by

$$\mu_m(\text{total}) = 2\mu_{Ni} + \mu_{Mn} + y\mu_{Mn} + (1-y)\mu_{In}, \tag{3.3}$$

where μ_{Ni}, μ_{Mn}, and μ_{In} are the magnetic moments of Ni, Mn, and In atoms, respectively. The line calculated on the basis of the theoretical data on the magnetic moments, $\mu_{Ni} = 0.28\mu_B$, $\mu_{Mn} = 3.72\mu_B$ and $\mu_{In} = -0.07\mu_B$, reported by Şaşıoğlu et al. [35] for the stoichiometric alloy, basically agrees with the experimental data in the whole concentration range [30]. The fully ordered $L2_1$ structures in the $Ni_2Mn_{1+y}In_{1-y}$ section are divided into two regions, $y > 0$ and $y < 0$, on number of Mn–Mn pairs in the second and third NNs. In $y > 0$, the number of the Mn(4b)–Mn(4d) pairs in the second NN linearly increases with increasing y, while that of the Mn(4b)–Mn(4b) pairs in the third NN is constant. On the other hand, in $y < 0$, the number of the Mn(4b)–Mn(4b) pairs in the third NN linearly decreases with decreasing y under the condition of no Mn(4b)–Mn(4d) pairs in the second NN. The fact that the concentration dependence of the μ_m in $y > 0$ is hardly different from that in $y \leq 0$ means that the magnetic moment of Mn atoms at the In(4d) sites is almost equivalent to that at the Mn(4b) sites as suggested by Kanomata et al. [33].

By a similar consideration, it is apparent that the T_C is basically determined by the Mn concentration located at the regular Mn(4b) sites, being independent of that at the In(4d) sites. According to the calculation on pressure dependence of exchange interaction in Ni_2MnSn reported by Şaşıoğlu et al. [36], the ferromagnetic interaction, $J^{(3)}_{MnMn}$, between two Mn atoms at the Mn($4b$) sites in the third NN increases with increasing pressure for interatomic Mn($4b$)–Mn($4b$) distance r_{tnn} ranging from 0.42 nm to about 0.37 nm, while in the region below 0.37 nm, $J^{(3)}_{MnMn}$ decreases abruptly with decreasing r_{tnn} and reaches $J^{(3)}_{MnMn} = 0$ at around 0.31 nm. Furthermore, the ferromagnetic Ni–Mn interaction is almost completely independent of pressure. If the pressure dependence of the exchange interaction in Ni_2MnIn is similar to that in Ni_2MnSn, since the ferromagnetic interaction of the Mn($4b$)–Mn($4d$) pairs in the second NN, $J^{(2)}_{MnMn}$, with $r_{snn} = 0.304$ nm may be negligibly small in comparison to the $J^{(3)}_{MnMn}$ with $r_{tnn} = 0.429$ nm, the T_C may be basically determined by the Mn concentration located at the regular Mn($4b$) sites.

Finally, it should be pointed out that for Ni_2MnIn, the magnetic moment of the Mn atom at the Mn($4b$) sites is ferromagnetically coupled with that at the $4d$ sites in the second NN, although being antiferromagnetically coupled for the Ni_2Mn(Ga, Al). The origin of this discrepancy may be brought about by difference in lattice parameter between these alloys. If the pressure dependence of the exchange interaction in Ni_2Mn(Ga, Al) is also similar to that in Ni_2MnSn, the interaction $J^{(2)}_{MnMn}$ of Mn($4b$)–Mn($4d$) pairs for $Ni_2MnGa_{0.5}Al_{0.5}$ with $r_{snn} = 0.291$ nm [15] may be negative, because it is significantly smaller than the r_{snn} ($= 0.31$ nm) showing $J^{(3)}_{MnMn} = 0$, while that for the Ni_2MnIn with $r_{snn} = 0.304$ nm may be negligibly small. This difference in distance of Mn($4b$)–Mn($4d$) pairs between Ni_2Mn(Ga, Al) and Ni_2MnIn may cause the difference in the magnetic coupling. Actually, it has been reported by Kanomata et al. [37] that for $Ni_2Mn_{1+y}Sn_{1-y}$ Heusler-type alloys possessing a smaller lattice parameter the magnetic moment of the Mn atom at the Mn($4b$) sites is antiferromagnetically coupled with that at the Sn($4d$) sites. Dependence of the degree of long-range order on magnetic properties for the stoichiometric Ni_2MnIn specimens annealed under some different conditions should also be investigated in order to confirm the feature of magnetic coupling of Mn($4b$)–Mn($4d$) pairs in the second NN, as carried out in the Ni_2Mn(Ga, Al) alloys.

3.4 Martensitic Transformation and Magnetic Properties in NiMnIn Alloy [38]

As mentioned in Sect. 3.1, in the NiMnIn and NiMnSn alloys, the spontaneous magnetization of the martensite phase is extremely smaller than that of austenite phase. Figure 3.11 shows the thermomagnetization $M(T)$ curves measured in a magnetic field $H = 0.05$ T under zero field cooling (ZFC) and field cooling (FC) protocols for $Ni_2Mn_{1.372}{}^{(57)}Fe_{0.02}In_{0.608}$ alloy. In the FC process, after a sharp rise at $T^A_C = 307$ K corresponding to the ferromagnetic transformation in austenite phase, the magnetization drastically drops due to martensitic transformation at $T_{Ms} = 302$ K and then becomes almost zero. With further cooling, the magnetic

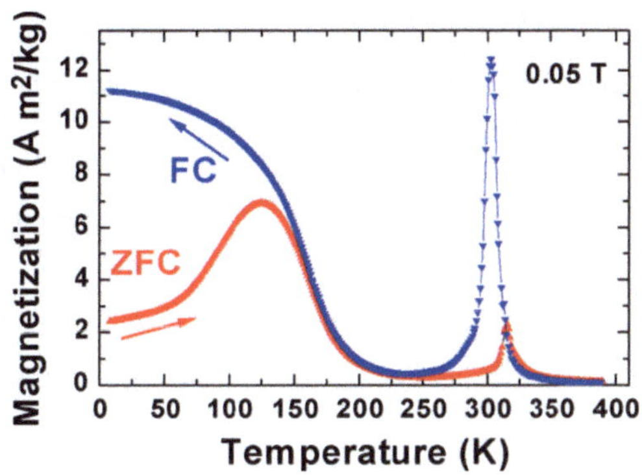

Fig. 3.8 Thermomagnetization, $M(T)$, curves measured in a magnetic field $H = 0.05\,\mathrm{T}$ under zero field cooling (ZFC) and field cooling (FC) protocols for $Ni_2Mn_{1.372}{}^{(57)}Fe_{0.02}In_{0.608}$ alloy [38]

transition again appears at $T_C^M = 162\,\mathrm{K}$, far below the martensitic transformation temperature. By the recent Mössbauer examination [38], the martensite phase with low magnetization appearing in the temperature region above 162 K is realized to be basically paramagnetic. On the other hand, the Mössbauer spectra obtained at temperatures below 162 K are very complicated, while including some kinds of magnetic components. Splitting of the ZFC and FC curves observed at low temperatures in Fig. 3.8 suggests a nontrivial magnetic ordering with coexisting antiferromagnetic and ferromagnetic interactions [32, 39–41]. It is very interesting to note that the ferromagnetic austenite phase transforms to the paramagnetic martensite phase because the magnetic entropy of the martensite phase with paramagnetism at lower temperatures is apparently higher than that of the austenite phase with ferromagnetism in contrast to ordinary ferromagnetic transformations. This abnormal behavior is clearly brought about by the lattice distortion due to martensitic transformation accompanying a lattice vibration entropy change larger than the magnetic entropy change with an opposite sign.

It has been reported that a mixture of 10- and 14-layered monoclinic martensite phases, i.e., the 10M and the 14M phases, appears in NiMnIn and NiCoMnIn alloys [42, 43]. The layered structures, such as the 10M- and 14M-type structures [$(3\bar{2})_2$ and $(5\bar{2})_2$ in Zhdanov notation, respectively], are sometimes considered to be tetragonal structure with a high density of ordered nanotwins [44]. If the 10M- and 14M-type structures have the same stacking unit composed of the distorted $L2_1$ phase denoted as a face-centered tetragonal (fct) structure (Fig. 3.9), the lattice parameters for the 10M- and 14M-type structures can be easily evaluated on the basis of the 2M structure possessing a lattice correspondence with the fct structure as $a_{2M} = c_{2M} = \left(\sqrt{a_{fct}^2 + c_{fct}^2}\right)/2$ and $b_{2M} = a_{fct}$. For the $Ni_2Mn_{1.372}{}^{(57)}Fe_{0.02}In_{0.608}$ alloy with the $M(T)$ curves shown in Fig. 3.8, the lattice parameters of the 2M structure, $a_{2M} = 0.4377\,\mathrm{nm}$, $b_{2M} = 0.5654\,\mathrm{nm}$, and $c_{2M} = 0.4377\,\mathrm{nm}$ and $\beta_{2M} = 99.54°$, which correspond to $a_{fct} = 0.5654$ and $c_{fct} = 0.6684\,\mathrm{nm}$ for the fct structure, were determined by using some peaks independent of the layered structures in the experimental X-ray diffraction pattern.

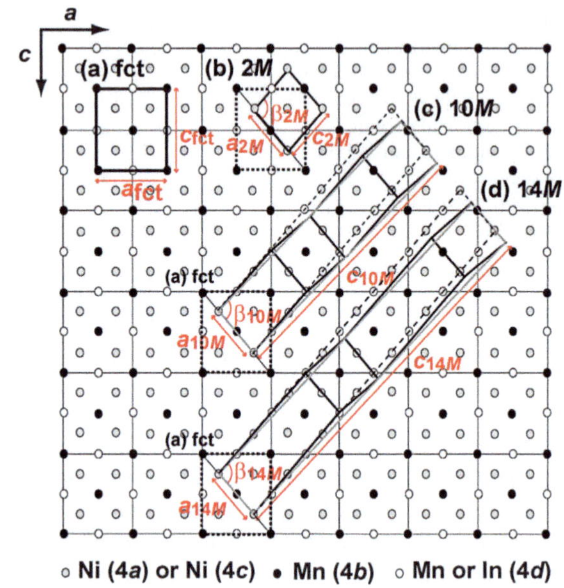

Fig. 3.9 Projections of the face-centered tetragonal (**a**), 2M (**b**), 10M (**c**), and 14M (**d**) structures on *b*-axis [38]

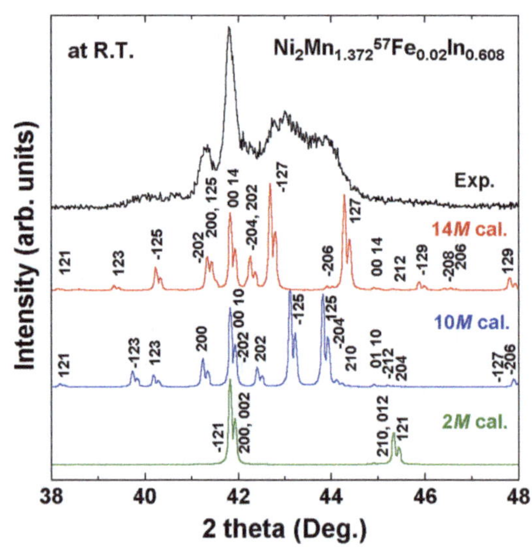

Fig. 3.10 X-ray powder diffraction patterns of $Ni_2Mn_{1.372}{}^{(57)}Fe_{0.02}In_{0.608}$ measured at room temperature, together with the calculated patterns as 2M, 10M, and 14M structures [38]

This result means that the basic tetragonal distortions from the $L2_1$ phase due to the martensitic transformation are about -6% along the a-axis and about $+11\%$ along the c-axis and that the volume change is about -1%. The X-ray diffraction patterns calculated for 2M, 10M and 14M together with the experimental one are shown in Fig. 3.10. The experimental pattern can be indexed as a mixture of the 10M- and 14M-type structures with the lattice parameters, $a_{10M} = 0.4377$ nm, $b_{10M} = 0.5654$ nm, $c_{10M} = 2.1594$ nm, $\beta_{10M} = 91.93°$, and

Table 3.1 Coordination number and distances between Ni–Mn and Mn–Mn pairs in the high-temperature $L2_1$ phase and the low-temperature fct phase of $Ni_2Mn_{1.372}{}^{(57)}Fe_{0.02}In_{0.608}$ [38]

Pair	$L2_1$ ($a = 0.5998$ nm) Distance, r^{A}_{inn} (nm)	Coordination number	fct ($a = 0.5654$ nm, $c = 0.6684$ nm) Distance r^{M}_{inn} (nm)	Coordination number	Ratio $(d_M - d_A)/d_A$ (%)
Ni–Mn(4*b*) in first NN	0.2597	4	0.2605	4	0.3
Ni–Mn(4*d*) in first NN	0.2597	4	0.2605	4	0.3
Mn(4*b*)–Mn(4*d*) second NN in	0.2999	6	0.2827	4	−5.7
			0.3342	2	11.4
Mn(4*b*)–Mn(4*b*) in third NN	0.4241	12	0.3998	4	−5.7
			0.4377	8	3.2

$a_{14M} = 0.4377$ nm, $b_{14M} = 0.5654$ nm, $c_{14M} = 3.0302$ nm, $\beta_{14M} = 94.35°$, respectively, obtained based on the calculated data for the 2M structure.

As listed in Table 3.1, in the $L2_1$ austenite phase of $Ni_2Mn_{1.372}{}^{(57)}Fe_{0.02}In_{0.608}$ alloy at $T = 320$ K, the distances between Ni and Mn(4*b* and 4*d*) positions in the first NN, between Mn(4*b*) and Mn(4*d*) positions in the second NN, and between Mn(4*b*) and Mn(4*b*) positions in the third NN are $r^{A}_{fnn} = 0.2597$ nm, $r^{A}_{snn} = 0.2999$ nm, and $r^{A}_{tnn} = 0.4241$ nm, respectively. If one can neglect the presence of the nanotwin boundaries in the layered structures, these atomic distances are affected by the simple tetragonal distortion from the cubic structure. As compared with the austenitic phase, the r^{A}_{fnn} in the first NN are only 0.3% larger than those (r^{M}_{fnn}) in the martensite phase, whereas both r^{A}_{snn} and r^{A}_{tnn} are split by the structural distortions into the two inequivalent distances (r^{M}_{snn} and r^{M}_{tnn}) listed in Table 3.1.

The observed abnormal magnetic phase transition sequence from paramagnetic martensite to ferromagnetic austenite seems to be explained as being due to a strong dependence of the exchange interactions on interatomic distances [33–36, 45] as well as to the magnetic coupling of Mn(4*b*)–Mn(4*d*) pairs in the austenite phase as discussed in the previous section. In the martensite phase, two sets of Mn–Mn distances [i.e., Mn(4*b*)–Mn(4*b*) and Mn(4*b*)–Mn(4*d*)] in the austenitic state are split into four sets due to the structural distortions of the cubic lattice (Table 3.1). Taking into account the coordination number, the effective changes of the exchange interactions may result from the decrease (from $r^{A}_{snn} = 0.2999$ to $r^{M}_{snn} = 0.2827$ nm) of the Mn(4*b*)–Mn(4*d*) distance and the increase (from $r^{A}_{tnn} = 0.4241$ to $r^{M}_{tnn} = 0.4377$ nm) of the Mn(4*b*)–Mn(4*b*) distance of the major pairs in the martensite phase. Assuming that the Mn–Mn exchange interaction curve for the martensite phase is the same as that for the austenite phase, it is strongly suggested that at least the Mn(4*b*)–Mn(4*d*) interactions in the *c* plane of the fct martensite phase, where their atomic distance decreases about 5.7% by martensitic transformation, must be antiferromagnetic as schematically illustrated in Fig. 3.11.

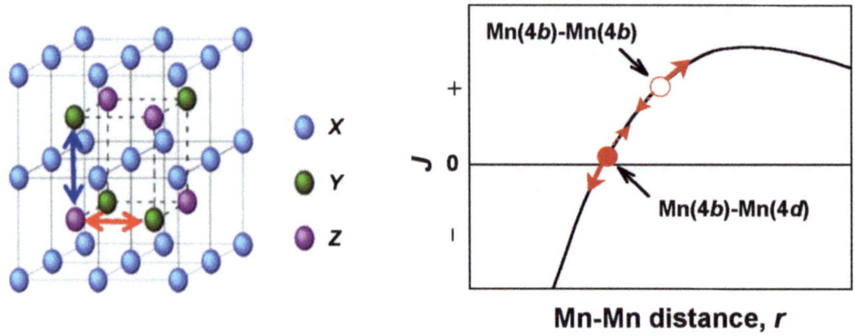

Fig. 3.11 Sketch Mn–Mn exchange interaction curve in Ni_2MnIn-based Heusler alloys [38]

3.5 Concluding Remarks

In this chapter, the order–disorder phase transformation from the $B2$ to the $L2_1$-type phase and magnetic properties of the $Ni_2Mn(Ga_xAl_{1-x})$ and the $Ni_2Mn_{1+y}In_{1-y}$ alloys were reported, and the influence of the martensitic transformation on the magnetic properties in the NiMnIn-based alloy was discussed.

1. The $T_t^{B2/L2_1}$ of the $Ni_2Mn(Ga_xAl_{1-x})$ alloys linearly increases with increasing Ga content. On the other hand, the T_C of the $L2_1$ phase in the $Ni_2Mn(Ga_xAl_{1-x})$ alloys of $x \geq 0.5$ and the T_N of the $B2$ phase of $x \leq 0.5$ are insensitive to Ga content, suggesting that the magnetic transition temperatures of the Ni_2MnAl alloy are comparable to those of the Ni_2MnGa alloy if the degree of long-range order is the same.
2. The saturation magnetization M_s and the T_C of the $Ni_2Mn(Ga_{0.5}Al_{0.5})$ alloy increase with decreasing annealing temperature. It can be concluded that these magnetic properties are brought about by the degree of long-range order between Mn($4b$) and Ga or Al($4d$) sites for the specimens.
3. In the $Ni_2Mn_{1+y}In_{1-y}$ alloys with the $L2_1$ structure, while the total magnetic moment almost linearly increases with increasing y through $y = 0$, the Curie temperature shows a unique behavior. That is, being almost constant in the low In region of $y > 0$, the T_C suddenly starts to decrease linearly with decreasing y in the high In region of $y < 0$.
4. In the $Ni_2Mn_{1.372}{}^{(57)}Fe_{0.02}In_{0.608}$ alloy, the magnetic feature of the martensite phase with low magnetization in temperatures just below martensitic transformation temperature is "paramagnetic," but not antiferromagnetic.

All these results strongly suggest that the unique and complex behaviors of magnetic properties appearing in both the disordered austenite phase and the martensite phase in the Heusler alloys are due to the magnetic exchange interaction which is a function of the interatomic distance of Mn($4b$)–Mn($4d$) pairs in the second nearest neighbor.

Acknowledgments The studies presented in this chapter were supported by a Grant-in-Aid for Scientific Research from the Japan Society for the Promotion of Science (JSPS). The authors are very grateful to Drs. W. Ito, H. Ishikawa, T. Miyamoto, M. Nagasako, and Prof. K. Ishida (Tohoku University, Sendai), Dr. V.V. Khovaylo (National University of Science and Technology MISiS, Moscow), Prof. T. Kanomata (Tohoku Gakuin University, Tagajo), and Prof. Y. Amako (Shinshu University, Matsumoto) for their helpful discussions.

References

1. K. Ullakko, J.K. Huang, C. Kantner, R.C. O'Handley, V.V. Kokorin, Large magneticfieldinduced strains in Ni_2MnGa single crystals. Appl. Phys. Lett. **69**, 1966 (1996)
2. P.J. Webster, K.R.A. Ziebeck, S.L. Town, M.S. Peak, Magnetic order and phase-transformation in Ni_2MnGa. Philos. Mag. B **49**, 295 (1984)
3. R.C. O'Handley, Model for strain and magnetization in magnetic shape-memory alloys. J. Appl. Phys. **83**, 3263 (1998)
4. A. Sozinov, A.A. Likhachev, K. Ullakko, Crystal structures and magnetic anisotropy properties of Ni–Mn–Ga martensitic phases with giant magnetic-field-induced strain IEEE Trans. Magn. **38**, 2814 (2002)
5. R. Kainuma, Y. Imano, W. Ito, Y. Sutou, H. Morito, S. Okamoto, O. Kitakami, K. Oikawa, A. Fujita, T. Kanomata, K. Ishida, Magnetic-field-induced shape recovery by reverse phase transformation. Nature **439**, 957 (2006)
6. R. Kainuma, Y. Imano, W. Ito, H. Morito, Y. Sutou, K. Oikawa, A. Fujita, K. Ishida, S. Okamoto, O. Kitakami, T. Kanomata, Metamagnetic shape memory effect in a Heusler-type $Ni_{43}Co_7Mn_{39}Sn_{11}$ polycrystalline alloy. Appl. Phys. Lett. **88**, 192513 (2006)
7. T. Krenke, E. Duman, M. Acet, E.F. Wassermann, X. Moya, L. Mañosa, A. Planes, Inverse magnetocaloric effect in ferromagnetic Ni–Mn–Sn alloys. Nat. Mater. **4**, 450 (2005)
8. Z.D. Han, D.H. Wang, C.L. Zhang, S.L. Tang, B.X. Gu, Y.W. Du, Large magnetic entropy changes in the $Ni_{45.4}Mn_{41.5}In_{13.1}$ ferromagnetic shape memory alloy. Appl. Phys. Lett. **89**, 182507 (2006)
9. S.Y. Yu, Z.H. Liu, G.D. Liu, J.L. Chen, Z.X. Cao, G.H. Wu, B. Zhang, X.X. Zhang, Large magnetoresistance in single-crystalline $Ni_{50}Mn_{50-x}In_x$ alloys ($x = 14$–16) upon martensitic transformation. Appl. Phys. Lett. **89**, 162503 (2006)
10. B. Zhang, X.X. Zhang, S.Y. Yu, J.L. Chen, Z.X. Cao, G.H. Wu, Giant magnetothermal conductivity in the Ni–Mn–In ferromagnetic shape memory alloys. Appl. Phys. Lett.**91**, 012510 (2007)
11. M. Khan, I. Dubenko, S. Stadler, N. Ali, Appl. Phys. Lett. **91**, 072510 (2007)
12. A. Planes, L. Mañosa, M. Acet, Magnetocaloric effect and its relation to shape–memory properties in ferromagnetic Heusler alloys. J. Phys. Condens. Matter **21**, 233201 (2009)
13. H. Ishikawa, R.Y. Umetsu, K. Kobayashi, A. Fujita, R. Kainuma, K. Ishida, Atomic ordering and magnetic properties in $Ni_2Mn(Ga_xAl_{1-x})$ Heusler alloys. Acta Mater.**56**, 4789 (2008)
14. R. Kainuma, N. Satoh, X.J. Liu, I. Ohnuma, K. Ishida, Phase equilibria and Heusler phase stability in the Cu-rich portion of the Cu–Al–Mn system. J. Alloys Compd. **266**, 191 (1998)
15. K. Ishikawa, R. Kainuma, I. Ohnuma, K. Aoki, K. Ishida, Phase stability of the X_2AlTi (X: Fe, Co, Ni and Cu) Heusler and B2-type intermetallic compounds. Acta Mater. **50**, 2233 (2002)
16. Y. Sutou, I. Ohnuma, R. Kainuma, K. Ishida, Ordering and martensitic transformations of Ni_2AlMn heusler alloys. Metall. Mater. Trans. A **29**, 2225 (1998)
17. K.R.A. Ziebeck, P.J. Webster, Helical magnetic order in Ni_2MnAl. J. Phys. F Met. Phys. **5**, 1756 (1975)
18. V.V. Godlevsky, K.M. Rabe, Soft tetragonal distortions in ferromagnetic Ni_2MnGa and related materials from first principles. Phys. Rev. B **63**, 134407 (2001)

19. J. Enkovaara, A. Ayuela, J. Jalkanen, L. Nordström, R.M Nieminen, First-principles calculations of spin spirals in Ni_2MnGa and Ni_2MnAl. Phys. Rev. B **67**, 054417 (2003)
20. A. Ayuela, J. Enkovaara, K. Ullakko, R.M Nieminen, Structural properties of magnetic Heusler alloys. J. Phys. Condens. Matter **11**, 2017 (1999)
21. R. Kainuma, F. Gejima, Y. Sutou, I. Ohnuma, K. Ishida, Ordering, martensitic and ferromagnetic transformations in Ni–Mn–Al Heusler shape memory alloys. Mater. Trans. JIM **41**, 943 (2000)
22. A. Fujita, K. Fukamichi, F. Gejima, R. Kainuma, K. Ishida, Magnetic properties and large magnetic-field-induced strains in off-stoichiometric Ni–Mn–Al Heusler alloys. Appl. Phys. Lett. **77**, 3054 (2000)
23. G. Inden, Determination of chemical and magnetic interchange energy in BCC alloys 1. General treatment. Z. Metallkde **66**, 577 (1975)
24. R. Kainuma, K. Urushiyama, C.C. Jia, I. Ohnuma, K. Ishida, Ordering and phase separation in b.c.c. aluminides of the Ni–Fe–Al–Ti system. Mater. Sci. Eng. A **240**, 240235 (1997)
25. R.W. Overholser, M. Wuttig, D.A. Neumann, Chemical ordering in Ni–Mn–Ga Heusler alloys. Scripta Mater. **40**, 1095 (1999)
26. F. Albertini, L. Pareti, A. Paoluzi, L. Morellon, P.A. Algarabel, M.R. Ibarra, L. Righi, Composition and temperature dependence of the magnetocrystalline anisotropy in $Ni_{2+x}Mn_{1+y}Ga_{1+z}(x+y+z=0)$ Heusler alloys. Appl. Phys. Lett. **81**, 4032 (2002)
27. A.N. Vasil'ev, A.D. Bozhko, V.V. Khovailo, I.E. Dikshtein, V.G. Shavrov, V.D. Buchelnikov, M. Matsumoto, S. Suzuki, T. Takagi, J. Tani, Structural and magnetic phase transitions in shape-memory alloys $Ni_{2+x}Mn_{1-x}Ga$. Phys. Rev. B **59**, 1113 (1999)
28. V. Sánchez-Alarcos, V. Recarte, J.I. Pérez-Landazábal, G.J. Cuello Correlation between atomic order and the characteristics of the structural and magnetic transformations in Ni–Mn–Ga shape memory alloys. Acta Mater **55** 3883 (2007)
29. E. Şaşıoğlu, L.M. Sandratskii, P. Bruno, Role of conduction electrons in mediating exchange interactions in Mn-based Heusler alloys. Phys. Rev. B **77**, 064417 (2008)
30. T. Miyamoto, W. Ito, R.Y. Umetau, R. Kainuma, T. Kanomata, K. Ishida, Phase stability and magnetic properties of $Ni_{50}Mn_{50-x}In_x$ Heusler-type alloys Scripta Mater.**62**, 151 (2010)
31. Y. Sutou, Y. Imano, N. Koeda, T. Omori, R. Kainuma, K. Ishida, K. Oikawa, Magnetic and martensitic transformations of $NiMnX(X = In, Sn, Sb)$ ferromagnetic shape memory alloys. Appl. Phys. Lett. **85**, 4358 (2004)
32. T. Krenke, M. Acet, E.F. Wassermann, X. Moya, L. Mañosa, A. Planes, Martensitic transitions and the nature of ferromagnetism in the austenitic and martensitic states of Ni–Mn–Sn alloys. Phys. Rev. B **72**, 014412 (2005)
33. T. Kanomata, T. Yasuda, S. Sasaki, H. Nishihara, R. Kainuma, W. Ito, K. Oikawa, K. Ishida, K.-U. Neumann, K.R.A. Ziebeck, Magnetic properties on shape memory alloys $Ni_2Mn_{1+x}In_{1-x}$. J. Magn. Magn. Mater. **321**, 773 (2009)
34. R.Y. Umetsu, Y. Kusakari, T.T. Kanomata, K. Suga, Y. Sawai, K. Kindo, K. Oikawa, R. Kainuma, K. Ishida, Metamagnetic behaviour under high magnetic fields in $Ni_{50}Mn_{50-x}In_x$ ($x = 14.0$ and 15.6) shape memory alloys. J. Phys. D Appl. Phys. **42**, 075003 (2009)
35. E. Şaşıoğlu, L.M. Sandratskii, P. Bruno, First-principles calculation of the intersublattice exchange interactions and Curie temperatures of the full Heusler alloys $Ni_2MnX(X = Ga, In, Sn, Sb)$. Phys. Rev. B **70**, 024427 (2004)
36. E. Şaşıoğlu, L.M. Sandratskii, P. Bruno, Pressure dependence of the Curie temperature in Ni_2MnSn Heusler alloy: A first-principles study. Phys. Rev. B **71**, 214412 (2005)
37. T. Kanomata, K. Fukushima, H. Nishihara, R. Kainuma, W. Itoh, K. Oikawa, K. Ishida, K.U. Neumann, K.R.A. Ziebeck, Magnetic and crystallographic properties of shape memory alloys $Ni_2Mn_{1+x}Sn_{1-x}$. Mater. Sci. Forum **583**, 119 (2008)
38. V.V. Khovaylo, T. Kanomata, T. Tanaka, M. Nakashima, Y. Amako, R. Kainuma, R.Y. Umetsu, H. Morito, H. Miki Magnetic properties of $Ni_{50}Mn_{34.8}In_{15.2}$ probed by Mössbauer spectroscopy. Phys. Rev. B **80**, 144409 (2009)

39. P.J. Brown, A.P. Gandy, K. Ishida, R. Kainuma, T. Kanomata, K.-U. Neumann, K. Oikawa, B. Ouladdiaf, K.R.A. Ziebeck, The magnetic and structural properties of the magnetic shape memory compound $Ni_2Mn_{1.44}Sn_{0.56}$. J. Phys. Condens. Matter **18**, 2249 (2006)
40. P.A. Bhobe, K.R. Priolkar, A.K. Nigam, Magnetostructural phase transitions in $Ni_{50}Mn_{25+x}Sb_{25-x}$ Heusler alloys. J. Phys. D Appl. Phys. **41**, 235006 (2008)
41. M. Khan, I. Dubenko, S. Stadler, N. Ali, Magnetostructural phase transitions in $Ni_{50}Mn_{25+x}Sb_{25-x}$ Heusler alloys. J. Phys. Condens. Matter **20**, 235204 (2008)
42. K. Oikawa, W. Ito, Y. Imano, Y. Sutou, R. Kainuma, K. Ishida, S. Okamoto, O. Kitakami, T. Kanomata, Effect of magnetic field on martensitic transition of $Ni_{46}Mn_{41}In_{13}$ Heusler alloy. Appl. Phys. Lett. **88**, 122507 (2006)
43. W. Ito, Y. Imano, R. Kainuma, Y. Sutou, K. Oikawa, K. Ishida, Martensitic and magnetic transformation behaviors in Heusler-type NiMnIn and NiCoMnIn metamagnetic shape memory alloys Metal. Mater. Trans. A **38**, 759 (2007)
44. A.G. Khachaturyan, S.M. Shapiro, S. Semenovskaya, Adaptive phase formation in martensitic transformation. Phys. Rev. B **43**, 10832 (1991)
45. V.D. Bchelnikov, P. Entel, V. Taskaev, V.V. Sokolovskiy, A. Hucht, M. Ogura, H. Akai, M.E. Gruner, S.K. Nayak, MonteCarlo study of the influence of antiferromagnetic exchange interaction on the phase transition of ferromagnetic Ni–Mn–X alloys (X = In, Sn, Sb). Phys. Rev. B **78**, 184427 (2008)

Chapter 4
Magnetic Interactions Governing the Inverse Magnetocaloric Effect in Martensitic Ni–Mn-Based Shape-memory Alloys

S. Aksoy, M. Acet, T. Krenke, E.F. Wassermann, M. Gruner, P. Entel, L. Mañosa, A. Planes, and P.P. Deen

Abstract Ni–Mn–X Heusler-type alloys (X: group IIIB–VB elements) undergo martensitic transformations, and many of them exhibit magnetic shape-memory and field-induced effects, one of the most predominant being the inverse magnetocaloric effect. To understand the cause of the inverse magnetocaloric effect, which involves a magnetic entropy increase with applied field, it is necessary to understand the

M. Acet (✉) · E.F. Wassermann
Physics Department, University of Duisburg-Essen, 47048 Duisburg, Germany
e-mail: mehmet.acet@uni-due.de; eberhard.wassermann@uni-due.de

M. Gruner · P. Entel
Faculty of Physics & CeNIDE, University Duisburg-Essen, 47048 Duisburg, Germany
e-mail: Markus.Gruner@uni-due.de; entel@thp.uni-duisburg.de

S. Aksoy
Faculty of Engineering & Natural Sciences, Sabanci University, 34956 Istanbul, Turkey
e-mail: saksoy@sabanciuniv.edu

T. Krenke
Thyssen Krupp Electrical Steel GmbH, 45881 Gelsenkirchen, Germany
e-mail: thorsten.krenke@thyssenkrupp.com

L. Mañosa
Departament d'Estructura i Constituents de la Matèria, Facultat de Física, Universitat de Barcelona Diagonal 647, 08028 Barcelona, Catalonia (Spain)
e-mail: lluis@ecm.ub.es

A. Planes
Facultat de Física, Departament d'Estructura i Constituents de la Matèria, Universitat de Barcelona, Diagonal 647, 08028 Barcelona Catalonia, Spain

Institut de Nanociència i Nanotecnologia, Universitat de Barcelona, Catalonia, Spain
e-mail: toni@ecm.ub.es; antoniplanes@ub.edu

P.P. Deen
European Spallation Source ESS AB P.O Box 176, SE-221 00 Lund, Sweden
e-mail: pascale.deen@esss.se

T. Kakeshita et al. (eds.), *Disorder and Strain-Induced Complexity in Functional Materials*, Springer Series in Materials Science 148,
DOI 10.1007/978-3-642-20943-7_4,

nature of the magnetic coupling in the temperature vicinity of the martensitic transition.

We present results on neutron polarization analysis experiments on Ni–Mn-based martensitic Heusler systems, with which we show that around M_s, the magnetic short-range correlations at temperatures $T < M_s$ are antiferromagnetic. We discuss the relationship of the magnetic coupling and the inverse magnetocaloric effect.

4.1 Introduction

The martensitic transformation in Heusler alloys is the source of a rich variety of physical phenomena that have diverse potential application possibilities particularly in the fields of magnetic shape-memory, and magnetocaloric materials [1]. The discovery of the magnetic shape-memory effect in Ni–Mn–Ga alloys [2] triggered broad research, first on the improvement of the properties of this alloy and on the understanding of their fundamental properties and then on the search for new Heusler materials that would exhibit favourable properties related in particular to the martensitic transformation [3–5]. Many Heusler alloys are now known to undergo martensitic transitions, and, with a vast number of possible elemental combinations, new ones await discoveries. The results of research on magnetic shape memory is not limited only to the behaviour described by its own topic, but has provided further understanding on other physical phenomena observed in these materials such as large conventional and inverse magnetocaloric effects (MCE) [5–10], large field-induced strains related to the reverse martensitic transformation [11, 12], large magnetoresistance [13–15], austenite arrest [16, 17], and exchange bias [18–20], to name a few. Much effort is invested also in theoretical work aiming to provide thorough understanding of these effects [21, 22].

In the course of search for magnetic shape-memory materials, one is confronted with the question as to what the interplay between the magnetic and lattice degrees of freedom is and how the effects mentioned above are to be understood within such a framework. But, before one can even attempt to deal with such a problem, it is first necessary to provide a portrait of the nature of magnetic coupling around and beyond the martensitic transition in these materials. In this work, we undertake such a study confined to Ni–Mn-based Heusler alloys that undergo martensitic transformations to provide a closer understanding of the inverse MCE. We first discuss the conventional and the inverse MCE in the prototype system $Ni_{50}Mn_{34}In_{16}$ and, subsequently, the relationship between the magnetic structure in the martensitic state of Ni–Mn-based Heusler alloys and the inverse MCE.

4.2 The Inverse Magnetocaloric Effect Around a Structural Transitions in a Ferromagnetic System

MCEs at first-order transitions have been thoroughly discussed in [23], and we extend the case to the inverse MCE in martensitic Heusler alloys. Martensitic transitions in ferromagnetic (FM) Heusler alloys are usually accompanied by

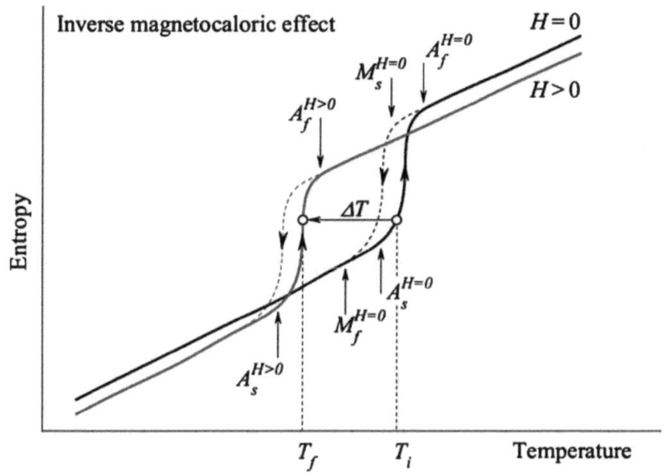

Fig. 4.1 Schematic representation of the temperature dependence of the entropy under zero field and applied field. The characteristic temperatures related to the transition are indicated by *vertical arrows* (some are omitted for clarity). The *dashed and solid curves* represent the forward and reverse transformations, respectively, as also indicated with *arrows*. ΔT is the temperature change associated with the inverse MCE

a thermal hysteresis, which is characterized by the martensite start and finish temperatures, M_s and M_f, and the austenite start and finish temperatures, A_s and A_f. These temperatures can shift to higher or lower values when a magnetic field is applied, whereby the field stabilizes the state with the higher magnetization. In many Ni–Mn-based martensitic Heusler alloys, the state with lower magnetization can be the low-temperature martensite state, and, at the same time, an externally applied magnetic field H can shift the characteristic temperatures to lower values [12]. In this case, the temperature dependence of the entropy $S(T)$ for $H = 0$ and $H > 0$ would show characteristic features in the transition region as shown in the schematic drawing in Fig. 4.1. The characteristic temperatures for $H = 0$ and $H > 0$ are indicated by arrows (some are omitted for clarity). The austenite-to-martensite transformation paths are shown with the dashed lines, and the reverse transformations are shown with the solid lines. At temperatures where the system is completely in austenite or martensite phases for both $H = 0$ and $H > 0$, $S_{H=0}(T) > S_{H>0}(T)$, whereas within the transition region, where austenite and martensite coexist, $S_{H=0}(T) < S_{H>0}(T)$. Although Fig. 4.1 describes basically the inverse MCE, the actual temperature dependence of the entropy with and without field in real systems can be more complex.

In the inverse MCE, the sample cools on applying a field adiabatically [24,25]. It owes its presence to the shift of the martensitic transformation to lower temperatures when a field is applied. Referring further to Fig. 4.1, if the sample is brought to a temperature T_i from a temperature $T < M_f^{H=0}$ such that $T_i > A_s^{H=0}$, it will be located on the reverse transformation branch with a certain proportion of martensite-to-austenite r_i. When a field is applied adiabatically, the temperature of the sample drops to T_f by an amount ΔT, and at the same time, the sample acquires a new proportion r_f such that $r_f < r_i$.

We discuss below the conventional and inverse MCE in $Ni_{50}Mn_{34}In_{16}$ and then give an experimental and theoretical account on the nature of magnetic coupling in the various crystallographic states of Ni–Mn-based Heusler alloys.

4.2.1 *Conventional and Inverse Magnetocaloric Effects in* $Ni_{50}Mn_{34}In_{16}$

The shift in the transformation temperature with applied field in $Ni_{50}Mn_{34}In_{16}$ can be as high as $\sim$ 10 KT^{-1} [12]. Figure 4.2a shows the temperature dependence of the measured temperature-change $\Delta T(T)$ and the entropy-change $\Delta S(T)$ calculated from magnetization isotherms [25]. The data show conventional MCE for 240 <

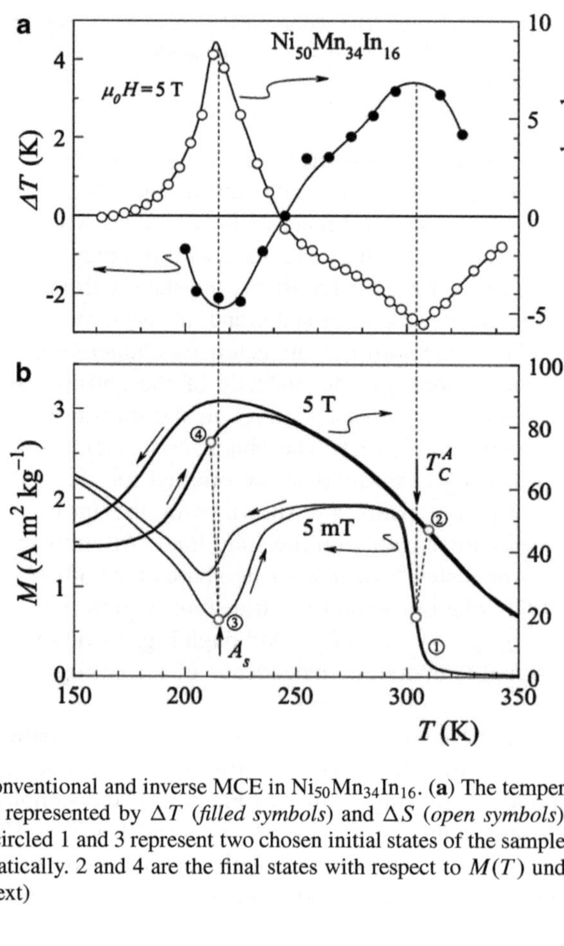

Fig. 4.2 The conventional and inverse MCE in $Ni_{50}Mn_{34}In_{16}$. (**a**) The temperature dependence of the MCE in 5 T represented by ΔT (*filled symbols*) and ΔS (*open symbols*). (**b**) $M(T)$ in 5 mT and 5 T. The encircled 1 and 3 represent two chosen initial states of the sample before a field of 5 T is applied adiabatically. 2 and 4 are the final states with respect to $M(T)$ under 5 T after the field is applied (see text)

$T \leq 350$ K, where the sample warms on applying a field ($\Delta S < 0$), and the inverse MCE for $180 < T \leq 240$ K, where the sample cools on applying a field ($\Delta S > 0$).

The temperature dependence of the cooling and warming magnetization data $M(T)$ (shown by arrows) in applied fields of 5 mT and 5 T are shown in Figs. 4.2b. The austenite Curie temperature T_C^A and the austenite start temperature A_s are shown with vertical arrows. The maximum in $\Delta T(T)$ (minimum in $\Delta S(T)$) in the conventional MCE occurs at about T_C^A, whereas in the inverse MCE, the minimum in $\Delta T(T)$ (maximum in $\Delta S(T)$) is located at about A_s on the $M(T)$ data taken under 5 mT, as shown with the vertical dashed lines.

A field of 5 T applied adiabatically at T_C^A (point 1) leads to a 4 K temperature rise, as read from Fig. 4.2a. The state of the sample with respect to the $M(T)$ curve in 5 T corresponds then to point 2. The crystallographic state of the sample does not change in this process, and it is austenite before and after applying the field. On the other hand, applying a 5 T-field at about A_s (point 3), where the sample is essentially 100% martensite, causes the temperature to drop about 2 K because of the inverse MCE. This carries the state of the sample to point 4 on the reverse transformation branch of the 5 T–$M(T)$ curve, with the sample now in a mixed state of martensite and austenite.

4.2.2 *Magnetic Coupling in Ni–Mn-Based Martensitic Heusler Alloys*

Many martensitic Heusler alloys show a strong drop in $M(T)$ below M_s [1]. However, at lower temperatures, $M(T)$ can recover and begin to increase with decreasing temperature as FM ordering sets in, as shown above in Fig. 4.2b. In order to understand the cause of this drop, we have carried out diffuse neutron scattering experiments on the D7 spectrometer at ILL-Grenoble on $Ni_{50}Mn_{37}Sn_{13}$ and $Ni_{50}Mn_{40}Sb_{10}$ using the XYZ polarization analysis technique. This method allows in paramagnets and antiferromagnets the separation of the coherent, incoherent, and magnetic contributions to the scattering solely from the geometrical scattering conditions [26]. The zero-field-cooled (ZFC), field-cooled (FC) and field-heated (FH) $M(T)$ curves for these samples are shown in Fig. 4.3a, b. $Ni_{50}Mn_{37}Sn_{13}$ ($M_s \approx 290$ K) orders ferromagnetically in both the austenite and martensite states at T_C^A and T_C^M, whereas $Ni_{50}Mn_{40}Sb_{10}$ ($M_s \approx 440$ K) orders ferromagnetically only in the martensitic state [27]. No FM ordering occurs in the austenite state of this alloy. The polarization analysis experiments for $Ni_{50}Mn_{37}Sn_{13}$ were carried out at temperatures $T > T_C^A$ and at the temperature corresponding to the local minimum in $M(T)$ at about 250 K. For $Ni_{50}Mn_{40}Sb_{10}$, they were carried out at $T > M_s$ and $T_C < T < M_s$.

Figure 4.4 shows the magnetic scattering cross section $(d\sigma/d\Omega)_{mag}$ as a function of the wave vector for $Ni_{50}Mn_{37}Sn_{13}$. As seen in Fig. 4.4a, for $T > M_s$, strong forward scattering is found at all measured temperatures, and finite scattering is

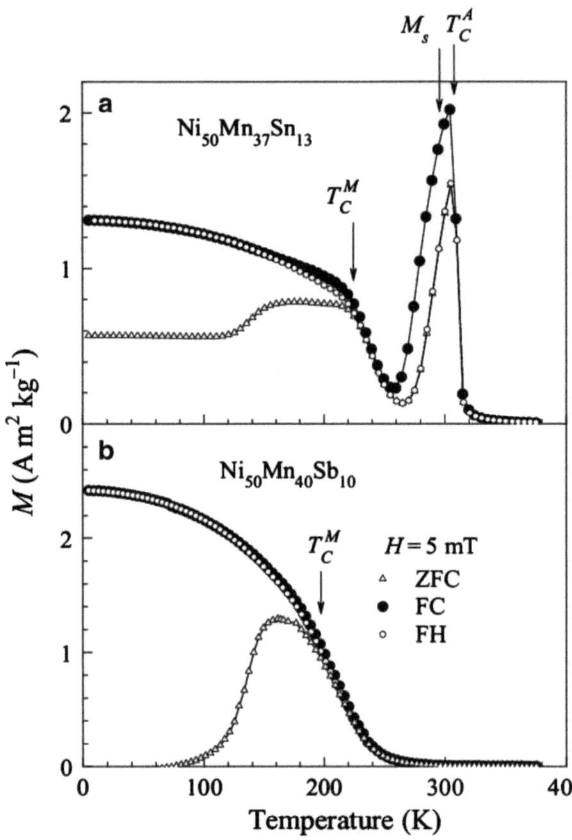

Fig. 4.3 Magnetization vs. temperature in 50 Oe for (**a**) $Ni_{50}Mn_{37}Sn_{13}$ and (**b**) $Ni_{50}Mn_{40}Sb_{10}$

observed at higher q-values. The data for 330 and 405 K are shifted vertically for clarity, and the spikes at about 2.2 $Å^{-1}$ are due to uncertainties in the separation process of the cross-section components with the XYZ-technique.

The forward scattering in Fig. 4.4 is an indication of the presence of strong FM correlations well in the paramagnetic state of the sample ($T_C^A \approx 300$ K). As seen in the inset, where the data are not shifted and are plotted in the low q-range, the forward scattering strengthens with decreasing temperature as expected. However, at 250 K (Fig. 4.4b), which corresponds nearly to the temperature of the minimum in $M(T)$ (Fig. 4.3b), the forward scattering is absent indicating the absence of FM exchange (the 500 K-data are plotted here for comparison). Instead what remains is a broad peak centred around 1.6 $Å^{-1}$, indicating the presence of short-range AF correlations. We note here that only Z-polarization analysis could be performed for the measurements at 250 K, at which the flipping ratio of the instrument decreases due to some coexisting FM rest-austenite.

The situation for $Ni_{50}Mn_{40}Sb_{10}$ is not much different as seen in Fig. 4.5, and FM correlations are also found at temperatures as high as 500 K well in the

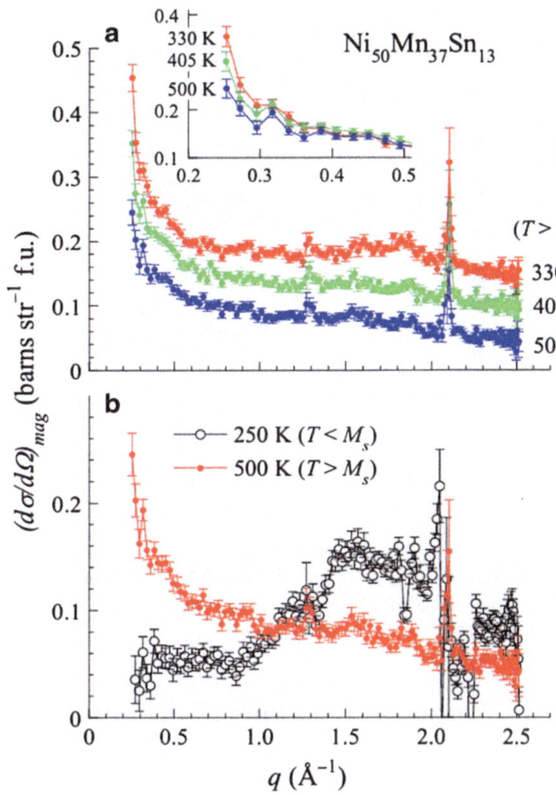

Fig. 4.4 The wave vector dependence of the magnetic differential cross section for $Ni_{50}Mn_{37}Sn_{13}$; (**a**) $(d\sigma/d\Omega)_{mag}$ for $T > M_s$; the spectra are shifted by 0.05 and 0.10 units for 330 and 405 K, respectively. The *inset* shows the same data in the range of the forward scattering. The forward scattering strengthens with decreasing temperature. (**b**) $(d\sigma/d\Omega)_{mag}$ at 250 K $T < M_s$ compared to that at 500 K ($T > M_s$). Forward scattering is not found at 250 K

paramagnetic regime. However, $Ni_{50}Mn_{40}Sb_{10}$ does not order ferromagnetically in the austenitic state as $Ni_{50}Mn_{37}Sn_{13}$ does. Nevertheless, we find practically the same behaviour for $T < M_s$ for both alloys. Namely, FM correlations disappear, and only AF correlations that appear as a broad peak centred around 1.6 Å^{-1} are found. In fact, as seen in the inset, the q-dependence of $(d\sigma/d\Omega)_{mag}$ for $Ni_{50}Mn_{37}Sn_{13}$ and $Ni_{50}Mn_{40}Sb_{10}$ is practically identical, suggesting that the spin configuration of the antiferromagnetism should be similar.

Additionally, the 320 K and the 500 K-data in Fig. 4.5 show similar q-dependence for $q > 1.6$ Å^{-1}, meaning that any profile similar to that of the 320 K-data for $q < 1.6$ Å^{-1} could be well hidden underneath the forward scattering of the 500 K-data. This then would not exclude the fact that FM and AF correlations would be coexisting at high temperatures.

The inverse MCE relies on the shift of M_s to lower temperatures when an external field is applied, and the rate of shift is different for different X species in Ni–Mn–X [12]. According to Fig. 4.1, the greater the shift, the larger ΔT would be. The size of the shift is expected to be related to the total energy difference of the martensite and austenite phases, which can be affected by type and strength of the magnetic coupling in each of the phases.

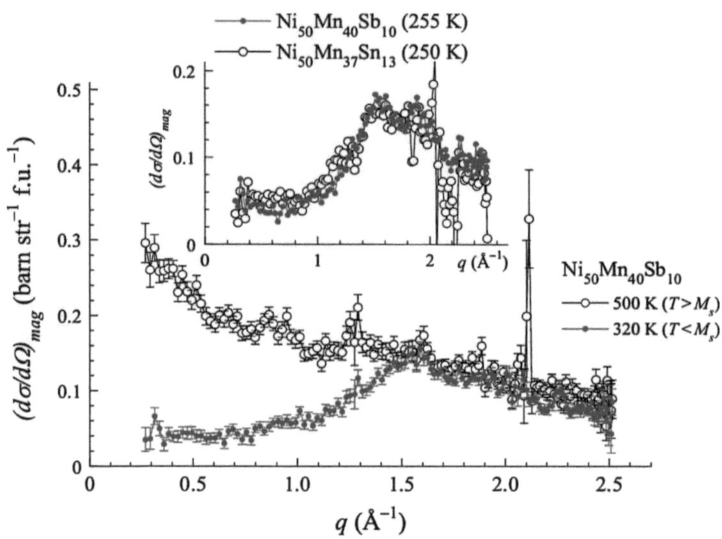

Fig. 4.5 $(d\sigma/d\Omega)_{mag}$ vs. q for $Ni_{50}Mn_{40}Sb_{10}$ at 320 and 500 K. The *inset* compares the magnetic scattering of $Ni_{50}Mn_{37}Sn_{13}$ and $Ni_{50}Mn_{40}Sb_{10}$ at similar temperatures corresponding to the martensitic state

4.2.3 Magnetic Exchange Constants in Ni–Mn-Based Martensitic Heusler Alloys

To examine closer the possibility of mixed magnetic coupling at high temperatures and below M_s, we have calculated the effective magnetic exchange constants J_{ij} between pairs of individual atoms for $Ni_{50}Mn_{40}Sb_{10}$ ($Ni_2Mn_{1.6}Sb_{0.4}$). The calculation was performed within the density functional theory (DFT) by means of the Liechtenstein approach [28] using the Munich SPR-KKR code (version 5.4) [29,30]. We have evaluated an FM cubic $L2_1$ Heusler configuration for the experimental lattice constant of $a = 5.967$ Å and a tetragonally distorted configuration with $(c/a) = 0.92$ at the same volume as the cubic phase. Excess Mn was assumed to be randomly distributed on the Sb sites describing disorder within the single-site coherent potential approximation. For the Brillouin-zone integration, we employed 624 irreducible k-points for the $L2_1$ structure and 1,183 for the tetragonal case. Levels up to the f-states are included in the angular momentum expansion.

A comparison of the exchange parameters for the cubic and tetragonal case is shown in Fig. 4.6. The most striking feature is the coexistence of strong FM Mn–Ni interactions and even stronger AF nearest neighbour Mn–Mn interactions between atoms located on the original Mn sites (Mn_1) and the Sb sites (Mn_2) of the Heusler structure. As the nearest neighbour Mn–Ni interactions are greater in number, ferromagnetism may prevail. However, this would suggest a close competition between FM, ferrimagnetic FI (with flipped Mn spins on the Sb

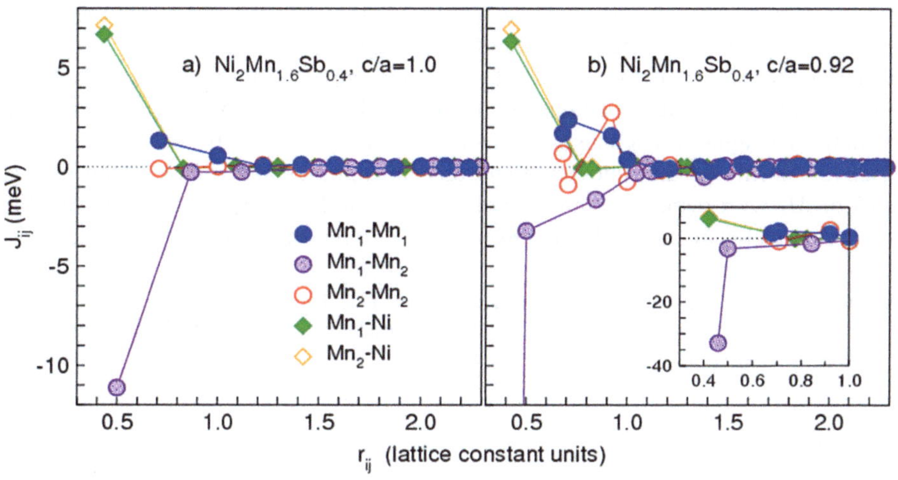

Fig. 4.6 Comparison of the exchange parameters J_{ij} between pairs of atoms i and j for different coordination shells obtained by DFT calculations for **(a)** the cubic $L2_1$ austenitic phase and **(b)** a tetragonally distorted structure, with its c-axis reduced by 8% compared to the a and b axes. The coordination shells are characterized by their interatomic distance r_{ij} given in units of the cubic lattice constant. The Ni–Ni contributions and the interactions with Sb atoms are small and are thus omitted for clarity. The results indicate a close competition between FM Mn–Ni contributions (positive values) and AF contributions (negative values) between nearest neighbour Mn atoms. The tetragonal distortion leads to a significant difference between the nearest neighbour Mn_1–Mn_2 J_{ij} within the a–b plane and with perpendicular components

sites) and, eventually, AF configurations. This is further corroborated by exchange constants calculated for an FI configuration (not shown), which exhibits a significant decrease of the FM Mn–Ni contributions caused by the breakdown of the induced moments on the Ni sites.

Tetragonal distortion leads to an anisotropy of the exchange constants: While the Mn_1–Mn_2 parameter nearly vanishes in the plane perpendicular to the shortened c-axis, it nearly triples its negative value for other directions. This situation can stabilize different FI or AF spin configurations than the cubic phase can. A detailed discussion of the exact magnetic ground state of the respective phases is, however, beyond the scope of the current paper and will be the subject of future research.

4.3 Conclusion

Neutron polarization analysis in Ni–Mn-based martensitic Heusler alloys demonstrates unequivocally the presence of FM and AF interactions in these alloys at temperatures within the austenite and the martensite states. The presence of AF interactions in such alloys has been justifiably speculated in the past, not only on the basis of the assumption that the Mn–Mn interaction would be oscillatory with distance but also on the basis of the observation of effects such as exchange bias, kinetic arrest, and the smeared feature of $M(T)$ in the FM transition in the martensite state. The q-dependence of the magnetic cross section shown here is a clear indication that the martensite state incorporates antiferromagnetism, the presence of which is also sustained by DFT calculations. The presence and strength of antiferromagnetic exchange in the martensite state is expected to affect the relative stability of the martensite and austenite phases and influence the rate of shift of M_s with respect to applied field. This would directly govern the available temperature change caused by the inverse MCE in Ni–Mn-based Heusler alloys.

Acknowledgment We would like to thank J. Minar and H.C. Herper for helpful discussions. This work was supported by the Deutsche Forschungsgemeinschaft (SPP 1239) and ILL-Grenoble.

References

1. A. Planes, L. Mañosa, M. Acet, Magnetocaloric effect and its relation to shape-memory properties in ferromagnetic Heusler alloys. J. Phys.: Condens. Matter **21**, 233201 (2009)
2. K. Ullakko, J.K. Huang, C. Kantner, R.C. O'Handley, V.V. Kokorin, Large magnetic-field-induced strains in Ni_2MnGa single crystals. Appl. Phys. Lett. **69**, 1966 (1996)
3. O. Soderberg, A. Sozinov, Y. Ge, S.-P. Hannula, V.K. Lindroos, *Giant Magnetostrictive Materials*, vol. 16, ed. by K.H.J. Buschow Handbook of Magnetic Materials (Elsevier, Amsterdam, 2006) p. 1
4. Y. Sutou, Y. Imano, N. Koeda, T. Omori, R. Kainuma, K. Ishida, K. Oikawa, Magnetic and martensitic transformations of NiMnX (X = In,Sn,Sb) ferromagnetic shape memory alloys. Appl. Phys. Lett. **85**, 4358 (2004)
5. T. Krenke, E. Duman, M. Acet, E.F. Wassermann, X. Moya, L. Mañosa, A. Planes, Inverse magnetocaloric effect in ferromagnetic Ni–Mn–Sn alloys. Nat. Mat. **4**, 450 (2005)
6. J. Marcos, L. Mañosa, A. Planes, F. Casanova, X. Batlle, A. Labarta, Multiscale origin of the magnetocaloric effect in Ni–Mn–Ga shape-memory alloys. Phys. Rev. B **68**, 094401 (2003)
7. L. Pareti, M. Solzi, F. Albertini, A. Paoluzi, Giant entropy change at the co-occurrence of structural and magnetic transitions in the $Ni_{2.19}Mn_{0.81}Ga$ Heusler alloy, Eur. Phys. J. B **32**, 303 (2003)
8. I. Dubenko, M. Khan, A.K. Pathak, B.R. Gautam, S. Stadler, N. Ali, Magnetocaloric effects in Ni–Mn–X based Heusler alloys with X = Ga, Sb,In. J. Magn. Magn. Mater. **321**, 754 (2009)
9. Z.D. Han, D.H. Wang, C.L. Zhang, H.C. Xuan, J.R. Zhang, B.X. Gu, Y.W. Du, The phase transitions, magnetocaloric effect, and magnetoresistance in Co doped Ni–Mn–Sb ferromagnetic shape memory alloys. J. Appl. Phys. **104**, 053906 (2008)
10. C. Jing, Z. Li, H.L. Zhang, J.P. Chen, Y.F. Qiao, S.X. Cao, J.C. Zhang, Martensitic transition and inverse magnetocaloric effect in Co doping Ni–Mn–Sn Heulser alloy. Eur. Phys. J. B **67**, 193 (2009)
11. R. Kainuma, Y. Imano, W. Ito, Y. Sutou, H. Morito, S. Okamoto, O. Kitakami, K. Oikawa, A. Fujita, T. Kanomata, K. Ishida: Magnetic-field-induced shape recovery by reverse phase transition. Nature **439**, 957 (2006)
12. T. Krenke, E. Duman, M. Acet, E.F. Wassermann, X. Moya, L. Mañosa, A. Planes, E. Suard, B. Ouladdiaf, Magnetic superelasticity and inverse magnetocaloric effect in Ni–Mn–In. Phys. Rev. B **75**, 104414 (2007)
13. K. Koyama, H. Okada, K. Watanabe, T. Kanomata, R. Kainuma, W. Ito, K. Oikawa, K. Ishida, Observation of large magnetoresistance of magnetic Heusler alloy $Ni_{50}Mn_{36}Sn_{14}$ in high magnetic fields. Appl. Phys. Lett. **89**, 182510 (2006)

14. V.K. Sharma, M.K. Chattopadhyay, K.H.B. Shaeb, A. Chouhan, S.B. Roy, Large magnetoresistance in $Ni_{50}Mn_{34}In_{16}$ alloy. Appl. Phys. Lett. **89**, 222509 (2006)
15. S. Chatterjee, S. Giri, S. Majumdar, S.K. De, Giant magnetoresistance and large inverse magnetocaloric effect in $Ni_2Mn_{1.36}Sn_{0.64}$alloy. J. Phys. D: Appl. Phys. **42**, 065001 (2009)
16. V.K. Sharma, M.K. Chattopadhyay, S.B. Roy, Kinetic arrest of the first order austenite to martensite phase transition in $Ni_{50}Mn_{34}In_{16}$: dc magnetization studies. Phys. Rev. B **76**, 140401R (2007)
17. W. Ito, K. Ito, R.Y. Umetsu, R. Kainuma, K. Koyama, K. Watanabe, A. Fujita, K. Oikawa, K. Ishida, T. Kanomata, Kinetic arrest of martensitic transformation in the NiCoMnIn metamagnetic shape memory alloy. Appl. Phys. Lett. **92**, 021908 (2008)
18. M. Khan, I. Dubenko, S. Stadler, N. Ali, Exchange bias behavior in Ni–Mn–Sb Heusler alloys. Appl. Phys. Lett. **91**, 072510 (2007)
19. Z. Li, C. Jing, J. Chen, S. Yuan, S. Cao, J. Zhang, Observation of exchange bias in the martensitic state of $Ni_{50}Mn_{36}Sn_{14}$ Heusler alloy. Appl. Phys. Lett. **91**, 112505 (2007)
20. A.K. Nayak, K.G. Suresh, A.K. Nigam, Observation of enhanced exchange bias behaviour in NiCoMnSb Heusler alloys. J. Phys. D: Appl. Phys. **42**, 115004 (2009)
21. V.D. Buchelnikov, P. Entel, S.V. Taskaev, V.V. Sokolovskiy, A. Hucht, M. Ogura, H. Akai, M.E. Gruner, S.K. Nayak, Monte Carlo study of the influence of antiferromagnetic exchange interactions on the phase transitions of ferromagnetic Ni–Mn–X alloys (X = In, Sn, Sb). Phys. Rev. B **78**, 184427 (2008)
22. M.A. Uijttewaal, T. Hickel, J. Neugebauer, M.E. Gruner, P. Entel, Understanding the phase transitions of the Ni_2MnGa magnetic shape memory system from first principles. Phys. Rev. Lett. **102**, 035702 (2009)
23. A.M. Tishin, Y.I. Spichkin, *The Magnetocaloric Effect and Its Applications* (Institute of Physics Publishing, Bristol, 2003)
24. X. Moya, L. Mañosa, A. Planes, S. Aksoy, M. Acet, E.F. Wassermann, T. Krenke, Cooling and heating by adiabatic magnetization in the $Ni_{50}Mn_{34}In_{16}$ magnetic shape-memory alloy. Phys. Rev. B **75**, 184412 (2007)
25. S. Aksoy, T. Krenke, M. Acet, E.F. Wassermann, X. Moya, L. Mañosa, A. Planes, Tailoring magnetic and magnetocaloric properties of martensitic transitions in ferromagnetic Heusler alloys. Appl. Phys. Lett. **91**, 241916 (2007)
26. J.R. Stewart, P.P. Deen, K.H. Andersen, H. Schober, J.-F. Barthélémy, J.M. Hillier, A.P. Murani, T. Hayes, B. Lindenau, Disordered materials studied using neutron polarization analysis on the multi-detector spectrometer, D7. J. Appl. Cryst. **42**, 69 (2009)
27. S. Aksoy, M. Acet, P.P. Deen, L. Mañosa, A. Planes, Magnetic correlations in martensitic Ni–Mn-based Heusler shape-memory alloys: Neutron polarization analysis. Phys. Rev. B **79**, 212401 (2009)
28. A.I. Liechtenstein, M.I. Katsnelson, V.A. Antropov, V.P. Gubanov, Local spin density functional approach to the theory of exchange interactions in ferromagnetic metals and alloys. J. Magn. Magn. Mater. **67**, 65 (1987)
29. Ebert H, in *Fully relativistic band structure calculations for magnetic solids – Formalism and Application*, ed. by H. Dreysse. Electronic Structure and Physical Properties of Solids, (Springer, Berlin, 1999) p. 191
30. http://olymp.cup.uni-muenchen.de/ak/ebert/SPRKKR

Chapter 5
Magnetic Field-Induced Strain in Ferromagnetic Shape Memory Alloys Fe-31.2Pd, Fe_3Pt, and Ni_2MnGa

Takashi Fukuda and Tomoyuki Kakeshita

Abstract Ferromagnetic shape memory alloys are a kind of multiferroic materials, in which a ferroelastic domain (variant in the martensite phase) and a magnetic domain are closely correlated. One interesting phenomenon caused by this correlation is the rearrangement of martensite variants by magnetic field, which is associated with a large magnetic field-induced strain of several percent. Typical alloys exhibiting such behavior are disordered Fe-31.2Pd (at.%), Fe_3Pt (degree of order $\sim$0.8), and Ni_2MnGa. In this chapter, we show martensitic transformation behavior and the magnetic field-induced strain in these three alloys. Then, we will derive the condition for realizing a large magnetic field-induced strain in ferromagnetic shape memory alloys. In addition, the origin of a weak first-order martensitic transformation in Fe_3Pt is discussed from its electronic structure.

5.1 Introduction

Multiferroic materials have two or more ferroic properties, such as ferromagnetic, ferroelectric, and ferroelastic properties, in one phase. A perovskite-type $BiMnO_3$ is a prototype of multiferroic material, which is simultaneously ferroelectric and ferromagnetic [1, 2]. Multiferroic materials are of interest because one of their intensive variables can be controlled not only by its conjugate extensive variable but also by another extensive variable. A ferromagnetic shape memory alloy (FMSMA) could be one of such multiferroic materials because ferromagnetic and ferroelastic properties coexist in its martensite phase. Due to such multiferroic nature of

T. Fukuda (✉) · T. Kakeshita
Graduate School of Engineering, Osaka University, Suita, Osaka 565-0871, Japan
e-mail: fukuda@mat.eng.osaka-u.ac.jp; kakesita@mat.eng.osaka-u.ac.jp

T. Kakeshita et al. (eds.), *Disorder and Strain-Induced Complexity in Functional Materials*, Springer Series in Materials Science 148,
DOI 10.1007/978-3-642-20943-7_5,

FMSMAs, a large strain of several percent is induced by a magnetic field in some FMAMAs.

The magnetic field-induced strain (MFIS) in FMSMAs was first discovered in Ni_2MnGa by Ullakko et al. [3]. Since then, the research in this field has attracted much interest, and as a result, a large MFIS of several percent was then found in Ni_2MnGa [4–6], Fe–Pd [7, 8] and Fe_3Pt [9, 10]. According to previous reports [3–6], the mechanism of MFIS has been explained as follows by considering large magnetic anisotropy in the martensite phase. That is, a thermally induced martensite is usually composed of several ferroelastic domains or variants separated by twinning planes as shown in Fig. 5.1b. Each variant corresponds to a magnetic domain, and the magnetic moment of each variant lies in the direction of magnetization easy axis. When a magnetic field is applied along the easy axis of one variant, the magnetic moment of the other variants, whose easy axes are not parallel to the field direction, will rotate toward the field direction. Therefore, the magnetocrystalline anisotropy energy of these variants becomes larger than that of the variant whose easy axis is parallel to the field direction. Then, the rearrangement of martensite variant (RMV) occurs to reduce the magnetic energy, as shown in Fig. 5.1c, if the shear stress for the twinning plane movement is comparatively low.

In this chapter, experimental results concerning martensitic transformation behavior and MFIS in three typical ferromagnetic shape memory alloys of Fe-31.2Pd (at.%), Fe_3Pt, and Ni_2MnGa are presented. Then we show the condition for realizing RMV by magnetic field in FMSMAs. In addition, we discuss the reason of a weak first-order martensitic transformation in Fe_3Pt from its electronic structure.

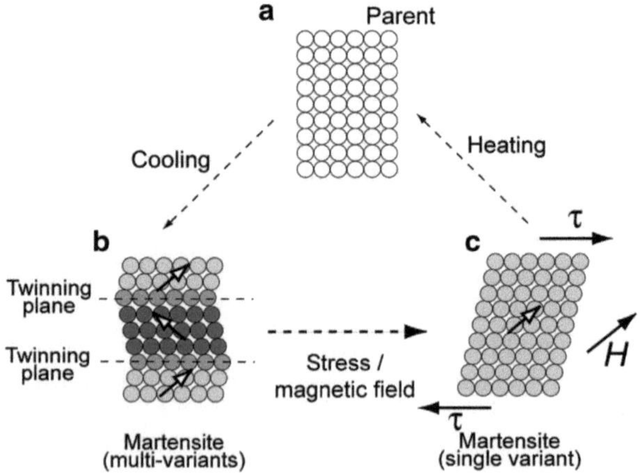

Fig. 5.1 Schematic illustration showing thermally induced martensitic transformation (**a**–**b**), and rearrangement of martensite variants by stress τ or by magnetic field H (**b**–**c**). The *open arrows* indicate the direction of magnetic moment

5.2 Martensitic Transformation in Fe-31.2Pd, Fe_3Pt, and Ni_2MnGa

Three alloys, disordered Fe-31.2Pd (at.%), ordered Fe_3Pt (degree of order $\simeq$ 0.8), and ordered Ni_2MnGa, exhibit a thermoelastic martensitic transformation from a cubic phase to a tetragonal phase (strictly, pseudotetragonal phase in Ni_2MnGa) [11–13]. The martensitic transformation in these alloys is detected as a sharp decrease in magnetic susceptibility as shown in Fig. 5.2. That is, in the cooling process, the magnetic susceptibility starts to decrease sharply as indicated by an arrow in each panel due to the fact that the ferromagnetic martensite phase has a larger magnetocrystalline anisotropy compared with the ferromagnetic parent phase. Therefore, this temperature corresponds to the martensitic transformation temperature, T_M, which is 230 K in Fe-31.2Pd alloy, 85 K in Fe_3Pt, and 202 K in Ni_2MnGa. Details of the anisotropy are described later.

We notice in Fig. 5.2c that there is an obvious hysteresis between the cooling and heating processes in the magnetic susceptibility of Ni_2MnGa, being a characteristic feature of a first-order martensitic transformation. On the other hand, the hysteresis is very small in Fe-31.2Pd and Fe_3Pt as shown in Fig. 5.2a, b, suggesting that they are weak first-order transformations.

The difference in transformation behavior between the three alloys is also seen in the temperature dependence of lattice parameters, which is shown in Fig. 5.3. That is, the lattice parameters clearly show a discontinuity at T_M in Ni_2MnGa while not in Fe-31.2Pd and Fe_3Pt. In the latter alloys, the lattice parameters change gradually below T_M. Nevertheless, there is one common feature among the three alloys:

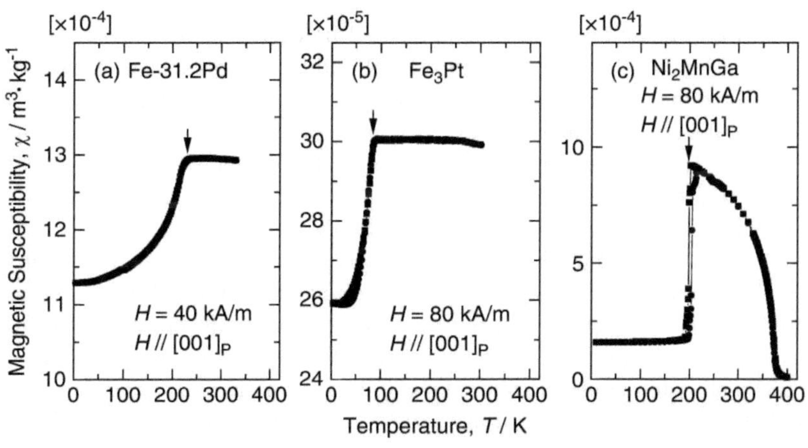

Fig. 5.2 Temperature dependence of magnetic susceptibility in the cooling and heating processes of (**a**) Fe-31.2Pd, (**b**) Fe_3Pt, and (**c**) Ni_2MnGa. Measurements are made using single crystals under a weak magnetic field applied in the $[001]_P$ direction. *Arrows* indicate a martensitic transformation temperature T_M

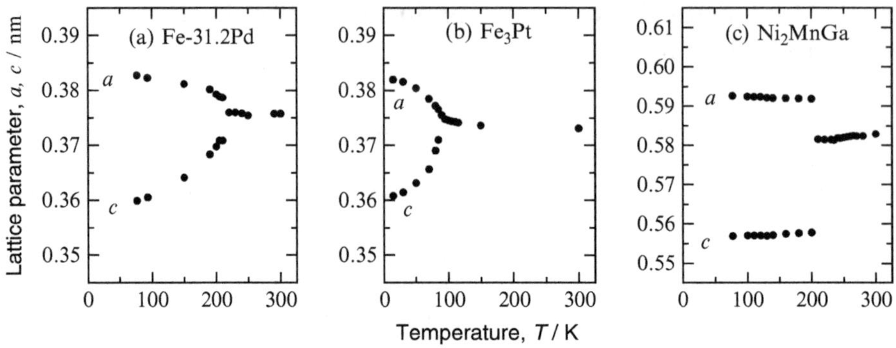

Fig. 5.3 Temperature dependence of lattice parameters obtained by X-ray diffraction of (**a**) Fe-31.2Pd, (**b**) Fe_3Pt, and (**c**) Ni_2MnGa. In case of Ni_2MnGa, a pseudotetragonal structure, which is reported by Webster et al. [21], is applied for convenience

the tetragonality c/a at temperatures sufficiently below T_M is almost the same among the three alloys and is about 0.94 at 77 K (at 13 K for Fe_3Pt). Incidentally, the martensites of the present three alloys come to have three variants (ferroelastic domains) because the c axis of the martensite phase corresponds to one of three equivalent $<001>_P$ directions (P stands for the parent phase).

5.3 Magnetic Field-Induced Strain in Fe-31.2Pd, Fe_3Pt, and Ni_2MnGa

In this section, characteristics of magnetic field-induced strain (MFIS) in the martensite state of Fe-31.2Pd (at.%), Fe_3Pt, and Ni_2MnGa are presented. Measurements of MFIS have been made by applying magnetic field in the $[001]_P$ direction and monitoring the strain in the same direction by a three-terminal capacitance method. Here, the $[001]_P$ direction corresponds to either the magnetization easy axis or the magnetization hard axis depending on variants, and the easy axis is the a axis in Fe-31.2Pd alloy, while it is the c axis in Fe_3Pt and Ni_2MnGa.

Before applying a magnetic field, each specimen has been cooled down below its T_M without applying a magnetic field, and the strain in the $[001]_P$ direction has been measured in the process. Figure 5.4 shows the strain monitored in the ZFC (zero-field-cooling) process. (The FC curves in the figure are described later.) The Fe-31.2Pd alloy has contracted below T_M (230 K). On the other hand, the Fe_3Pt has expanded after a small contraction below T_M (85 K). In the case of Ni_2MnGa, no significant shape change is observed below T_M (202 K).

After the ZFC process described above, a magnetic field of up to 3.2 MA/m has been applied to each specimen and then removed. The magnetic field-induced strain obtained is shown in Fig. 5.5. In the field applying process, the specimen expands by 3.1% in the Fe-31.2Pd, while it contracts by 2.3% in Fe_3Pt and 4.3% in Ni_2MnGa.

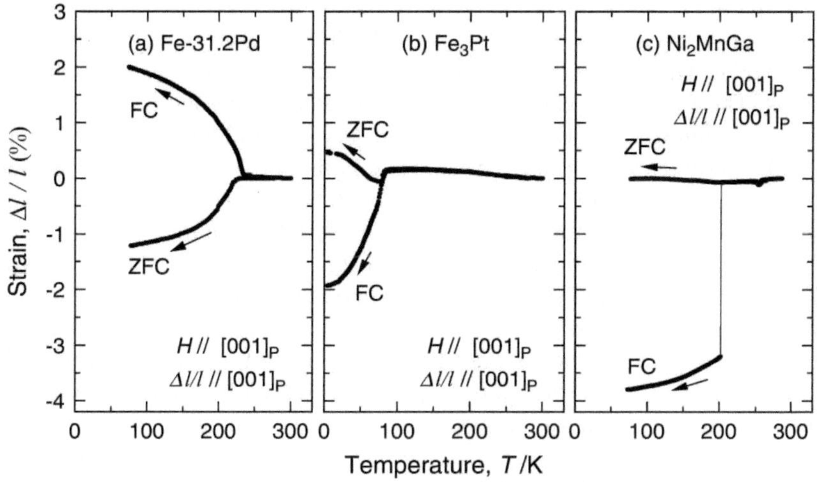

Fig. 5.4 Strain induced in the zero-field-cooling process (ZFC) and field-cooling (FC) process of (**a**) Fe-31.2Pd, (**b**) Fe_3Pt, and (**c**) Ni_2MnGa. The strain is measured in the $[001]_P$ direction, and the magnetic field of the FC process is applied in the $[001]_P$ direction

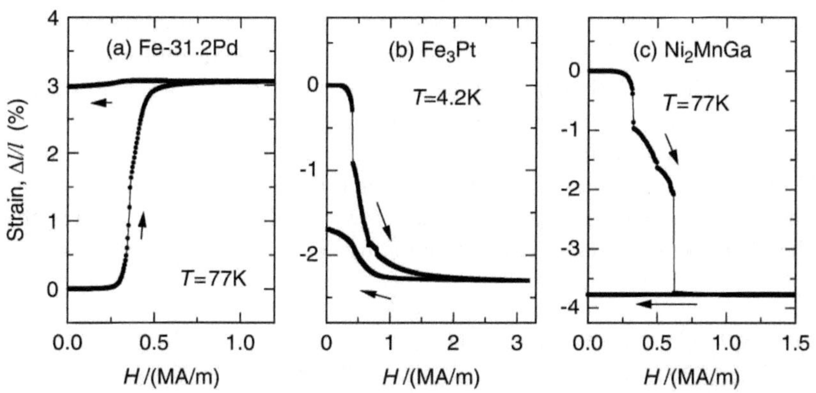

Fig. 5.5 Magnetic field-induced strain of (**a**) a Fe-31.2Pd at 77 K, (**b**) a Fe_3Pt at 4.2 K, and (**c**) a Ni_2MnGa at 77 K. Measurements are made after the ZFC process shown in Fig. 5.4. Magnetic field is applied along the $[001]_P$ direction and the strain is measured along the same direction

In the field removing process, the field-induced strain essentially does not recover in Fe-31.2Pd and Ni_2MnGa. On the other hand, a part of the strain (about 0.6%) recovers in Fe_3Pt. The origin of the recovery of the strain is unknown at present, but possibly caused by the elastic energy accumulated due to some defects during the field applying process.

In order to understand quantitatively the rearrangement of martensite variants (RMV) by magnetic field, we focus our attention on the fraction of the most preferable variant under a magnetic field, f_P. Under the magnetic field applied in the

$[001]_P$ direction, f_P is the fraction of the variant whose easy axis lies in the $[001]_P$ direction. It is given as $f_P = (1 - f_c)$ for Fe-31.2Pd and $f_P = f_c$ for the Fe_3Pt and Ni_2MnGa, where f_c is the fraction of the variant whose c axis lies in the $[001]_P$ direction. The fraction f_c can be easily calculated by using lattice parameters and the strain monitored in the $[001]_P$ direction, because the length of the specimen in the $[001]_P$ direction in a multivariant state is proportional to the average lattice parameter in this direction, which is given by $L = f_c c + (1 - f_c)a$, and we can directly obtain L by monitoring the strain in the $[001]_P$ direction.

We first calculate the fraction f_c after ZFC (77 K for Fe-31.2Pd and Ni_2MnGa, 4.2 K for Fe_3Pt) shown in Fig. 5.4. The calculated value is about 50% in Fe-31.2Pd, about 30% in Fe_3Pt, and about 30% in Ni_2MnGa. If the three variants were equivalently formed by the martensitic transformation, f_c would be one third. The deviation of f_c from one third is probably caused by some internal defects introduced through preparing the specimen. We next calculate f_c reached under the maximum magnetic field shown in Fig. 5.5. The value of f_c is approximately 0% for Fe-31.2Pd, 70% for Fe_3Pt, and 100% for Ni_2MnGa. Thus, f_P under the magnetic field is about 100% for Fe-31.2Pd and Ni_2MnGa. The result is quite natural because the total energy is the lowest in this state. On the other hand, f_P is 70% and does not reach 100% in Fe_3Pt. The reason for imperfect rearrangement in Fe_3Pt is unclear, but one possibility is due to the presence of some internal defects.

Confirmation of the RMV by magnetic field has also been made by an optical microscopy observation and the X-ray diffraction, and the result for the Fe-31.2Pd alloy is shown in Fig. 5.6. In the ZFC process, a banded surface relief appears below T_M, and the surface relief at 81 K is shown in Fig. 5.6a. Then we apply a magnetic field in the $[001]_P$ direction. When the magnetic field exceeds about 0.28 MA/m, RMV initiates as shown in Fig. 5.6b, and when the field exceeds about 0.4 MA/m, the whole area becomes a single variant as shown Fig. 5.6c. In this way, it is obvious that the MFIS occurs in association with the RMV.

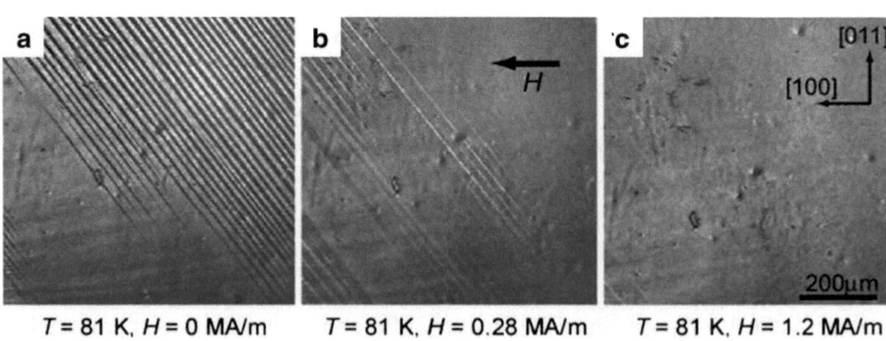

Fig. 5.6 A series of optical micrographs showing rearrangement of variants by magnetic field in a Fe-31.2Pd alloy single crystal. (**a**) The martensite phase with zero magnetic field; banded contrast appears due to the self-accommodation of martensite variants. (**b**) Magnetic field of 0.28 MA/m is applied along the $[001]_P$ direction. (**c**) Magnetic field increases to 1.2 MA/m; banded contrast completely disappears

In order to understand the effects of temperature on the RMV by magnetic field, we have examined RMV in a field-cooling process. The FC curves in Fig. 5.4 show the strain measured along the $[001]_P$ direction in the cooling process under the magnetic field of 3.2 MA/m applied in the $[001]_P$ direction. For all the alloys, the specimen starts to expand or contract at T_M. In the case of Fe-31.2Pd and Ni_2MnGa, the fraction f_P reaches nearly 100% at T_M. This means that the shape change below T_M in the FC curve is essentially caused by the change in lattice parameters. In the case of Fe_3Pt, however, the fraction f_P is about 50% at T_M, and it increases with decreasing temperature down to 40 K, and then maintains a constant value (75%) below it. This result means that the RMV by magnetic field basically depends on temperature. Also, we have confirmed that the RMV by magnetic field occurs at all temperatures below T_M in the three alloys after ZFC process.

Incidentally, the recovery of MFIS observed in Fe_3Pt (Fig. 5.5b) is of technological importance because we can repeatedly induce the strain by applying and removing a magnetic field. Thus, we have examined temperature dependence of the recoverable strain in Fe_3Pt, and some of the results are shown in Fig. 5.7. As seen in the figure, a large strain of about 1% with a small hysteresis is obtained at 20 K. This strain is nearly five times as large as that in the magnetostriction of Terfenol-D.

In the above experiments, the magnetic field has been applied in the $[001]_P$ direction. Then, we have examined the RMV by applying magnetic field in the $[110]_P$ and $[111]_P$ directions. As a result, we have confirmed in the present three alloys that the RMV occurs when the magnetic field is applied in the $[110]_P$ direction, although the fraction of the preferable variant does not reach 100% in this case. We have also confirmed in the three alloys that the RMV does not occur when the field is applied in the $[111]_P$ direction. This result is quite reasonable because the magnetic anisotropy energy is the same among the three variants when the magnetic

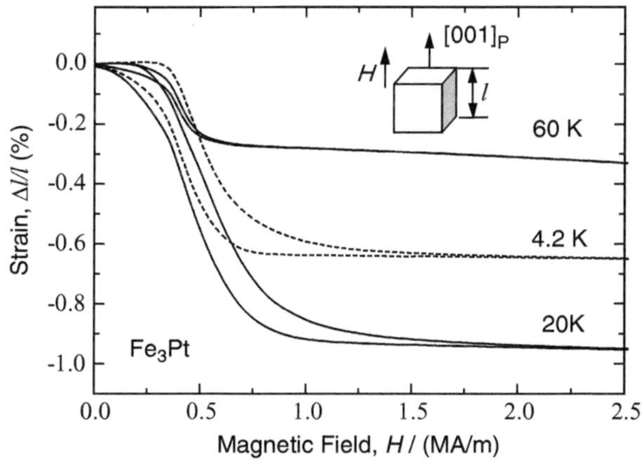

Fig. 5.7 Reversible magnetic field-induced strain of Fe_3Pt measured at 60, 20, and 4.2 K. The strain repeatedly appears when application and removal of magnetic field is repeated

field is applied in the $[111]_P$ direction, and strongly ensures that the RMV is certainly related to magnetocrystalline anisotropy.

To make use of FMSMAs as actuator materials, a quick response of strain to magnetic field is desirable. Recently, it has been confirmed by Sakon et al. using Fe-31.2Pd [14] and Fe_3Pt [15] that almost the same results of MFIS shown in Fig. 5.5 can be induced by a pulsed magnetic field with a pulse duration of 5 ms. This result suggests that these materials are possibly actuated under a frequency of 400 Hz and higher. Moreover, RMV under a pulsed magnetic field was confirmed to occur in Fe_3Pt by a synchrotron X-ray study [16].

5.4 Condition for Rearrangement of Martensite Variants by Magnetic Field

As described in the introduction, the rearrangement of martensite variants (RMV) by magnetic field is related to the magnetocrystalline anisotropy. Ullakko et al. proposed an energy condition for realizing the RMV by magnetic field [17]. They considered that the RMV occurs when the uniaxial magnetocrystalline anisotropy is larger than the energy dissipated during the RMV. Essentially, the same condition can be quantitatively described in terms of shear stress as described below.

Since the twinning plane moves under a magnetic field, we consider that the magnetic field exerts a shear stress across the twinning plane. We referred to this shear stress as magnetic shear stress, τ_{mag}. In order to realize the RMV by magnetic field, τ_{mag} should be larger than the stress required for the twinning plane movement, τ_{req}. That is, the RMV by magnetic field is realized when $\tau_{mag} \geq \tau_{req}$ is satisfied. In the following, we will quantitatively demonstrate that this condition is actually satisfied.

The value of τ_{mag} is expressed as $\Delta U_{mag}/s$, where ΔU_{mag} is the magnetic energy difference per unit volume between the two variants separated by the twinning plane considered, and s is the corresponding twinning shear. Since the magnetic energy is mainly composed of magnetocrystalline anisotropy energy and Zeeman energy, the maximum of ΔU_{mag} is equal to the uniaxial magnetocrystalline anisotropy constant $|K_u|$ under the $[001]_P$ field.

The value of $|K_u|$ is obtained from the area enclosed by the two magnetization curves: one is measured along the hard magnetization axis and the other is measured along the easy magnetization axis. In order to obtain these curves, the magnetization of a single variant state (nearly a single variant state in the case of Fe_3Pt) has been measured in the martensite phase. Here, the single variant state is realized by applying a compressive stress along one of $<001>_P$ directions. Examples of the magnetization curves along the easy and hard axes thus obtained for the Fe-31.2Pd, Fe_3Pt, and Ni_2MnGa at 4.2 K are shown in Fig. 5.8. A characteristic feature of the magnetization curves is that the saturated magnetization of Ni_2MnGa is small

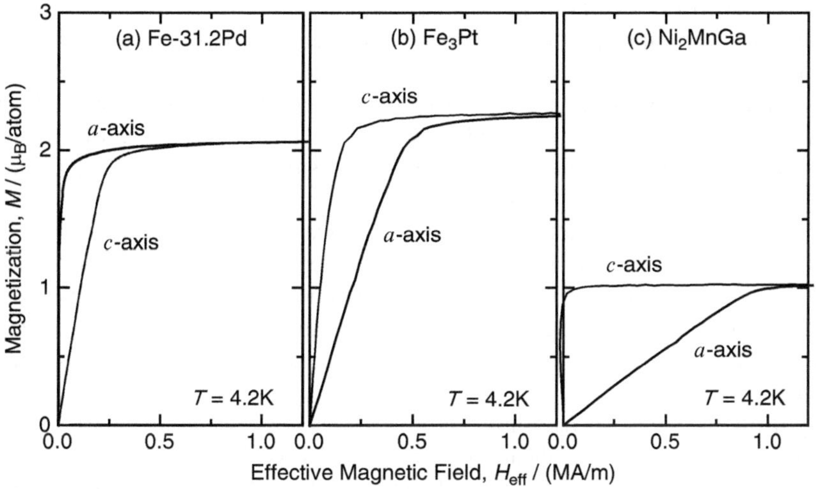

Fig. 5.8 Magnetization curves measured at 4.2 K along the *a* axis and *c* axis of (**a**) Fe-31.2Pd, (**b**) Fe_3Pt, and (**c**) Ni_2MnGa. The enclosed area corresponds to the uniaxial magnetocrystalline anisotropy constant

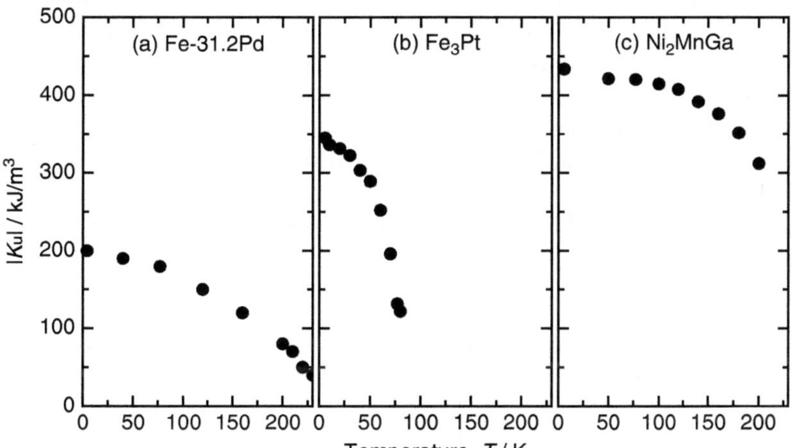

Fig. 5.9 Temperature dependence of magnetocrystalline anisotropy constant $|K_u|$ for (**a**) Fe-31.2Pd, (**b**) Fe_3Pt, and (**c**) Ni_2MnGa

compared with those of other alloys, while the anisotropy field of Ni_2MnGa is higher than those of the other alloys.

From the area enclosed by two magnetization curves, we obtain $|K_u|$ and its temperature dependence as shown in Fig. 5.9. The value of $|K_u|$ is the largest for Ni_2MnGa and smallest for Fe-31.2Pd at 4.2 K. The value of $|K_u|$ is nearly proportional to $(1 - c/a)$ in Fe-31.2Pd and Fe_3Pt, while it is not so simple for Ni_2MnGa. There appears to be a fundamental relation between the tetragonality (c/a) and $|K_u|$, although we are not sure of the formula. The value of K_u of Fe-31.2Pd shown in Fig. 5.9a is one order of magnitude larger than that of Fe-30Pd alloy at room temperature reported by Cui et al. [18]. In the case of Ni_2MnGa, the value of K_u is the same order as that of off-stoichiometric Ni–Mn–Ga alloys [19, 20].

The value of s can be calculated from the lattice parameters shown in Fig. 5.3 as $s = \{1 - (c/a)^2\}/(c/a)$, and its temperature dependence is shown in Fig. 5.10. The value of s is about 0.12 at the lowest temperature in this study as seen in Fig. 5.10. It decreases significantly as temperature increases in Fe-31.2Pd and Fe_3Pt, while it is almost independent of temperature in Ni_2MnGa as seen in Fig. 5.10.

By using the value of $|K_u|$ shown in Fig. 5.9 and s shown in Fig. 5.10, we obtain the maximum of magnetic shear stress τ^m_{mag}, which is shown in Fig. 5.11 by solid squares. The value of τ^m_{mag} does not change significantly in Fe-31.2Pd, whereas it decreases with increasing temperature in Ni_2MnGa.

In order to obtain the value of τ_{req}, tensile tests of Fe-31.Pd and compressive tests of Fe_3Pt and Ni_2MnGa have been carried out. Figure 5.12 shows examples of stress–strain curves obtained at 80 K for Fe-31.2Pd (a), 20 K for Fe_3Pt (b), and 77 K for Ni_2MnGa (c). The direction of applied stress is $[001]_P$ for Fe-31.2Pd and Ni_2MnGa, and $[114]_P$ for Fe_3Pt. Each stress–strain curve has a stage related to RMV. The stress level is below 4 MPa for all alloys. The corresponding τ_{req} is obtained by multiplying by the Schmidt factor, which is 0.5 for Fe-31.2Pd and Ni_2MnGa, and 0.42 for Fe_3Pt. The obtained values of τ_{req} are about 0.8 MPa for Fe-31.2Pd at 80 K, about 1.4 MPa for Fe_3Pt at 20 K, and about 1.6 MPa for Ni_2MnGa at 77 K. Similar results have been obtained at different temperatures below each T_M, and the result of τ_{req} for Fe-31.2Pd and Ni_2MnGa is shown by solid squares in Fig. 5.10a and b, respectively.

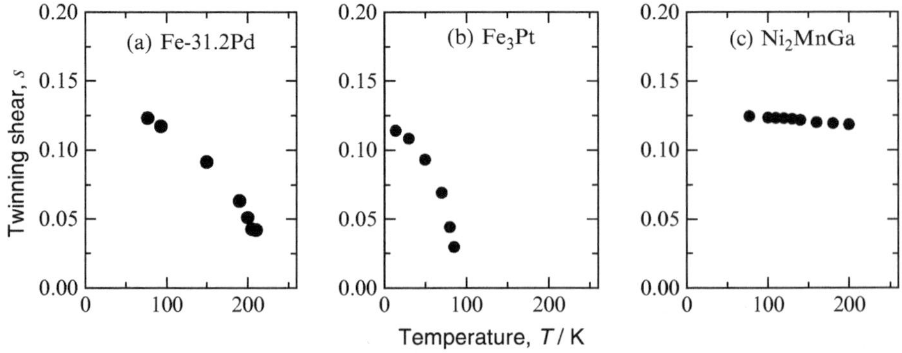

Fig. 5.10 Temperature dependence of twinning shear s for (**a**) Fe-31.2 Pd, (**b**) Fe_3Pt, and (**c**) Ni_2MnGa

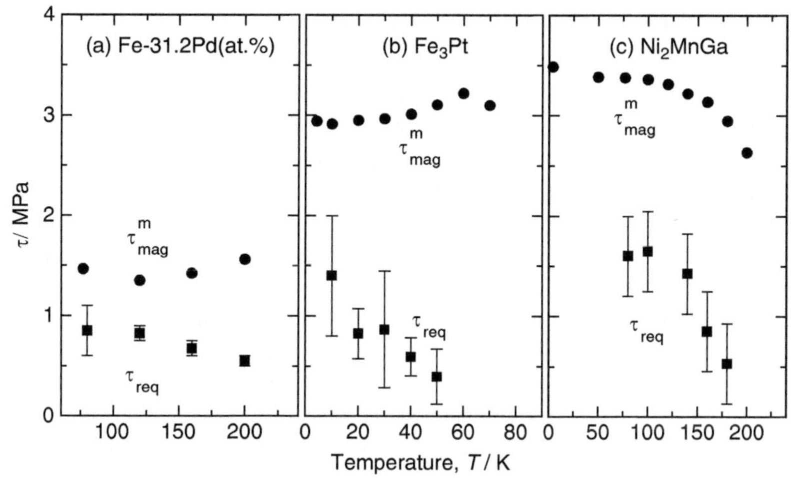

Fig. 5.11 The maximum of magnetic shear stress $\tau^{\mathrm{m}}_{\mathrm{mag}}$ (*solid squares*), and the stress required for twinning plane movement τ_{req} (*solid squares*)

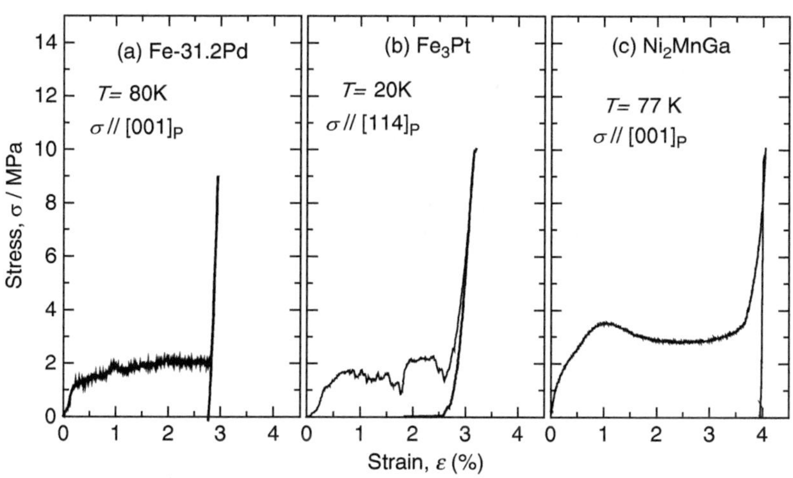

Fig. 5.12 Stress–strain curves showing a stage caused by rearrangement of martensite variants for (**a**) Fe-31.2Pd, (**b**) Fe_3Pt, and (**c**) Ni_2MnGa

Comparing $\tau^{\mathrm{m}}_{\mathrm{mag}}$ and τ_{req} shown in Fig. 5.11, it is evident that the value of $\tau^{\mathrm{m}}_{\mathrm{mag}}$ is larger than τ_{req} at any temperature below T_{M} for both Fe-31.2Pd and Ni_2MnGa. In this way, we have confirmed that the condition for RMV by magnetic field discussed earlier is quantitatively satisfied below T_{M} in the Fe-31.2 Pd and Ni_2MnGa. A similar result is expected for Fe_3Pt. To confirm this, more accurate values of K_{u} and τ_{req} for Fe_3Pt are required.

5.5 Origin of Martensitic Transformation in Fe_3Pt

As known from Fig. 5.3, the $L1_2$ – FCT transformation in Fe_3Pt is weak first order and could be directly related to the instability of the $L1_2$-type structure. In the following, therefore, we investigate the origin of the $L1_2$ – FCT transformation observed in Fe_3Pt alloys from the electronic structure of the parent phase.

The electronic structure of perfectly ordered Fe_3Pt has been calculated by the density functional theory using the all-electron full potential (linearized) augmented plane wave plus local orbitals (L/APW + lo) method. The generalized gradient approximation (GGA) is employed for exchange–correlation interactions.

Figure 5.12 shows the total energy plotted in a contour map. We know from the figure that the total energy is minimum at $c/a = 1$ and $V = V_0 = 352.9\,\text{a.u.}^3/\text{cell}$. This result means that the ground state of the perfectly ordered Fe_3Pt is cubic. The equilibrium lattice parameter obtained from V_0 is 0.374 nm, and the Bulk modulus is calculated to be 176 GPa. These values are in good agreement with experimentally obtained values of 0.375 nm and 179.2 GPa, respectively, for a highly ordered Fe_3Pt.

The total energy is plotted as a function of c/a in Fig. 5.14 with a fixed volume of $340\,\text{a.u.}^3/\text{cell}$, $352.9\,\text{a.u.}^3/\text{cell}$ ($= V_0$), and/or $360\,\text{a.u.}^3/\text{cell}$. Obviously, the total energy is the lowest for $c/a = 1$ with $V = V_0$, as described above. However, we notice in Fig. 5.14 that the energy increase by tetragonal distortion is quite small, especially when $c/a < 1$. We also notice in Fig. 5.14 that the total energy curve for the fixed volume of $V = 360\,\text{a.u.}^3/\text{cell}$ has two local minima located at $c/a = 1$ and $c/a \simeq 0.85$. Such a small increase of energy by tetragonal distortion and the existence of two local minima in the energy curve suggest that a tetragonal structure could be stabilized by giving a slight change in the present system such as introducing defects or changing composition. From the c/a dependence of total

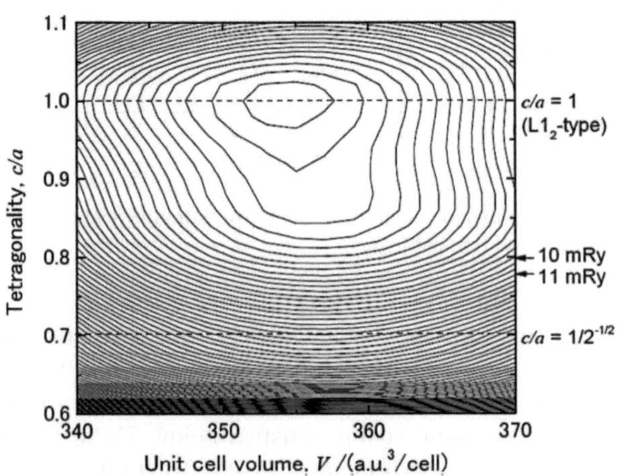

Fig. 5.13 Contour map of total energy plotted as a function of tetragonality c/a and the volume of the unit cell

energy shown in Fig. 5.14, the elastic modulus $c'(=1/2(c_{11}-c_{12}))$ is calculated to be 11.6 GPa. This value of c' is significantly small compared with that of normal metals (for example, 48 GPa for BCC-Fe and 52 GPa for FCC-Pt).

In order to understand the origin of the instability for the tetragonal distortion observed experimentally in Fe_3Pt, we have calculated the density of states (DOS) of the perfectly ordered Fe_3Pt, and the result for the cubic phase ($c/a = 1$) is shown in Fig. 5.15. In the figure, the majority and minority spin bands are shown separately and the Fermi Energy E_F is shown by a horizontal dashed line. Incidentally, the total magnetic moment calculated from the DOS profile is 2.13 μ_B/atom, being in good

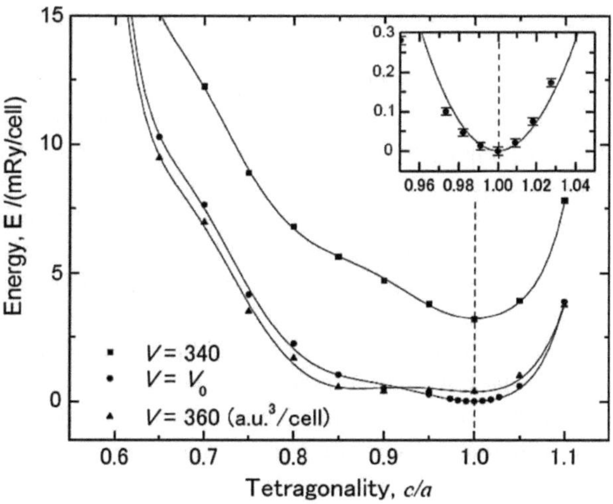

Fig. 5.14 Total energy plotted as a function of tetragonality c/a for three fixed volumes. The *inset* shows the magnification with the fixed volume of $V_0 = 352.9\,a.u.^3$/cell

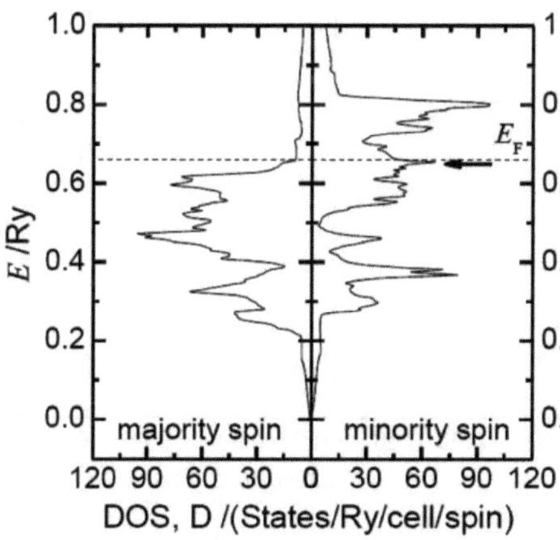

Fig. 5.15 Total density of states of ordered Fe_3Pt. The *dotted line* indicates the Fermi energy

agreement with experimentally obtained value of 2.15 μ_B/atom. A characteristic feature in Fig. 5.15 is the existence of a sharp peak in the DOS of minority spin band slightly below E_F as indicated with an arrow. As seen from the magnification of this peak shown in Fig. 5.16a, the peak top is located at about 0.01 Ry below E_F.

Then, we have examined how the peak shown in Fig. 5.16a changes by giving a tetragonal distortion while fixing the volume of the unit cell to V_0, and the result is shown in Fig. 5.16b–d. Obviously, the single peak seen in the DOS profile of $c/a = 1$(Fig. 5.16a) splits to two peaks as c/a decreases to 0.99 (Fig. 5.15b), and the separation becomes larger with decreasing c/a. This result suggests that the band Jahn–Teller effect appears strongly in Fe_3Pt That is, the split of the peak by the tetragonal distortion shifts the energy of states near E_F lower. However, the decrease in electronic energy by the shift is not large enough to stabilize the so-called FCT structure as seen in Figs. 5.13 and 5.14 in the case of the perfectly ordered Fe_3Pt.

Considering the fact that the FCT martensite appears in a partly disordered Fe_3Pt while not in the perfectly ordered Fe_3Pt, it is speculated that the band Jahn–Teller effect will be enhanced by partial disordering (i.e., by decreasing S). This speculation is explained qualitatively from the local band structure of Fe_3Pt shown in Fig. 5.17. As seen in the local DOS profile, a significant hybridization exists between d-band of Fe (solid curve) and d-band of Pt (dotted curve) in the $L1_2$-type Fe_3Pt. With decreasing S, a part of the Fe–Pt pairs in the first nearest neighbor is replaced by the Fe–Fe pairs and Pt–Pt pairs. This replacement will reduce a part of the hybridization between the d-bands of Fe and Pt. At the same time, the number of Pt d-states lying in the low energy region is expected to increase because of the increase in the number of the Pt–Pt pairs. Thus, we can speculate that the number of low energy states increases by introducing partial disorder. If such a change in

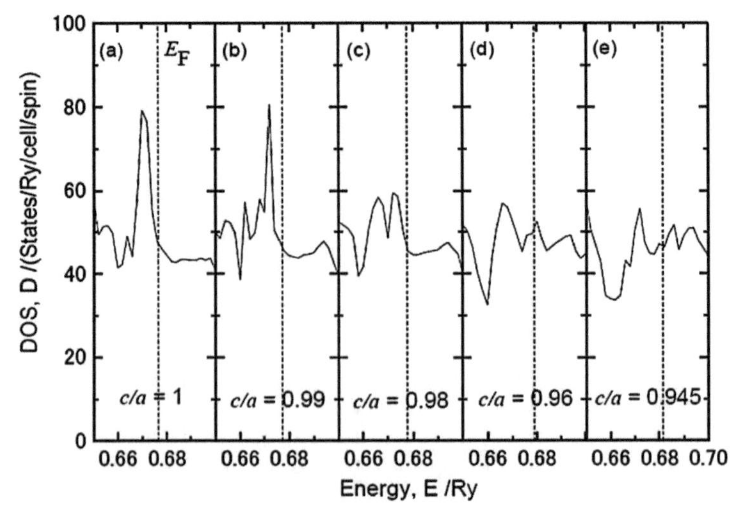

Fig. 5.16 Total density of states of minority spin of Fe_3Pt in the vicinity of Fermi energy EF. The tetragonality is changed form $c/a = 1$ to $c/a = 0.945$

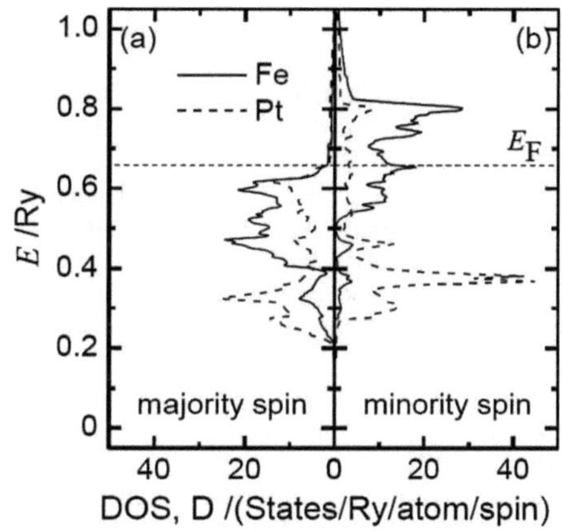

Fig. 5.17 Local density of states for Fe and Pt atoms of Fe_3Pt with $L1_2$-type structure

DOS profile is realized in the cubic phase, E_F should approach the peak indicated with an arrow in Fig. 5.15 and enhance the band Jahn–Teller effect. In this way, the introduction of partial disorder will stabilize the FCT martensite in Fe_3Pt.

5.6 Summary

Three ferromagnetic shape memory alloys, Fe-31.2Pd, Fe_3Pt, and Ni_2MnGa, exhibit a large magnetic field-induced strain of several percent, in association with rearrangement of martensite variants at any temperature below each T_M. The uniaxial magnetocrystalline anisotropy constant $|K_u|$, twinning shear s, and the stress required for twinning plane movement τ_{req} of the three alloys are almost in the same order. That is, $|K_u|$ is in the order of several hundred kJ/m^3, s is in the order of 0.1, and τ_{req} is in the order of 1 MPa. We have quantitatively confirmed that the condition, $\tau_{mag} \geq \tau_{req}$, is satisfied when the rearrangement of martensite variants by magnetic field is realized, where τ_{mag} is a magnetic shear stress, and its maximum value under the $[001]_P$ field is given by $|K_u|/s$. The formation of FCT martensite in Fe_3Pt could be caused by band Jahn–Teller effect.

References

1. T. Kimura, S. Kawamoto, I. Yamada, M. Azuma, M. Takano, Y. Tokura, Magnetocapacitance effect in multiferroic $BiMnO_3$. Phys. Rev. B **67**, 180401(R) (2003)
2. W. Prellier, M.P. Singh, P. Murugavel, The single-phase multiferroic oxides: From bulk to thin film J. Phys.: Condens. Matter **17**, R803 (2005)
3. K. Ullakko, J.K. Huang, C. Kantner, R.C. O'Handley, V.V. Kokorin, Large magnetic-field-induced strains in Ni_2 MnGa single crystals Appl. Phys. Lett. **69**, 1966 (1996)

4. R. Tickel, R.D. James, Magnetic and magnetomechanical properties of Ni_2 MnGa J. Magn. Magn. Mater. **195**, 627 (1999)
5. S.J. Murray, M.A. Marioni, A.M. Kukla, J. Robinson, R.C. O'Handley, S.M. Allen, Large field induced strain in single crystalline Ni-Mn-Ga ferromagnetic shape memory alloy J. Appl. Phys. **87**, 5774 (2000)
6. A. Sozinov, A.A. Likhachev, N. Lanska, K. Ullakko, Giant magnetic-field-induced strain in NiMnGa seven-layered martensitic phase Appl. Phys. Lett. **80**, 1746 (2002)
7. R.D. James, M. Wuttig, Magnetostriction of martensite Philos. Mag. A **77**, 1273 (1998)
8. J. Koeda, Y. Nakamura, T. Fukuda, T. Kakeshita, T. Takeuchi, K. Kishio, Giant magnetostriction of Fe-Pd alloy single crystal exhibiting martensitic transformation Trans. Mater. Res. Soc. Jpn. **26**, 215 (2001)
9. T. Kakeshita, T. Takeuchi, T. Fukuda, M. Tsujiguchi, T. Saburi, R. Oshima, S. Muto, Giant magnetostriction in an ordered Fe3Pt single crystal exhibiting a martensitic transformation Appl. Phys. Lett. **77**, 1502 (2000)
10. T. Sakamoto, T. Fukuda, T. Kakeshita, T. Takeuchi, K. Kishio, Magnetic field-induced strain in iron-based ferromagnetic shape memory alloys J. Appl. Phys. **93**, 8647 (2003)
11. M. Sugiyama, R. Oshima, F.E. Fujita, Mechanism of FCC-FCT Thermoelastic Martensite Transformation in Fe-Pd Alloys Trans. Jpn. Inst. Metals **27**, 719 (1986)
12. S. Muto, R. Oshima, F.E. Fujita, Electron microscope study on martensitic transformations in Fe- Pt alloys: General features of internal structure Metall. Trans. A **19**, 2723 (1988)
13. J. Pons, V.A. Chernenko, R. Santamarta, E. Cesari, Crystal structure of martensitic phases in Ni-Mn-Ga shape memory alloys Acta Mater. **48**, 3027 (2000)
14. T. Sakon, A. Takaha, Y. Matsuoka, K. Obara, T. Saito, M. Motokawa, T. Fukuda, T. Kakeshita, Field-induced strain of shape memory alloy Fe-31.2%Pd using a capacitance method in a pulsed magnetic field Jpn. J. Appl. Phys. **43**, 7467 (2004)
15. T. Sakon, A. Takaha, K. Obara, K. Dejima, H. Nojiri, M. Motokawa, T. Fukuda, T. Kakeshita, Magnetic-field-induced strain of shape-memory alloy Fe_3Pt studied by a capacitance method in a pulsed magnetic field Jpn. J. Appl. Phys. **46**, 146 (2007)
16. Z.W. Ouyang, Y.H. Matsuda, H. Nojiri, T. Inami, K. Ohwada, M. Tsubota, T. Sakon, T. Fukuda, T. Kakeshita, Direct observation of field-induced variant transformation in Fe_3Pt using pulsed magnetic field x-ray diffraction. J. Appl. Phys. **102**, 113917 (2007)
17. K. Ullakko, J.K. Huang, V.V. Kokorin, R.C. O'Handley, Magnetically controlled shape memory effect in Ni_2 MnGa intermetallics Scr. Metall. **36**, 1133 (1997)
18. J. Cui, T.W. Shield, R.D. James, Phase transformation and magnetic anisotropy of an iron-palladium ferromagnetic shape-memory alloy Acta Mater. **52**, 35 (2004)
19. O. Heczko, L. Straka, N. Lanska, K. Ullakko, J. Enkovaara, Temperature dependence of magnetic anisotropy in Ni-Mn-Ga alloys exhibiting giant field-induced strain J. Appl. Phys. **91**, 8228 (2002)
20. R. Tickle, R.D. James, Magnetic and magnetomechanical properties of Ni_2MnGa J. Magn. Magn. Mater. **195**, 627 (1999)
21. P.J. Webster, K.R.A. Ziebeck, S.L. Town, M.S. Peak, Magnetic order and phasde transformation in Ni_2 MnGa Philos. Mag. B **49**, 295 (1984)

Chapter 6
Soft Electronic Matter: Inhomogeneneous Phases in Strongly Correlated Condensed Matter

Peter B. Littlewood

Abstract The physics of strong correlations has at its core a competition between the delocalizing effects of the kinetic energy, and the localizing Coulomb potential. The classic competition is thus the Mott transition between paramagnetic metal and insulating antiferromagnet, but physical systems often add many complexities, via chemical doping, multiple orbitals, and coupling to the lattice degrees of freedom. Combined with this extra interplay, the result is quite frequently the appearance of various types of spatially modulated phases, with simple examples being charge- and spin-density waves, but most importantly new phases of coupled modulated order – where several broken symmetry states coexist in complex spatial patterns. This chapter views this topic via an examination of three different physical systems: the doped Mott insulator $La_2NiO_{4+\delta}$; the magnetic and/or charge-ordered states of the doped manganites based on $LaMnO_3$; and the heavy fermion superconductor $CeCoIn_5$.

6.1 Introduction

The physics of strong correlations has at its core a competition between the delocalizing effects of the kinetic energy, and the localizing Coulomb potential. The classic competition is thus the Mott transition between paramagnetic metal and insulating antiferromagnet, but physical systems often add many complexities, via chemical doping, multiple orbitals, and coupling to the lattice degrees of freedom. Combined with this extra interplay, the result is quite frequently the appearance of various types of spatially modulated phases, with simple examples being charge- and spin-density waves, but most importantly new phases of coupled modulated

P.B. Littlewood (✉)
Cavendish Laboratory, Cambridge University, JJ Thomson Ave, Cambridge CB3 0HE, UK
e-mail: pbl21@cam.ac.uk

T. Kakeshita et al. (eds.), *Disorder and Strain-Induced Complexity in Functional Materials*, Springer Series in Materials Science 148,
DOI 10.1007/978-3-642-20943-7_6, © Springer-Verlag Berlin Heidelberg 2012

order – where several broken symmetry states coexist in complex spatial patterns. This chapter views this topic via an examination of three different physical systems: the doped Mott insulator $La_2NiO_{4+\delta}$; the magnetic and/or charge-ordered states of the doped manganites based on $LaMnO_3$; and the heavy fermion superconductor $CeCoIn_5$.

6.2 A Microscopic View

Relatively little in this chapter is concerned with microscopics and therefore too unwieldy for most purposes. However, this introduction sets the scene with an elementary exposition of the underlying microscopic ideas. It was Peierls [1] who asked the obvious question about why crystal structures are in general so complex, and a glimmer of the answer comes from considering the response function of an electron gas. The principle is that a complex structure may be seen to emerge as a result of the instability of a simpler one.

The (spin or charge) density response function measures the response of the electronic charge density to an external potential V

$$\delta\rho(\mathbf{q},\omega) = \chi(\mathbf{q}\omega)\; V(\mathbf{q},\omega).$$

For a solid with a bandstructure (with n the band index and k the momentum), this is given in the (bad) approximation of noninteracting electronic quasiparticles [2]

$$\chi_o(\mathbf{q},\omega) = \frac{\delta\rho(\mathbf{q},w)}{V(\mathbf{q},w)} = 2\sum_{knn'} \frac{f\left(\varepsilon_n'(k)\right) - f(\varepsilon_n(k+\mathrm{q}))}{\varepsilon_n(k+\mathrm{q}) - \varepsilon_n'(k) - \hbar\omega}.$$

The static response is sketched for a free electron band in Fig. 6.1. In one dimension, this function is (logarithmically) singular at $q = 2k_F$, whereas in higher dimensions is well behaved. The 1D singularity is, of course, a signature of instability: this will lead to either a charge- or a spin-density wave instability depending on which is the dominant channel.

The simplest treatment of an interacting theory is to calculate the self-consistent response in mean-field theory – for example for the simplest model case of a

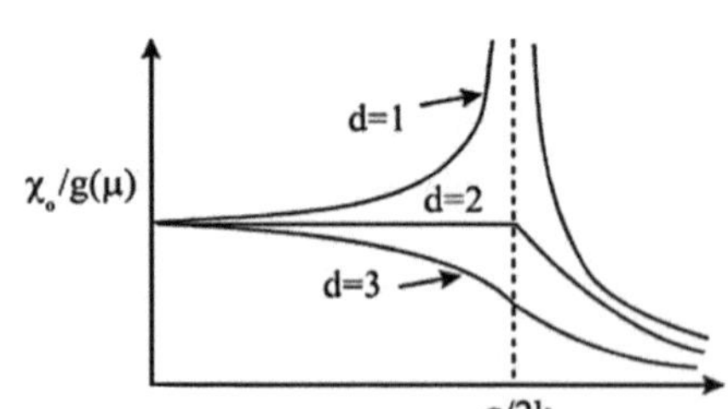

Fig. 6.1 Sketch of the static response function as a function of momentum in different dimensions

strong local interaction U, the self-consistent magnetic response function $M(q)/H(q)$ becomes [3]

$$\chi_\sigma(\mathbf{q},\omega) = \frac{\mu_B^2 \chi_o(\mathbf{q},\omega)}{1 - \frac{1}{2} U \chi_o(\mathbf{q},w)}$$

and it is now evident that this predicts an instability for large enough U at the wavevector where the bare susceptibility is peaked.

For free electrons, we saw that in three dimensions, the maximum is indeed at $q = 0$, but in one dimension, the peak is at $2k_F$ (actually a singularity). In general, the value of the susceptibility depends on the details of the band structure, but as a general rule, any quasi-one-dimensionality in the band structure – i.e., strong anisotropy in the dispersion between different directions – leads to peaks at finite momentum.

In the case where the instability occurs at finite q, the ensuing magnetic order is periodic, and generally this is called a spin-density wave (SDW), or an antiferromagnet. (Spin-density wave as a term is usually reserved for cases when the magnetic period is not exactly a lattice vector, and where the amplitude of the magnetic order parameter is small. Examples include a number of quasi-one-dimensional organic metals, and metallic Cr.) Of course, the calculation needs to be generalized to minimize the energy with a now finite-order parameter, beyond the linear response theory implicit above.

Without doing the calculation explicitly, it is worth thinking through the answer in cartoon form in one dimension; the important parameter is the ratio of the interaction strength U to the bandwidth $4t$, where t is the hopping matrix element in a 1D lattice. We saw that in 1D the instability was likely to occur at $2k_F$, and therefore in the spin-density-wave state, there should be a periodic spin-density component at a wavevector $2k_F$ (see Fig. 6.2). If there are n electrons per unit cell, note that $k_F = n\pi/2a$, where a is the lattice periodicity. Furthermore, the instability in a 1D model will occur for infinitesimal U, because the response function is singular at $2k_F$.

Now let us imagine increasing U from zero. When it is small, the spin-density-wave state will be of small amplitude: It is best to think of it as two periodic charge-density waves, each of period $2\pi/2k_F = 2a/n$ but opposite in phase. For definiteness, let us take $n = 1$, although the argument works for other densities too. Because there is a new periodicity in the structure (reciprocal lattice vector $2k_F$), then there is a new Brillouin zone plane (at wavevector k_F), and hence, a gap in the single particle bands – the onset of the spin-density wave is also a metal–insulator transition. Scattering from this periodic structure self-consistently regenerates the spin-density modulation with the correct period. When the amplitude is weak ($U/t \ll 1$), the spin-density modulation is very close to being sinusoidal (i.e., a small gap means that the NFE approximation is good); so the charge density remains very nearly uniform. However, if we increase the value

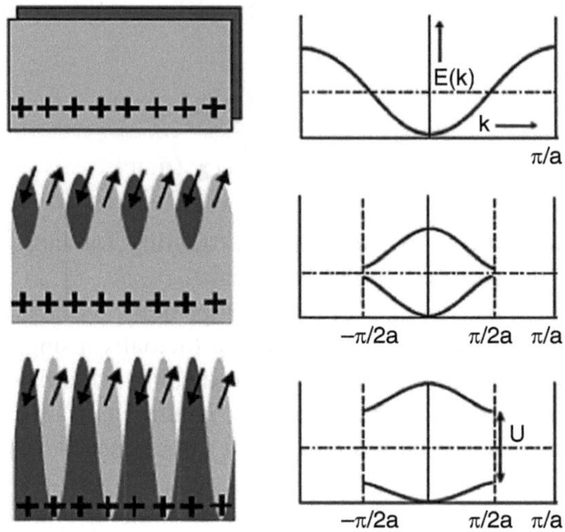

Fig. 6.2 Evolution of the charge and spin density in a 1D SDW instability from weak to strong coupling

of U, we expect the amplitude of the modulation to grow, and therefore, the gap to increase.

Eventually the picture (for $U/t \gg 1$) will surely become that shown in the lower panels of Fig. 6.2. The wave can no longer be sinusoidal, because the charge density must remain always positive, for each spin. It eventually localizes so that each period of the spin-density modulation contains precisely one electron, and there is very little overlap from one electron to its neighbor – but, of course, this is still an antiferromagnet. The charge gap is of order U, because the excitation of a carrier involves moving it from one site onto the neighboring charge. However, although the ground state is still antiferromagnetic, the (superexchange) interaction which determines the magnetic transition temperature must be quite small, because it will depend on the overlap of wavefunctions from one electron to its neighbor: it is not difficult to argue that in fact the exchange interaction $J = t^2/U$. This is, of course, a Mott insulator.

Note the distinction between the two regimes: in "weak coupling" ($U/t \ll 1$), the instability that produces the antiferromagnetic order also opens a gap at the Fermi surface – but the magnetic interaction is the driving force. In "strong coupling" ($U/t \gg 1$), it is fundamentally the interaction between charges that produces the Mott transition, and subsequently magnetic order appears on a low energy scale. These two regimes are, however, smoothly connected.

The picture shown in the figure has been carefully constructed for the conventional Mott insulating case when there is an average of one electron for each site. What happens if the system is doped away from this point? Obviously in the weak coupling – sinusoidal – limit, nothing much changes, since there is no particular connection between the Fermi wavevector and the lattice spacing. In the

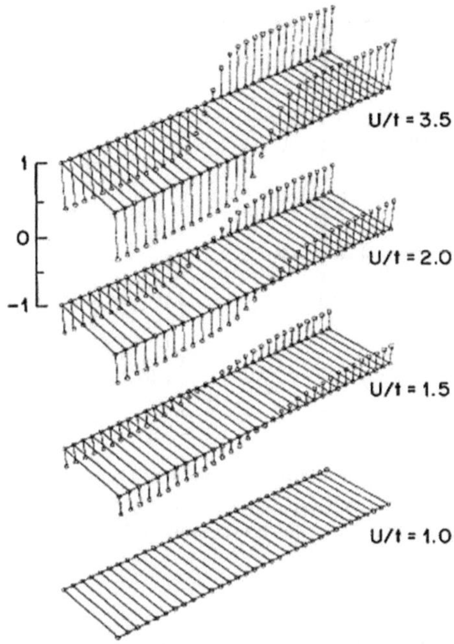

Fig. 6.3 Mean-field solution of the square lattice Hubbard model showing a vertical domain wall for varying $U = t$ on a $(33, 0) \times (0, 2)$ supercell. t is the nearest neighbor hopping matrix element on a square lattice. Plotted is the staggered magnetization – spins on alternate sites have been flipped. For large $U = t$, it is clear that the excess charge (holes in this case) is localized to a boundary between two regions of undoped antiferromagnet. The system has developed stripes

strong coupling limit, the lattice is visible to the added carriers, since there is a natural tendency for the pattern to be commensurate with the underlying lattice. The evolution to the strong coupling limit is thus natural – there will be a tendency for the excess doped carriers (either electrons or holes) to localize. Explicit solution of the 2D Hubbard model demonstrates this, as shown in Fig. 6.3 [4].

6.3 Example 1: La_2NiO_4

Real systems have additional complexities, often multiple bands, and frequently also strong electron–phonon coupling. The latter cannot be ignored when Coulomb interactions are strong, and there are also situations where the leading order instability is a charge-density wave where the effective interaction is mediated by the lattice.

It is worth noticing that the lattice effects act to enhance the well-known tendency of strong electronic correlations to produce crystalline order. While the importance of such physics of some systems is still hotly debated (e.g., for the high-temperature

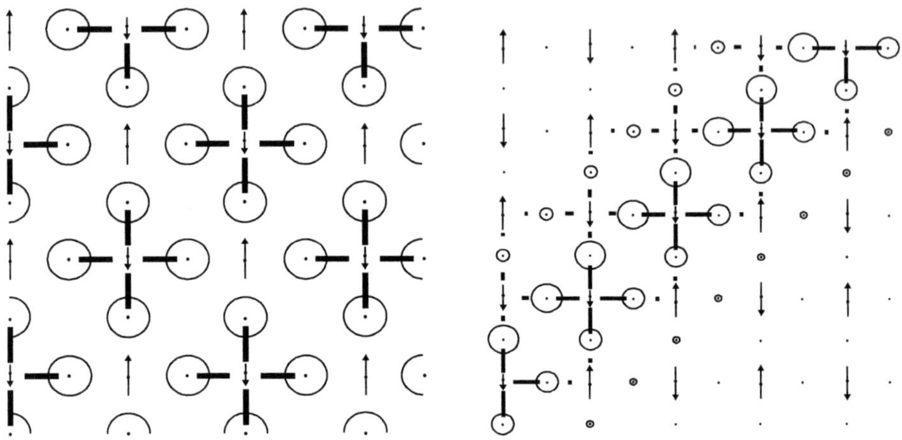

Fig. 6.4 Left figure shows the polaron lattice obtained at a doping concentration of $1/2$ hole per unit cell. *Circles* indicate the excess charge density on oxygen, the *arrows* the spins on Ni, and the *bars* indicate bond deformations. The right figure shows a fragment of a domain wall that forms the striped phase of $La2NiO_{4+\delta}$

superconducting cuprates based on La_2CuO_4), there can be no doubt that stripes are present in the analogous $La_{2-x}Sr_xNiO_4$ compound, which is insulating whatever the hole doping, and shows ordered incommensurate phases [5–7]. It is believed that here strong electron–phonon coupling stabilizes the striped phases, and Fig. 6.4 shows the calculation of a stripe within an extended Hubbard model including electron–phonon coupling [8]. Notice that the order is extremely complex, with a combination of spin, charge, and lattice order. For a real material, it is not always productive to try and assign the origin of a particular ordered structure to a specific mechanism.

6.4 Example 2: Colossal Magnetoresistance in Manganites

6.4.1 The Basics: Double Exchange and Jahn–Teller

$La1 - xSrxMnO_3$ is a prototype for the broad class of cubic perovskite manganites, where with replacement of a trivalent rare earth by a divalent alkaline earth, the *nominal* valence of *Mn* can be continuously tuned between 3+ (corresponding to the Mn(III) configuration found at $x = 0$) and 4+ (corresponding to the Mn(IV) configuration found at $x = 1$). The important physical ingredients can be seen by reference to Fig. 6.5. In a cubic environment, the *Mn d* levels are crystal-field split into a low-lying triplet of t_{2g} symmetry and a doublet of e_g symmetry. *Mn* is a strongly correlated ion, whereby double occupancy of the tightly bound d orbitals is suppressed by Coulomb repulsion, and the direct on-site exchange interaction aligns

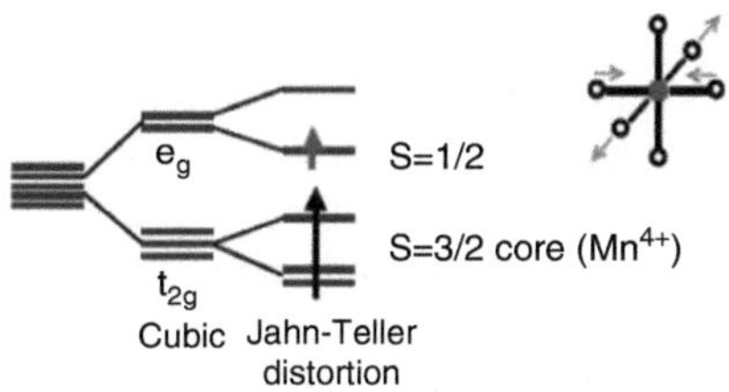

Fig. 6.5 The five degenerate d-levels on an isolated atom are split into a triplet t_{2g} and a doublet e_g in the cubic environment of the perfect octahedral perovskite. The octahedral symmetry may be broken by a distortion of the Jahn–Teller variety (e.g., the motion of the O atoms in the inset) which splits the degeneracy further

the spins in the different d orbitals. The t_{2g} levels are strongly localized, whereas an electron in the higher lying e_g state is potentially itinerant. At $x = 0$, each e_g level is singly occupied: double occupancy of these highest occupied e_g levels is suppressed by Coulomb repulsion, and $LaMnO_3$ is an antiferromagnetic Mott insulator [9, 10].

At finite doping ($x > 0$), there are some empty e_g levels, and hence, hopping is possible. The exchange (Hund's rule) coupling J between the spin of an itinerant carrier and each core spin is rather larger than the hopping matrix element t between neighboring e_g levels. Consequently, each conduction electron is forced to align with the core spin texture (this may be viewed as a strong coupling version of the RKKY interaction). Clearly, the kinetic energy of the conduction electrons is minimized (maximum bandwidth) if the core spins are parallel to one another, and this so-called *double exchange* is the fundamental mechanism [11–13] of metallic ferromagnetism at low temperatures in the doped manganites. At a sufficiently high temperature, the energy of this ferromagnetic configuration is overwhelmed by the entropy gain available from a randomization of the manganese spin system. Thus, the system lowers its free energy by entering the *paramagnetic* state.

Another important feature of the manganites arises from the doubly degenerate e_g level in Fig. 6.5. This degeneracy may be broken by a distortion of the oxygen cage away from cubic symmetry that lowers the energy of the occupied e_g level on the Mn_{3+} ion. If the level is singly occupied, there is a gain in energy that is *linear* in the amplitude of the distortion – whereas any energy cost from elastic deformation is quadratic in the distortion. Consequently, in an isolated molecular state (where there is no kinetic energy change due to hopping) this distortion is mandatory at low temperatures – a local broken symmetry called the *Jahn–Teller* effect. Things are more complicated in a solid: one effect is geometric, because the octahedral share oxygen atoms at their corners. So an elongation of the Mn–O bond length on one unit cell typically corresponds to a contraction in the neighbor. So in pure $LaMnO_3$, there is an antiferrodistortive arrangement of the distorted octahedra [14], leading to a "doubled" unit cell.

This is to be distinguished from an equivalent and simultaneous source of doubling of the unit cell due to rigid rotations of the octahedra, which are promoted by the small size of the A-site cations. (These rotations mainly affect O–Mn–O bond angles, whereas the Jahn–Teller distortions mainly affect O–Mn–O bond lengths.) However, these rotations and distortions are clearly coupled: since it is easier to accommodate the Jahn–Teller mandated bond length variations by a combination of rotations and distortions, the Jahn–Teller ordered structures are more evident when the A-site cations are "too small" to fit in the cage (measured by the so-called tolerance factor) [15, 16]. Another difference in the solid is, of course, hopping and the presence of bands. This means that the assumption that the energy lowering from the JT effect is *linear* in the distortion is no longer true, because we may have some kinetic energy cost. However, if we can arrange the pattern of distortions so that they open up a gap at the fermi energy (e.g., in a 1D band model, by choosing a periodicity $2\pi/Q = \pi/k_F$), we will have the classic Peierls distortion [1] – with an energy lowering $\propto u^2 \ln u$, which is still singular though not nearly as much. The message is just that JT is no longer mandatory and also that we expect naturally to get periodic distortions that have a wavevector depending on doping. Whether one should think about these charge-ordered structures from a local (strong coupling) or an extended (weak coupling) perspective depends on context. Since the e_g level is progressively depopulated with increasing x, the tendency toward JT distortion is suppressed; the long-range-ordered antiferrodistortive phase disappears near $x = 0.2$. It is important to note that the lattice displacements associated with the JT distortions are large, and therefore, the disappearance of the long-range order does not mean that (static or dynamic) JT fluctuations may not be pronounced. Such fluctuations have been discovered to be very prominent in the manganites [10, 17].

Double exchange and Jahn–Teller physics are two competing effects in the solid that control much of the basic physics of the manganites. Double exchange favors itinerancy for the carriers, and a ferromagnetic metal ground state. The Jahn–Teller effect acts to localize carriers: with doping the e_g level is emptied, and one can expect that there is a preference for gaining the maximum Jahn–Teller splitting for each filled site Mn_{3+}, while having no symmetry breaking distortion associated with the empty sites Mn_{4+}. This is, of course, a recipe for inhomogeneity: dynamic or static, short- or long-range ordered, and there are a set of closely related theoretical models that have been applied to the further discussion [18–22].

The movement of atoms from their equilibrium position generates a potential from which the electrons scatter, and this interaction is generically called the electron–phonon interaction. Jahn–Teller is a rather special case of this (because the potential breaks a local symmetry). Electron–phonon interaction is characterized by the induced potential due to a lattice displacement $u(\mathbf{r}) = u_{\mathbf{q}} e^{i\mathbf{q}\cdot\mathbf{r}}$ of the form

$$V(\mathbf{q}, \omega) = g_{\mathrm{q}}\, u_{\mathrm{qw}},$$

where g_{q} is the *electron–phonon coupling constant*, a function that can in principle be calculated within any model. We treat this separately from the bare stiffness of the phonon modes K the contribution to the elastic constants from all but the highest valence bands.

With appropriate scaling, we can write g_q and K_q so they both have the dimension of energy, and then the natural *dimensionless* coupling constant for the electron–phonon coupling problem is

$$\lambda = \frac{g^2}{Kt} = \frac{g^2}{M\Omega^2 t},$$

where $K \approx M\Omega^2$ sets the characteristic phonon frequency Ω and t is the electronic bandwidth. Notice that both K and t appear in the denominator – this is because if we have either very soft phonons or very narrow band electrons, the kinetic energy cost of making distortions in the lattice (coupled to fluctuations in the electron density) is small. The electron–phonon problem is a complicated one – especially in strongly interacting systems with a high density of carriers. We do know, however, that for a single carrier there is a very rapid crossover in the behavior for $\lambda > 1$, corresponding to the appearance of a self-trapped "polaron." Here, the carrier deforms the lattice strongly enough to trap in the accompanying potential, and thereby, becomes much less mobile; the crossover is in practice rather abrupt.

6.4.2 *Competing and Cooperating Phases in Manganites*

Armed with this parameter, we can now put together a "theorist's" phase diagram (at some moderate doping), shown in Fig. 6.6. If λ is small, then we expect to have a transition between a ferromagnetic and paramagnetic metal as we raise the temperature. As we mentioned above, there is not a very large change in the bandwidth, so we do not expect a huge change in the resistance, nor in consequence, do we expect a "colossal" magnetoresistance. If we increase the electron–phonon coupling at low temperatures, at some point we expect an abrupt transition where

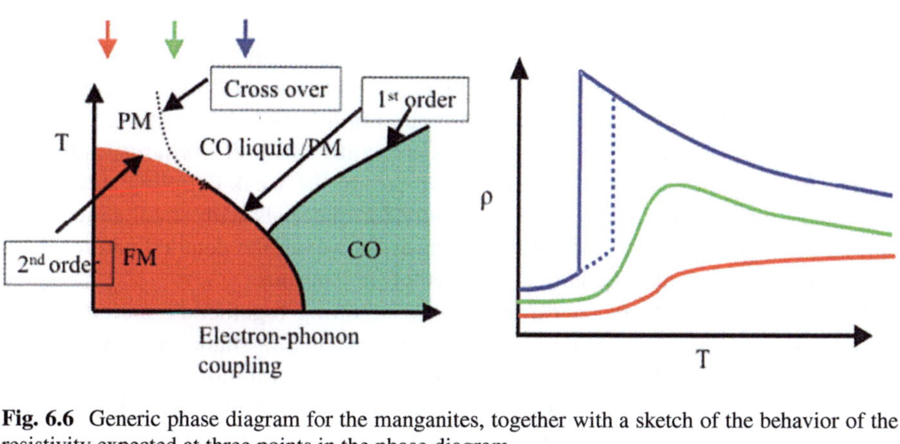

Fig. 6.6 Generic phase diagram for the manganites, together with a sketch of the behavior of the resistivity expected at three points in the phase diagram

the carriers crystallize into an array of corresponding lattice distortions; this is often a metal-to-insulator transition and is likely discontinuous (first order, with a latent heat). Of course, this transition is not driven by magnetism, though it has magnetic consequences – the metallic phase is generally ferromagnetic (DE), while the insulating phase likely has a small antiferromagnetic superexchange coupling.

If one raises the temperature of this charge-ordered system, eventually the lattice distortions thermally disorder at a melting transition – but not until very high temperatures should one expect the distortions to be locally small. So the disordered high-temperature paramagnet thus formed is not expected to have a high conductivity – we shall call it a polaronic liquid (though usually the density of carriers is so high that no individual polaronic carrier can be identified).

It is in the intermediate range of coupling constant and temperature that the magnetism and electron–phonon coupling show an interesting interplay. Since the bandwidth – crudely estimated – in the ferromagnet is larger than that in the paramagnet, the effective electron phonon coupling is *larger* in the high-temperature paramagnetic phase than in the low-temperature metal. Thus for intermediate values of the bare coupling λ, one may by raising the temperature cross over from weak to strong coupling behavior (even though λeff will not change more than a factor of two at most, the effects may be large). Thus even if the FM/PM phase boundary remains continuous, there will be a large increase in scattering above the transition and perhaps a substantially enhanced resistivity; such a resistivity can, of course, be suppressed by a magnetic field (which aligns the spins, increases t, and thus reduces λ. Such can be the origin of bulk CMR in a homogeneous phase.

However, for slightly larger bare coupling, the cooperative magnetic and lattice effects can lead to a first-order liquid–gas transition, with unusually the *fermi gas* phase being the low temperature one (since the kinetic energy is dominated by quantum mechanical motion, unlike the usual classical gas). Here, one has the possible two-phase coexistence, and with interactions and disorder, the possibility of hysteretic and percolative effects on the resistivity. Since a magnetic field can change the relative proportions of two phases in coexistence and give rise to bulk CMR effects arising from inhomogeneous phases.

This three-phase picture is more a cartoon than a theory, of course, but the general principles embodied by these ideas seem to be correct. In *LSMO*, at large doping the CMR is small, because the electron–phonon coupling is fairly weak. *LSMO* has a well-matched A-site cation, so rotations are quite stiff, "K" is large, and once the electron kinetic energy is substantial we are on the weak coupling side of the phase diagram. But for *LCMO* and especially for materials with small cations, we tend to the strong coupling side of the diagram. However, the detailed nature of the magnetic, and especially the lattice order, is left mysterious.

We alluded above to the *packing problem* of fitting together JT distorted sites with nominally undistorted ones, given that the octahedra share oxygen atoms at their corners. Figure 6.7 shows this graphically. One straightforward solution to the packing problem is to arrange the distorted sites in rows, sometimes called "stripes." This can neatly solve the frustration issue (but it will necessarily introduce

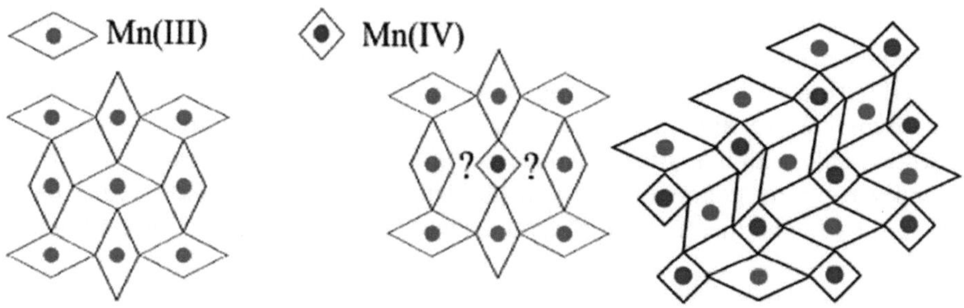

Fig. 6.7 The left figure shows a cartoon of the structure of a single layer of $LaMnO_3$; the diamond shapes represent the distorted octahedral, which share oxygen atoms at the vertices. Adding a hole to the perfect *Mn(III)* array produces geometrical frustration (center figure). But if the holes line up in diagonal lines (as shown here for a 50% concentration, the geometrical frustration can be relieved). Notice that accompanying this charge ordering is an accompanying orbital ordering, and also that the system will naturally develop a macroscopic strain

a macroscopic strain) and can also accommodate any concentration of charge by varying the separation between the stripes.

6.4.3 Ginzburg–Landau Theory for Manganites

We have now outlined the microscopic complexities of competition between the localizing effects of the Coulomb interaction and the kinetic-energy-induced itinerant ferromagnetism. Complexities abound because of coupling to the lattice, and induced orbital order. While it is indeed possible to attack this subject with microscopic models – and much has been done in that direction [18–22] – it is also of interest to consider the competing phenomena in a phenomenological way, using Ginzburg–Landau theory. The advantage of such phenomenology is that it exposes potential mesoscale phenomena that are difficult to attack ab initio.

Ginzburg–Landau theory [23] allows the study phase transitions in a phenomenological way and it consists in expressing the free energy as a power expansion of the order parameters and their gradients. In this problem, we should consider the magnetization, the modulated charge and orbital order, and antiferromagnetic order. Even restricting to these few degrees of freedom, this is a multicomponent theory. However, to understand some general principles, we make a number of simplifications: antiferromagnetism is assumed to be a slave to the charge and orbital order, and is neglected; and we will approximate both magnetism and charge order by two scalar fields.

In our approach [24], order parameters are thus the magnetization $M(r)$ and the charge-orbital modulation $\psi(r) = \rho(r)e^{i(Q_c.r+\phi(r))}$. Here, $\mathbf{r}$ is the spatial coordinate, ρ is the amplitude of the modulation, $\mathbf{Q_c} = \mathbf{a}^*/n$ is a wavevector commensurate with the lattice ($\mathbf{a}^*$ being a reciprocal lattice vector and n an integer), and ϕ is a (possibly spatially varying) phase that allows us to incorporate structures with incommensurate periodicities. $n = 4$ gives the correct periodicity for the lattice distortions measured in $x = 0.5$; though the charge modulation has period 2, the orbital order follows a zig-zag pattern with period 4. To further simplify the discussion, we study a 1D modulation since charge modulation occurs in one direction only within a domain. Notice that if $\nabla\phi = 0$, ψ is a wave of amplitude ρ and wavevector commensurate with the lattice. If $\nabla\phi \neq 0$, the wavevector is $\mathbf{Qc} + <\nabla\phi>$ and therefore, in general, cannot be expressed as a rational fraction of $\mathbf{a}^*$.

The free energy density can be written in three parts: magnetization, charge modulation, and coupling terms. The first two are

$$\mathcal{F}_M = \frac{1}{2}a_M(T - T_c)M^2 + \frac{1}{4}b_M M^4 + \frac{1}{2}\xi_M^2(\nabla M)^2,$$

$$\mathcal{F}_\psi = \frac{1}{2}a_\rho(T - T_{\mathrm{CO}})\rho^2 + \frac{1}{4}b_\rho\rho^4 + \frac{1}{2}\xi_\rho^2(\nabla\rho)^2 + \frac{1}{2}\xi_\rho^2\rho^2(\nabla\phi - q)^2 + \frac{1}{n}\eta\rho^n\cos(n\phi).$$

The magnetic energy taken alone describes a phase transition to homogenous magnetism below the Curie temperature T_c. It is conventional to scale the order parameters to be dimensionless; the parameters a, b, etc., are constants, and the coefficients of the gradient terms thus have an additional dimension of [Length]^2. $\mathcal{F}_\psi$ is the free energy extensively used to study commensurate–incommensurate phase transitions of charge-density waves, spin-density waves, or modulated lattice distortions [25,26]. $q_o = 1/2 - x$ is the predicted deviation from commensurability around $x = 0.5$. The two phase-dependent terms compete with each other. The first

$$\frac{1}{2}\xi_\rho^2\rho^2(\nabla\phi - q_o)^2$$

favors a uniform incommensurate modulation with $\nabla\phi = q_o$. Note that the definition of q guarantees that the ground state at $x = 0.5$ is commensurate. The last term

$$\frac{1}{n}\eta\rho^n\cos(n\phi)$$

is an Umklapp term that favors commensurability with $\phi = 2\pi j/n$, j integer (for $\eta < 0$) Taken alone, $\mathcal{F}_\psi$ describes two ordered phases. Upon cooling below T_{CO} the amplitude ρ of the charge-density wave is formed but provided $n > 2$ the Umklapp term is small, and the modulation is incommensurate. As temperature is lowered, ρ grows, the Umklapp term may become dominant, and a lock-in transition occurs.

We now discuss coupling between the two order parameters. The lowest order coupling term that arises is

$$d_1\rho^2M^2$$

with d_1 positive so that there is a free energy penalty for homogeneous coexistence of magnetism and charge modulation. Were this the only coupling term the free energy would be generally stabilized either by a homogenous magnetization or by charge modulation, depending on which transition temperature is the larger. Next one can, of course, introduce uniform coupling terms of higher powers of M and ρ, but they make no qualitative changes unless they have a negative magnitude. More interesting is that there is a leading order coupling term in the *gradient* of the form

$$d_2\rho^2 M^2(\nabla\phi - q_o).$$

That there is a term *linear* in the gradient is expected because there is no symmetry about $x = 1/2$; different signs of the gradient correspond physically to compression or extension of the CDW period, i.e., to extra 3+" or 4+" sites. One can also justify this term microscopically: if we consider the effect of charge modulation on the Fermi surface, then it is clear that if we choose a wavevector that does not match the chemical doping, one will be left with small pockets of carriers at the Fermi surface; these metallic electrons are then available to mediate double exchange and thereby promote ferromagnetism. The asymmetry around $x = 1/2$ is due to the asymmetry between electron and hole pockets. Now note that this gradient term can be incorporated into $\mathcal{F}_\psi$ by completing the square, and replacing q by

$$q_{\text{eff}} = q_o - \frac{d_2}{\xi_\rho^2}M^2 = \frac{1}{2} - x - \frac{d_2}{\xi_\rho^2}M^2.$$

The sign of d_2 is unknown *a priori* and we here choose it to be positive. Once this sign is fixed, however, this gradient coupling has profound consequences for the phase diagram. First, note that even if we are at commensurability ($x = 1/2$), if magnetism is present, then there is a tendency to *incommensurate* charge modulation. This reproduces the experiments of Chen et al. [27], on La1/2Ca1/2MnO3, where the onset of charge modulation is incommensurate, and accompanied by ferromagnetism – which is replaced by Neel order at the transition to the commensurate phase. The incommensurate phase of Pr1/2Ca1/2MnO3 is paramagnetic but accompanied by the onset of ferromagnetic spin fluctuations [28, 29]. The second feature of this term is that if $x < 1/2$, it is possible for coexisting magnetism to "cancel" the chemical tendency to incommensurability, consistent with the observation of canted magnetism [30] in the manganites that show *commensurate* charge modulation. No such cancellation is possible for $x > 1/2$, and so the system is naturally incommensurate. Calculations for a specific set of parameters [24] are shown in Fig. 6.8.

The important conclusion of this is that the competition between two phases is not invariably a choice between the one and the other – but may instead result in a third phase, which is an inhomogeneous mixture of the two. By approaching this on a macroscopic level, we can gain a more general insight: here, *homogeneous* coexistence of ferromagnetism and charge order is forbidden, but boundaries between the two phases can have a *nefative* free energy so the system is driven to a

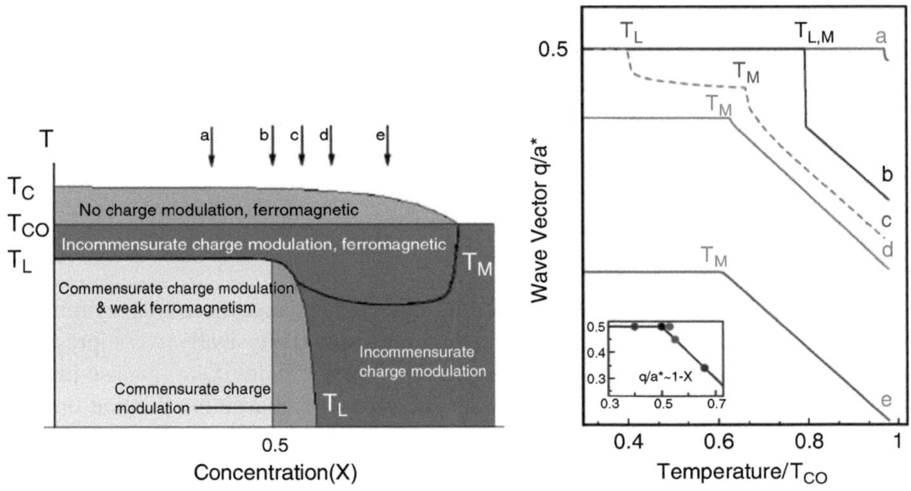

Fig. 6.8 The left figure is a schematic phase diagram which results from the minimization of the free energy. The scale of the axes depends on the particular parameters used. The complex phases arise provided $T_C > T_{CO}$, a condition that can be relaxed if the model is extended to account for discontinuous phase transitions at T_{CO}. The values of T_{CO} and T_C are parameters in the model, whereas the lock-in temperature T_L is a consequence of the competition between the Umklapp and incommensurate modulation terms. T_M is the temperature below which the magnetization is zero. The labels a–e correspond to the placement of curves in the right hand figure, showing the wavevector itself as a function of temperature: the *inset* shows the low-temperature saturation value. From [24]

new structure that stabilizes a finite density of interfaces. This is not an uncommon situation in soft matter, where the entropy from interfaces is frequently the source of mixing; in these inorganic materials, it seems that the cause is primarily energetic.

6.5 Example 3: Superconductivity and Magnetism in $CeCoIn_5$

As a final example, we choose something very different – the heavy fermion compound $CeCoIn_5$. This compound is a superconductor with a superconducting T_c at ambient pressure of 2.3 K [31]. Moreover, it lies close to a magnetic quantum critical point exhibiting strong AF spin fluctuations [32]. The zero-field SC gap symmetry is most likely to be of d_{x2-y2}-wave type [33]. At high magnetic fields close to the H_{c2} here superconductivity is suppressed, a distinct new state is observed [34], initially thought to be a realization of the Fulde–Ferrell–Larkin–Ovchinnikov state [35,36], which is an inhomogeneous superconducting state where the pairs have a finite momentum. Neutron diffraction data show instead that this state is an almost commensurate SDW at $\mathbf{Q} = (q, q, 0.5)$, which disappears at the same upper critical field with SC in a first-order transition [37]. Moreover, the

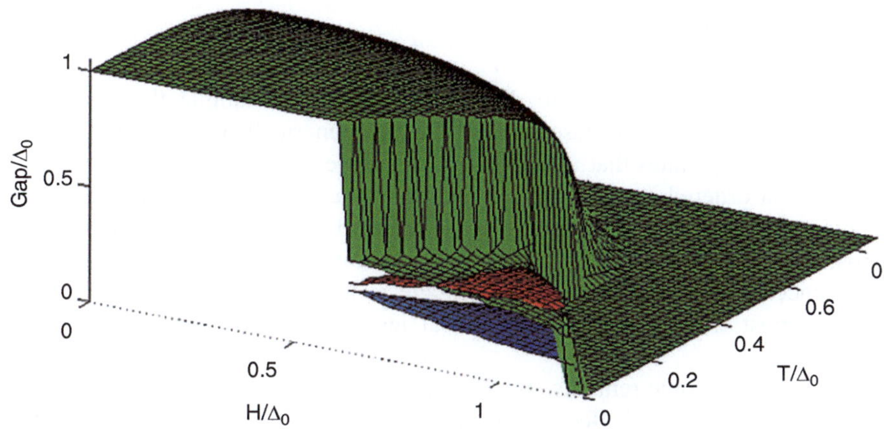

Fig. 6.9 The temperature and field dependence of the three order parameters in a model of $CeCoIn_5$. Here, the d-wave singlet gap is shown in *green*, the staggered triplet SC order parameter is *red*, and the SDW order parameter M is *blue*. In the low-T high field state all three order parameters coexist. From [40]

modulation wavevector *is not coupled to the magnitude of the external magnetic field* ruling out the FFLO mechanism that produces superfluid density modulations that scale with the field. The neutron results agree with previous NMR results [38, 39] that reported SDW ordering in the HFSC state.

Aperis et al. [40] considered a microscopic model of competition between a d-wave superconductor and an itinerant SDW. In contrast to the previous section, this is a full microscopic – albeit mean-field – theory. (However, a G-L-type approach by Agterberg and Sigrist leads to a similar conclusion [41]). Of course, both SDW and SC compete for the electronic fermi surface in this material and they are antagonistic. However, consider a situation in which they coexist: then *by symmetry* a third-order parameter must exist, corresponding to superconducting triplet correlations with the wavevector of the SDW. If a commensurate wavevector near $(0, 0, \pi)$ is involved this is called π-triplet, related to an old proposal for modulated singlet superconductivity in a different context known as η-pairing [42]. It turns out that energetically such a state is stabilized by applying a magnetic field, which suppresses the singlet order parameter – and near H_{c2} a first-order phase transition occurs from the d-wave superconductor to a triple order parameter state: singlet, π-triplet, and SDW. At a higher field, all are simultaneously suppressed (in a first-order transition). A remarkable feature is that we have a magnetic state *only* when superconductivity coexists (Fig. 6.9).

6.6 Concluding Remarks

This chapter has endeavored to discuss by means of example the underlying origins of inhomogeneous phases in what are now generically called "multiferroic" materials: namely states that arise because of competition or cooperation between two different ordered states of matter. The fundamental message is that the result of competition between two apparently incompatible states is not necessarily to choose between the one and the other – in several interesting cases, inhomogeneous coexistence is preferred. The other message is that often this behavior can be captured most readily at the semiclassical level, based on a Ginzburg–Landau theory.

In conclusion, some remarks on what has not been considered are given. There has been little mention of effects on the macroscale, which can arise through effects mediated by disorder, and by strain. Almost all order parameters explicitly couple to elastic strain, and the patterns chosen (e.g., for the charge-ordered phases in manganites) generally reflect the constraints of strain ordering, which are well discussed elsewhere [43–45]. The effects of disorder – intrinsic or strain-induced – on the development of heterogeneity can often be overwhelming, and explicit phase-separation phenomena are clearly influenced by that [22, 46, 47]. Perhaps more dangerously, the approach taken in this chapter is entirely classical, based only on the statistical physics of coupled order parameters, with even that usually approximated at the mean-field level. Quantum fluctuations of these fields are potentially not only responsible for new phases (as soft phonons can mediate superconductivity, for example) but perhaps they generate entirely new quantum states of matter.

Acknowledgments Many of the specific ideas presented here have been developed in conjunction with collaborators: the lattice coupling of stripe phases with J. Zaanen; the inhomogeneous magnetic order in manganites with G.C. Milward and M.J. Calderon; and $CeCoIn_5$ with G. Varelogiannis and A. Aperis. Others who have particularly influenced the development of these ideas include T. Lookman, A.J. Millis, N.D. Mathur, and A. Saxena.

References

1. See discussion in W.A. Harrison, *Electronic Structure and the Properties of Solids: The Physics of the Chemical Bond* (Dover Publications, USA, 1990)
2. N.W. Ashcroft, N.D. Mermin, *Solid State Physics* (Thomson Learning, Toronto, 1976)
3. D. Bohm, D. Pines, A collective description of electron-electron interactions. 3. Coulomb interactions in a degenerate electron gas. Phys. Rev. **92**, 609 (1953)
4. M. Inui, P.B. Littlewood, Hartree-Fock study of the magnetism in the single band Hubbard model. Phys. Rev. B **44**, 4415 (1991)
5. C.H. Chen, S.W. Cheong, A.S. Cooper, Charge modulations in $La_{2-x}Sr_xNiO_{4+y}$ – Ordering of polarons. Phys Rev. Lett. **71**, 2461 (1993)
6. S.M. Hayden et al., Incommensurate magnetic correlations in $La_{0.8}Sr_{0.2}NiO_4$. Phys. Rev. Lett. **68**, 1061 (1992)

7. E.D. Isaacs et al., Diffuse X-ray scattering from $La_{2-x}Sr_xNiO_4$ and $La_{2-x}Sr_xCuO_4$. Phys. Rev. Lett. **72**, 3421 (1994)
8. J. Zaanen, P.B. Littlewood, Freezing electronic correlations by polaronic instabilities in doped La_2NiO_4. Phys. Rev. B **50**, 7222 (1994)
9. J. Goodenough, Theory of the role of covalence in the Perovskite-type manganites [La, M(II)]MnO3, Phys. Rev. **100**, 564 (1955)
10. For a review, see Y. Tokura (ed.), *Colossal Magnetoresistance Oxides* (Gordon & Breach, New York, 2000)
11. C. Zener, Interaction between the D-shells in the transition metals.2. Ferromagnetic compounds of manganese with perovskite structure. Phys. Rev. **81**, 403 (1951)
12. P.W. Anderson, H. Hasegawa, Considerations on double exchange. Phys. Rev. **100**, 675 (1955)
13. P.G. de Gennes, Effects of double exchange in magnetic crystals. Phys. Rev. **118**, 141 (1960)
14. J. Kanamori, Crystal distortion in magnetic compounds. J. Appl. Phys. Suppl. **31**, 14S (1960)
15. H.Y. Hwang et al., Lattice effects on the magnetoresistance in doped $LaMnO_3$. Phys. Rev. Lett. **75**, 914 (1995)
16. L.M. Rodriguez-Martinez, J.P. Attfield, Cation disorder and the metal-insulator transition temperature in manganese oxide perovskites. Phys. Rev. B **58**, 2426 (1998)
17. P.G. Radaelli et al., Charge localization by static and dynamic distortions of the MnO_6 octahedra in perovskite manganites. Phys. Rev. B **54**, 8992 (1996)
18. A.J. Millis, P.B. Littlewood, B.I. Shraiman, Double exchange alone does not explain the resistivity of $La_{1-x}Sr_xMnO_3$. Phys. Rev. Lett. 74, 5144 (1995)
19. H. Roder, J. Zang, A.R. Bishop, Lattice effects in the colossal-magnetoresistance manganites. Phys. Rev. Lett. **76**, 1356 (1996)
20. A.J. Millis, B.I. Shraiman, R. Mueller, Dynamic Jahn-Teller effect and colossal magnetoresistance in $La_{1-x}Sr_xMnO_3$. Phys. Rev. Lett. **77**, 175 (1996)
21. T.V. Ramakrishnan et al., Theory of insulator metal transition and colossal magnetoresistance in doped manganites. Phys. Rev. Lett. **92**, 157203 (2004)
22. E. Dagotto, T. Hotta, A. Moreo, Colossal magnetoresistant materials: The key role of phase separation. Physics Reports **344**, 1 (2001)
23. J.C. Toledano, P. Toledano, *The Landau Theory of Phase Transitions*, (World Scientific, Singapore, 1987)
24. G.C. Milward, M.J. Calderon, P.B. Littlewood, Electronically soft phases in manganites. Nature. **433**, 607 (2005)
25. W.L. McMillan, Theory of discommensurations and commensurate-incommensurate charge-density-wave phase transition. Phys. Rev. **14**, 1496 (1976)
26. R.M. Fleming, D.E. Moncton, D.B. McWhan, F.J. DiSalvo, Broken hexagonal symmetry in the incommensurate charge-density wave structure of $2H-TaSe_2$. Phys. Rev. Lett. **45**, 576 (1980)
27. C.H. Chen, S.W. Cheong, Commensurate to incommensurate charge ordering and its real-space images in $La_{0.5}Ca_{0.5}MnO_3$. Phys. Rev. Lett. **76**, 4042 (1996)
28. R. Kajimoto, H. Yoshizawa, Y. Tomioka, Y. Tokura, Commensurate-incommensurate transition in the melting process of orbital ordering in $Pr_{0:5}Ca_{0:5}MnO_3$: A neutron diffraction study. Phys. Rev. B **63**, 212407 (2001)
29. R. Kajimoto et al., Anomalous ferromagnetic spin fluctuations in an antiferromagnetic insulator $Pr_{1-x}Ca_xMnO_3$. Phys. Rev. B **58**, R11837 (1998)
30. Z. Jirak, S. Krupicka, Z. Simsa, M. Dlouham, S. Vratislav, Neutron diffraction study of $Pr_{1-x}Ca_xMnO_3$ perovskites. J. Magn. Magn. Mater. **53**, 153 (1985)
31. C. Petrovic et al., Heavy-fermion superconductivity in $CeCoIn_5$ at 2.3 K. J. Phys. Condens. Matter **13**, L337 (2001)
32. A. Bianchi et al., Avoided antiferromagnetic order and quantum critical point in $CeCoIn_5$, Phys. Rev. Lett. **91**, 257001 (2003)
33. K. Izawa et al., Angular position of nodes in the superconducting gap of quasi-2D heavy-fermion superconductor $CeCoIn_5$. Phys. Rev. Lett. **87**, 057002 (2001)
34. A. Bianchi et al., Possible Fulde-Ferrell-Larkin-Ovchinnikov superconducting state in $CeCoIn_5$. Phys. Rev. Lett. **91**, 187004 (2003)

35. P. Fulde, R.A. Ferrell, Superconductivity in strong spin-exchange field. Phys. Rev. **135**, A550, (1964)
36. A.I. Larkin, Y.N. Ovchinnikov, Inhomogeneous state of superconductors. Sov. Phys. JETP **20**, 762 (1965)
37. M. Kenzelmann et al., Coupled superconducting and magnetic order in $CeCoIn_5$. Science **321**, 1652 (2008)
38. V.F. Mitrovic et al., Observation of spin susceptibility enhancement in the possible Fulde-Ferrell-Larkin-Ovchinnikov state of $CeCoIn_5$. Phys. Rev. Lett. **97**, 117002 (2006)
39. B.L. Young et al., Microscopic evidence for field-induced magnetism in $CeCoIn_5$. Phys. Rev. Lett. **98**, 036402 (2007)
40. A. Aperis, G. Varelogiannis, P.B. Littlewood, Magnetic-field-induced pattern of coexisting condensates in the superconducting state of $CeCoIn_5$. Phys. Rev. Lett. **104**, 216403 (2010)
41. D.F. Agterberg, M. Sigrist, H. Tsunetsugu, Order parameter and vortices in the superconducting Q phase of $CeCoIn_5$. Phys. Rev. Lett. **102**, 207004 (2009)
42. C.N. Yang, Eta-pairing and off-diagonal long range order in a Hubbard model. Phys. Rev. Lett. **63**, 2144 (1989)
43. N.D. Mathur, P.B. Littlewood, Mesoscopic textures in manganites. Phys. Today **56**, 25 (2003)
44. K.H. Ahn, T. Lookman, A.R. Bishop, Strain-induced metal-insulator phase coexistence in perovskite manganites. Nature **428**, 401 (2004)
45. T. Lookman, P.B. Littlewood, Nanoscale heterogeneity in functional materials. MRS Bull. **34**, 822 (2009)
46. J. Burgy, A. Moreo, E. Dagotto, Relevance of cooperative lattice effects and correlated disorder in phase separation theories for CMR manganites. Phys. Rev. Lett. **92**, 097202 (2004)
47. M.B. Salamon, M. Jaime, The physics of manganites: Structure and transport. Rev. Mod. Phys. **73**, 583 (2001)

Chapter 7
Defects in Ferroelectrics

Wenwu Cao

Abstract Ferroelectric materials are one of the smartest materials known to us, which have multiple functional properties, including piezoelectric, dielectric and pyroelectric characteristics. Since functional properties are usually associated with response agilities of materials to external stimuli, better functional properties may be created if one could make the crystal structure or mechanical structure of materials less inert As discussed in this chapter, various defects, including vacancies, aliovalent dopants, domain walls, grain boundaries, interstitial defects, surfaces, etc. have been introduced into ferroelectric materials to weaken the stability of crystal structure or domain structure so that some intended functional properties can be greatly enhanced. In fact, any ferroelectric material used as a functional material contains some types of defects. These defects may be chemically introduced through doping or are being physically created through thermal processes or domain engineering. Understanding the role of each type of defect can help us use defects properly to our advantage in designing better functional materials and in creating smaller and more advanced electric or electromechanical devices that can further facilitate our life.

7.1 Introduction

Ferroelectric materials are the most widely used functional materials today. They belong to the category of "smart" materials with three main functional properties, i.e., dielectric, piezoelectric, and pyroelectric properties. All three functional properties are very closely related to our daily life applications. For example, most of mutilayer ceramic capacitors (MLCC) are made of doped ferroelectric

W. Cao (✉)
Department of Mathematics, The Pennsylvania State University, University Park, PA 16802, USA
e-mail: cao@math.psu.edu

T. Kakeshita et al. (eds.), *Disorder and Strain-Induced Complexity in Functional Materials*, Springer Series in Materials Science 148,
DOI 10.1007/978-3-642-20943-7_7,

$BaTiO_3$ ceramic, which are widely used in electronic devices. In fact, these LMCCs are responsible for the miniaturization of today's electronic devices. Many electromechanical devices, such as ultrasonic medical imaging transducers, underwater SONARs, piezoelectric actuators, and stress sensors, are made of ferroelectric materials. Some infrared sensors and night vision devices are also made of ferroelectric materials utilizing their large pyroelectric coefficients near the phase transition temperature.

For practical applications, ferroelectric materials are often being modified chemically or physically to increase their functionalities or tune the phase transition temperature to suit particular targeted applications. Chemical doping is a commonly used method to increase the responsiveness of ferroelectric materials through introducing defects into the crystal structure. Such defects can be either on the lattice sites or interstitial between ideal ionic positions. There are two basic strategies in designing the doping elements as substitutional defects. One is to replace some ions by other types of ions that have larger or smaller atomic radii, so that the crystal structure becomes mechanically distorted near the doping site, which makes the crystal structure less stable or more responsive to external mechanical stimuli. The other strategy is to introduce aliovalent doping, i.e., doping with different valance ions so that the local charge balance is disturbed. In order for the system to reach electrical equilibrium for the latter case, vacancies must be formed so that local dipolar defects are produced in the crystal structure. Such dipolar defects will shift the ferroelectric–paraelectric phase transition temperature, or the Curie temperature T_c, and also modify the agility of the material in responding to external electric stimuli. Doping can manipulate the phase transition temperature T_c closer to room temperature so that anomaly at T_c can be explored for certain functional properties, such as the very large pyroelectric effects near T_c so that these materials can be used to make room temperature infrared sensors.

In general, the overall philosophy for enhancing functionalities in ferroelectric materials is to introduce defects into otherwise ordered crystal structure to create local disorder, which help push the ferroelectric system further away from energetically stable state so that they will become more responsive to external stimuli. To a large degree, defects are the magic ingredients for making ferroelectric materials smarter.

The most useful inorganic ferroelectric materials are oxides, and they often have the perovskite crystal structure with the general chemical formula of ABO_3. Using conventional processing techniques, oxygen vacancies are often produced in these ceramic or single crystal perovskite materials, which can produce interesting transport properties that can also induce structural changes through electromechanical coupling. Oxygen vacancies often make ferroelectric materials semiconducting because the hopping energy barrier between adjacent oxygen sites is rather low and the conduction increases quickly with the increase in temperature. In general, defect-free ferroelectric oxides are good insulators. The presence of oxygen vacancies can make the ferroelectric material a conductor, a semiconductor, or an insulator depending on the vacancy concentration and the hole-hopping energy barrier. By controlling the behavior of oxygen vacancies, one can tune the transport properties and other functional properties of ferroelectrics for targeted applications.

Surfaces and domain walls are natural physical defects that are present even in a ferroelectric material with perfect lattice structure. They can also generate new functional properties, or degrade existing functional properties, or be used to enhance certain functional properties. For example, by controlling the poling direction of ferroelectric single crystals, different domain patterns can be produced and one can utilize these domain patterns to produce or to change the macroscopic symmetry of a single crystal. Such methods of functional property enhancement by controlling domain structures is often termed "domain engineering" and the method has become one of the most important and successful methods in recent years for fabricating ferroelectric materials with ultrahigh piezoelectric and dielectric properties.

In this chapter, through some practical examples, we will provide a comprehensive review on the roles of different defects in ferroelectrics. Due to the page limitation, our main goal here is to analyze the basic physical principles and to provide some references for people to get an idea on the effects of different types of defects and on the role of these defects in controlling the functional properties of ferroelectric materials.

7.2 Vacancies in Perovskite Ferroelectric Materials

The most widely studied ferroelectric material is barium titanate ($BaTiO_3$), including its ceramic form and single crystal form, which has a perovskite structure ABO_3 as shown in Fig. 7.1 Its high-temperature paraelectric phase is cubic with $m\bar{3}m$ symmetry and it has three consecutive structural phase transitions on cooling to: tetragonal phase with $4mm$ symmetry at 130°C, orthorhombic phase with $mm2$ symmetry at 0°C, and rhombohedral phase with $3m$ symmetry at −90°C [1]. Since its discovery in the early 1940s, people are continuously fascinated by the complex characteristics of $BaTiO_3$ and are continuously trying more ways to explore its multifunctional properties. It is by far the most widely used ferroelectric materials today.

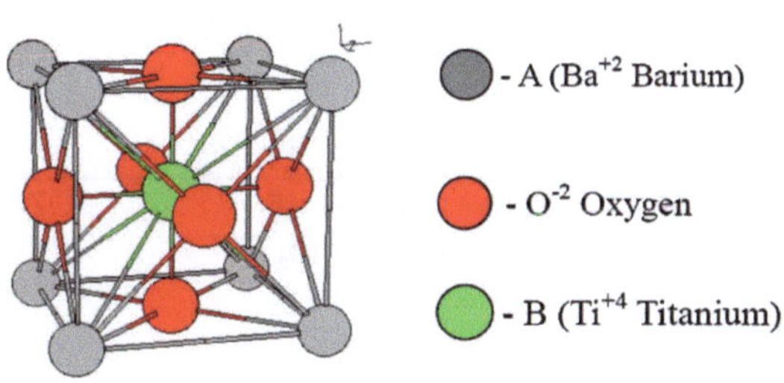

Fig. 7.1 Illustration of the crystal structure of barium titanate

Doped $BaTiO_3$ ceramic and single crystals have many different functional properties and are used widely used for electromechanical, dielectric, thermal-electric, and electro-optical applications. The largest markets for barium titanate ceramics are MLCC and positive temperature coefficient resistors (PTCR) [2]. The functional properties of $BaTiO_3$ can be tuned by introducing various dopants owing to the high defect solubility of its perovskite structure.

Barium titanate has been used as a model ferroelectric system in almost all ferroelectric related books, particularly when discussing ferroelectric phase transitions. It is a proper ferroelectric material with typical Curie–Weiss behavior and a distinct soft zone center optical mode whose softening drives the ferroelectric phase transition. The ability to accommodate different kinds of cation defects makes it an excellent model system to engineer desired functional properties through chemical doping method. It has been well documented that defect chemistry plays an important role in the formation of barium titanate structure. As shown in Fig. 7.2, the phase diagram of $BaO–TiO_2$ is rather complex. Earlier works on the phase diagram did not take into account the defective structures near the stoichiometric composition [3–5]. As shown in the dashed lines near the 50/50 perovskite $BaTiO_2$ composition, there is a small solubility region in which defective perovskite phases can form [6].

The solubility on the Ti-rich side is larger than that on the Ba-rich side due to the fact that the energies of forming a TiO_2 Schottky defect (2.9 eV/defect) are larger than that of forming a BaO Schottky defect (2.58 eV/defect) [7, 8]. Using the Kröker–Vink notation [9], one can write out the defect formation process on both sides of the stoichiometric composition line.

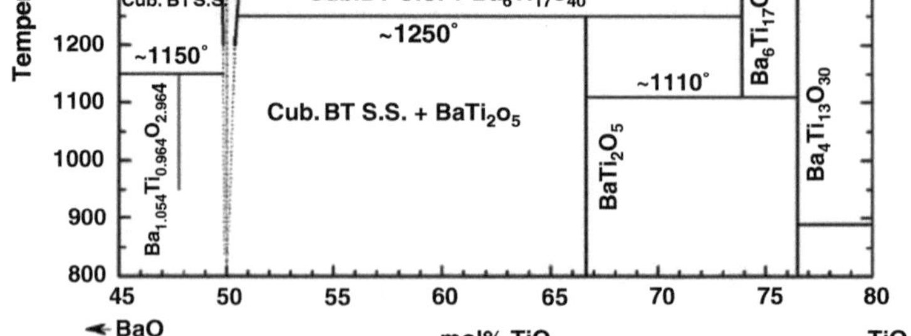

Fig. 7.2 Modified phase diagram from [6]

On the Ba-rich side,

$$\mathrm{BaO} \overset{\mathrm{BaTiO_3}}{\longrightarrow} \mathrm{Ba}_{\mathrm{Ba}}^{\mathrm{X}} + \mathrm{O}_{\mathrm{O}}^{\mathrm{X}} + V_{\mathrm{Ti}}^{''''} + 2V\ddot{\mathrm{o}}; \text{or} \tag{7.1}$$

$$\mathrm{Ba}_a\,\mathrm{Ti}_b\,\mathrm{O}_{a+2b} \overset{\mathrm{BaTiO_3}}{\longleftrightarrow} a\mathrm{Ba}_{\mathrm{Ba}}^{\mathrm{X}} + b\mathrm{Ti}_{\mathrm{Ti}}^{\mathrm{X}} + (a+2b)\mathrm{O}_{\mathrm{O}}^{\mathrm{X}} + kV_{\mathrm{Ti}}^{''''} + 2kV\ddot{\mathrm{o}}. \tag{7.2}$$

where $a = b + k$ ($k = 0$ at $a = b$). Ba_2TiO_4 ($a = 2$ and $b = 1$) and $Ba_{1.054}Ti_{0.946}O_{2.946}$ ($a = 1.054$ and $b = 0.946$) are the secondary phases around stoichiometric $BaTiO_3$ on the Ba-rich side.

On the Ti-rich side,

$$\mathrm{TiO_2} \overset{\mathrm{BaTiO_3}}{\longrightarrow} \mathrm{Ti}_{\mathrm{Ti}}^{\mathrm{X}} + 2\mathrm{O}_{\mathrm{O}}^{\mathrm{X}} + V_{\mathrm{Ba}}^{''} + V\ddot{\mathrm{o}}; \text{or} \tag{7.3}$$

$$\mathrm{Ba}_\alpha\,\mathrm{Ti}_\beta\,\mathrm{O}_{\alpha+2\beta} \overset{\mathrm{BaTiO_3}}{\longleftrightarrow} \alpha\mathrm{Ba}_{\mathrm{Ba}}^{\mathrm{X}} + \beta\mathrm{Ti}_{\mathrm{Ti}}^{\mathrm{X}} + (\alpha+2\beta)\mathrm{O}_{\mathrm{O}}^{\mathrm{X}} + mV_{\mathrm{Ba}}^{''} + mV\ddot{\mathrm{o}} \tag{7.4}$$

where $\beta = \alpha + m$ ($m = 0$ at $\beta = \alpha$). $BaTi_2O_5$ ($\alpha = 1$ and $\beta = 2$) and $Ba_6Ti_{17}O_{40}$ ($\alpha = 6$ and $\beta = 17$) are confirmed as secondary phases around stoichiometric $BaTiO_3$ on the Ti-rich side [10, 11].

Besides the two types of partial Schottky defects, there are also full Schottky defects that may form on both Ti- and Ba-rich sides; each full Schottky defect contains a Ba vacancy, a Ti vacancy, and three O vacancies:

$$\mathrm{BaTiO_3} \leftrightarrow V_{\mathrm{Ba}}^{''} + V_{\mathrm{Ti}}^{''''} + 3V\ddot{\mathrm{o}} \tag{7.5}$$

The formation energy for a full Schottky defect is higher compared to that of a partial Schottky defect. It amounts to 3.33 eV/defect on the Ti-rich side and 3.48 eV/defect on the Ba-rich side.

The existence of various defects can change the electric properties of $BaTiO_3$ drastically because oxygen vacancies can easily hop between their ideal ionic positions to produce ionic conduction and the conduction characteristics depend on the defect concentration.

7.3 Doping of Aliovalent Defects

Better functional materials should be more agile to external stimuli. For ferroelectric materials, the external stimuli usually refer to electric field, mechanical stresses or temperature, or the combination of different fields. The degree of electric response is reflected in the dielectric property, while the degree of mechanical response is reflected in the piezoelectric response. Due to the presence of oxygen vacancies, barium titanate can be electrically very lossy, even conductive when the vacancy concentration is high enough. Through cation doping and oxygen annealing, the electric properties of barium titanate can be tuned to fit different applications. A common way to improve the dielectric properties of $BaTiO_3$ is to dope the system by 3+ rare-earth cations. The reason to choose 3+ cations is because they have the possibility to go into either "A-" or "B-" site of the perovskite crystal structure to create either donor or acceptor doping situations. Local dipolar defects are created with such aliovalent doping, which can influence both the electric and mechanical responses as well as change the ferroelectric phase transition temperature T_c.

One of the key factors determining which site the doped ions will go is the ionic radius of the doping element. A common practice is to use the Goldschmidt tolerance factor to give a qualitative prediction, which, for perovskites, is defined by [12]

$$t = \frac{r_A + r_o}{\sqrt{2}(r_B + r_o)} \tag{7.6}$$

where $r_i (i = \text{A, B, O})$ are the ionic radii of "A-", "B-", and "O-" site ions in a perfect perovskite structure. An ideal cubic perovskite should satisfy the relation: $r_A + r_o = \sqrt{2}(r_B + r_o)$, i.e., with the tolerance factor equals 1. For pure barium titanate, $t \sim 1.06$ due to the fact that Ti^{4+} is smaller than its cavity and Ba^{2+} is larger than its cavity. In general, the preferential site for the dopant is where it can make the tolerance factor closer to 1. As shown in Fig. 7.3, the tolerance factor t_A and t_B for rare-earth 3+ ion R^{3+} doping has a crossover at the ionic radius $r \sim 0.9$ Å. Therefore, qualitatively, for rare-earth 3+ ion doping, when the ionic radius is smaller than 0.9, the doping has a tendency to go into the B-site, while for ions with ionic radius larger than 0.9, the doping has a tendency to go into the A-site.

From reported experimental observations, there seem to be three different regimes. For rare-earth dopants with ionic radius $r < 0.87$ Å, the dopants will occupy the B-site. The compensation mechanism is mainly via oxygen vacancies. For rare-earth dopant with ionic radius >0.94 Å, the dopants will occupy the A-site

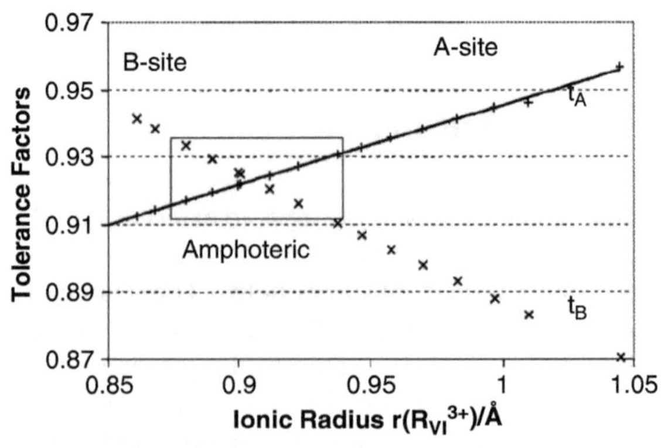

Fig. 7.3 The tolerance factors t_A and t_B for substitution of trivalent ions on the A- and B-sites, as functions of the ionic radius [16]

and the compensation mechanism will depend on the Ba/Ti ratio. For $Ba/Ti = 0.99$, the compensation is mainly via electrons [13], while for $Ba/Ti = 1.01$, a significant part of the compensation is by titanium vacancies [14]. When the ionic radii of the dopant ions are in the regime of $0.87\,Å < r < 0.94\,Å$, the dopants will choose their sites according to the Ba/Ti ratio. In general, one may find them in both sites with certain preference depending on the environmental situation and they will induce volume change accordingly based on the Vegard's law [15].

Interestingly, for some rare-earth dopings, the valence of the dopants may be changed by the local environment. When doping ions go into the A-sites, they tend to become 2+ ions; when they go into the B-sites, they tend to become 4+ ions. This valance change is also affected by the temperature. Usually, higher temperatures make the ions to have a tendency to change into higher valance ions.

7.4 Defects and Dielectric Properties

The rich defect chemistry in $BaTiO_3$ makes it possible to use different dopings to manipulate its dielectric properties. Undoped barium titanate behaves like a semiconductor with a band gap ~3.1 eV and a resistivity of $10^{10}\,\Omega\,cm$ [17]. The Curie temperature varies with lattice constant so that the one could use doping to change the lattice constant and hence adjust the Curie temperature [18]. Because the dielectric constant peaks at the Curie temperature, i.e., at the paraelectric–ferroelectric phase transition temperature T_c, pushing down the phase transition temperature of $BaTiO_3$ to near room temperature will drastically increase the dielectric constant at room temperature. On the other hand, for practical applications, temperature stability and field stability are also very important considerations. Therefore, an ideal doping should have two effects: (1) bring down the phase transition temperature closer to room temperature; and (2) to broaden the dielectric peak at the phase transition so that better temperature stability can be achieved. One solution to this problem is to combine the properties of normal ferroelectric with relaxor ferroelectric materials; the latter has defused phase transition, i.e., has a very broad dielectric peak near the dielectric maximum and the characteristic frequency-dependent dielectric maximum. Based on the work of [19], the ferroelectric lead barium zirconate titanate $(Pb_{0.65}Ba_{0.35})(Zr_{0.70}Ti_{0.30})O_3$ shows typical relaxor behavior as shown in Fig. 7.4a. Its dielectric constant is approximately 6,000 at room temperature compared to about 1,200 of $BaTiO_3$ ceramic, and remained almost constant under an electric field as high as 20 kV/cm, as shown in Fig. 7.4b.

It was found that Nb doping can decrease the dielectric maximum temperature to below room temperature so that the room temperature dielectric constant value becomes smaller (Fig. 7.5). On the other hand, proper amount of Nb doping may trigger faster grain growth, causing the dielectric constant to increase to a room temperature value of 7,000 for a lead barium zirconate titanate with 1% Nb doping, as shown in Fig. 7.5.

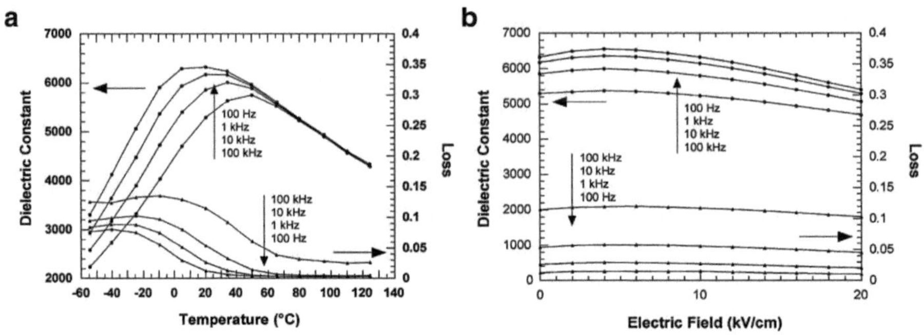

Fig. 7.4 The dielectric properties of the PBZT base composition showing (**a**) typical relaxor behavior measured at low field (1 V oscillation) and (**b**) a stable voltage dependence at room temperature [19]

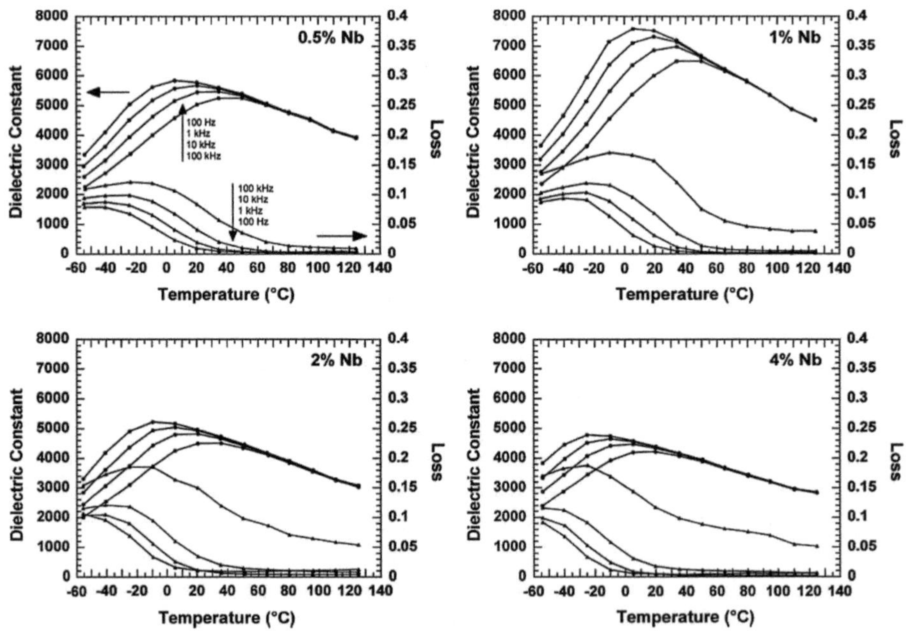

Fig. 7.5 Dielectric constant and loss versus temperature of lead barium zirconate titanate ceramic with different amount of Nb doping [19]

Although relaxor type ferroelectrics have broader dielectric maximum and can produce very large dielectric properties, the frequency dependence is often undesired and the temperature stability range is relatively small. For higher temperature applications, it is better to lift the dielectric constant value at room temperature while not to lower the Curie temperature.

Utilizing the high defect solubility of the perovskite $BaTiO_3$, people have tried many different doping elements to improve the dielectric performance, including

Fe, Mg, Yb, Er, Dy, and Sm [20, 21]. Rare-earth element doping is the most widely studied due to the rich phenomena produced by such dopings. As mentioned above, the rare-earth elements with intermediate atomic radii could enter either the A-site or the B-site of the perovskite structure, creating donor or acceptor doping situation. The peak value of the dielectric constant is usually increased by donor doping, while the room temperature dielectric constant could be either increased or decreased by such aliovalent substitutions.

It is also possible to introduce interstitial defects to produce donor-like doping. For example, B^{3+} can be introduced into $BaTiO_3$ through vapor doping [22]. In vapor doping, B_2O_3 can vaporize as a single molecule and be doped into the perovskite lattice.

A particular impurity in perovskite lattice may act as an acceptor, a donor, or an uncharged impurity, depending on the site it occupies. Because B^{3+}(0.23 Å) is too small to occupy the cation ($Ba^{2+} \sim 1.35$ Å and $Ti^{4+} \sim 0.68$ Å) sites of $BaTiO_3, B_2O_3$ usually acts as a glass-forming agent in $BaTiO_3$ ceramic. If B^{3+} ions are incorporated into the lattice, they form boron interstitials due to their ultrasmall ionic size. The defect chemistry equation can be written as follows:

$$B_2O_3 \rightarrow 2B_i^{\cdot\cdot\cdot} + 3O_O^x + 3V_{Ba}'', \tag{7.7}$$

where $B_i^{\cdot\cdot\cdot}$ stands for an interstitial boron with three positive charges, O_O^x stands for an oxygen ion on an oxygen site, and V_{Ba}'' stands for a barium vacancy with two negative charges. The observed lattice expansion can be explained by the formation of interstitial boron ions. It is believed that the preferred site should be the position labeled by "2" in Fig. 7.6 The interstitial boron ions caused both the a and c lattice constants in the tetragonal phase to increase from 4.0002 Å and 4.0426 Å to 4.0137 Å and 4.598 Å, respectively, so that the total volume of the lattice is increased in such interstitial doping. The ferroelectric phase transition temperature was also increased from 128 to 130°C, but the room temperature dielectric constant was suppressed by almost 20% [22].

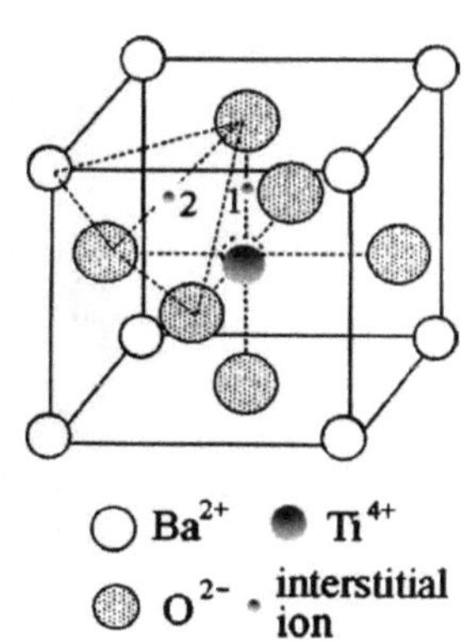

Fig. 7.6 Possible boron interstitial sites in the $BaTiO_3$ perovskite structure [22]

In general, some aliovalent dopings help stabilize oxygen vacancies that are usually present in $BaTiO_3$, which will help reduce the dielectric loss and increase the temperature stability. Other dopings may bring down the Curie point closer to room temperature so that the dielectric response can be drastically increased at room temperature. This is usually at the cost of temperature stability, and hence, cannot be used for high temperature capacitors.

7.5 Grain Boundary and Positive Temperature Coefficient Resistor

The main difference between a ceramic and a single crystal is the presence of grain boundaries in ceramic. Grain boundary is another type of defect that can be considered as a thin distorted crystal layer or amorphous layer on the crystalline surface. In ferroelectric materials, grain boundaries can produce very interesting phenomena when they are coupled with doping defects.

It was found that under heavy donor doping, $BaTiO_3$ ceramic exhibits an unusual increase in resistance by about six orders of magnitude above a critical temperature (between 100 and 200°C). People have utilized this phenomenon to make PTCRs, which are used widely as electric current limiters, temperature sensors, etc. Through different experimental studies and theoretical analyses, it is recognized that PTC phenomenon is caused by grain boundaries in ceramic materials [23–27]. Shown in Fig. 7.7 is the illustration of a $BaTiO_3$ ceramic structure. The grains are separated by grain boundaries. When the ceramic is heavily donor doped, a depletion layer is formed at the interface between grain and grain boundary, which produces a Schottky barrier in the paraelectric phase for which the dielectric constant is relatively small.

Under heavy Nb doping, the grains have a lot of donor states, while the grain boundaries do not have regular crystal structure and the doping elements do not go into them as easy as in the domains so that the grain boundaries are in acceptor

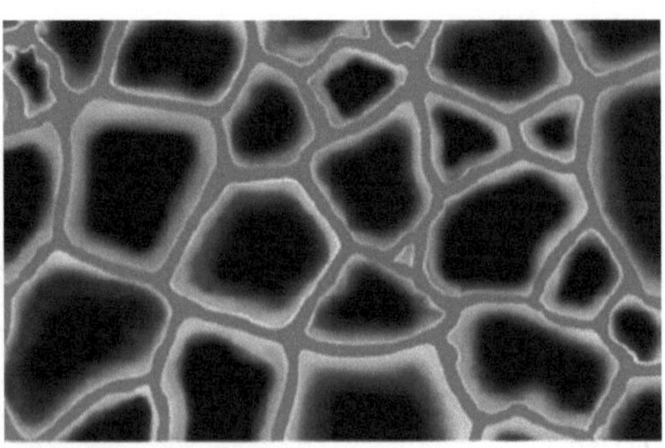

Fig. 7.7 Illustration of a ceramic structure

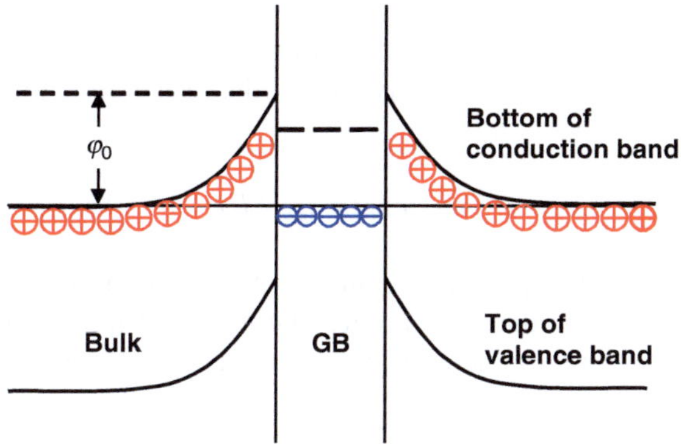

Fig. 7.8 Grain boundary and grains on both sides produced double Schottky barrier that prevents current flow so that the combined structure has very high resistance

states with some accumulations of electrons. Across a grain boundary, there are two interfaces between the grain and the grain boundary as shown in Fig. 7.7. Such a structure makes a pair of symmetric Schottky barriers as shown in Fig. 7.8 [28]. Current follow is blocked both ways by an energy barrier whose height depends on the density of the acceptor states in the grain boundary. This energy barrier will produce very high electrical resistivity.

As shown in Fig. 7.8, the barrier height ϕ_0 determines the resistivity value. It is also strongly coupled to the dielectric properties. Assuming the depletion layer formed at the grain boundary has a width b, then this width depends on the concentration of the acceptor states N_A at the grain surface and the concentration of bulk donor states n_D:

$$b = N_A/n_D \tag{7.8}$$

According to Poisson's equation, the height of the potential barrier ϕ_0 is given by

$$\phi_0 = e^2/\varepsilon \times b^2/2 \tag{7.9}$$

and the barrier height ϕ_0 controls the resistivity $\rho \sim \exp(\phi_0/kT)$.

Below T_c, the barrier height is very low due to the large dielectric permittivity. Above T_c, ϕ_0 increases quickly with a drastic decrease in the dielectric constant since the barrier height is inversely proportional to the dielectric permittivity:

$$\phi_0 = A/\varepsilon = A/C \times (T - T_c) \tag{7.10}$$

Thus, the resistivity increases exponentially with temperature above T_c

$$\rho \sim \exp(A/Ck \times (1 - T_c/T)) \tag{7.11}$$

This relation (7.11) holds until the Fermi level is reached. When the acceptor states reached the Fermi level, they reemit electrons into the bulk and ϕ_0 reaches its maximum value $\phi_{0\,\max}$. Above the Fermi level, the resistivity will start to decrease exponentially.

$$\rho \sim \exp(\phi_{0\,\max}/T) \tag{7.12}$$

The characteristic resistivity vs. temperature is illustrated in Fig. 7.9.

The fast decrease of the permittivity with temperature immediately above T_c drives the resistivity to increase exponentially, producing several orders of

Fig. 7.9 Typical resistivity vs. temperature plot for a PTC material.

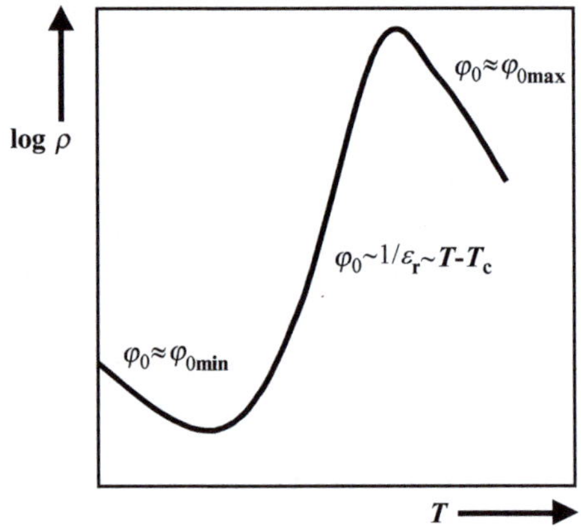

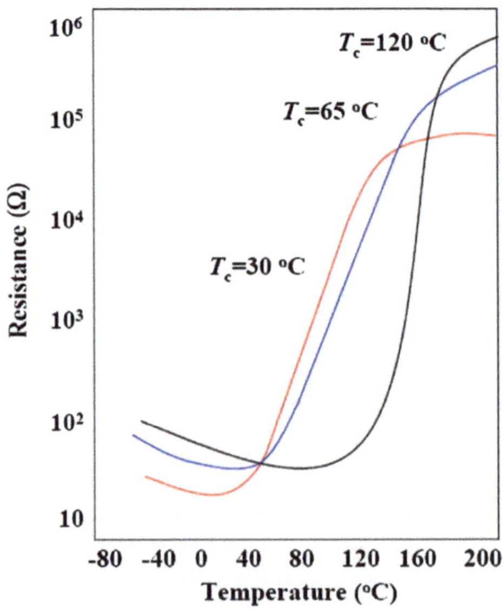

Fig. 7.10 Typical PTC resistance versus temperature curves for different T_c

magnitude changes to the resistivity in a temperature range of a few tens of degrees. In other words, the resistivity is super sensitive to temperature in this temperature range. Therefore, they can be used to make temperature control switches.

Since the PTC effect is directly coupled to the ferroelectric phase transition temperature T_c, one could tune the PTC characteristics by changing the T_c of $BaTiO_3$ through defect doping. There are already commercial PTC thermistors made of donor-doped $BaTiO_3$ ceramic with T_c ranging from 30 to 120°C as shown in Fig. 7.10.

7.6 Domain Walls as a Type of Mobile Defects

One of the characteristics of ferroelectric materials is the presence of domain structures in the ferroelectric phase. Domains refer to a region in which all dipoles are aligned in the same direction. The formation of domain patterns during a ferroelectric phase transition from a high symmetry phase to a low symmetry phase is a reflection of the system trying to recover those lost symmetries. The number of domain states or variants in the low-temperature phase is equal to the ratio of the number of operations in the high symmetry group over that of the low symmetry group. Although macroscopically we often treat many systems, such as ceramic materials, as isotropic, i.e., having a spherical symmetry, the highest symmetry in terms of crystal structure is only cubic $m\bar{3}m$. There are in total 32 point groups describing the crystal symmetries [29]. We must distinguish the macroscopic symmetries and the lattice symmetries of crystal structures. The number of domain states in the ferroelectric phase may be predicted if the low-temperature phase is a subgroup of the high-temperature symmetry group. For example, there are 48 operations in the cubic $m\bar{3}m$ point group, and if the ferroelectric phase transition is from cubic to a tetragonal $4mm$ symmetry phase, like in the case of $BaTiO_3$, there will be six domain states because the symmetry operation in the $4mm$ symmetry group is eight. Phase transitions may also happen between subgroup symmetries of the same parent group, although they may not have direct group–subgroup relationship, such as between the tetragonal and rhombohedral symmetry phases in ferroelectric $BaTiO_3$ and $Pb(Zr_xTi_{1-x})O_3$ (PZT) solid solutions. In general, because all domain states are energetically the same, all of them may appear in the same crystal and form the so-called domain structures.

The consequence of multidomain formation is the creation of domain walls, which refer to the spatial transition region between two domains of different orientations. Domain walls are also defective structures of the crystal, in which the ionic ordering is distorted. However, different from vacancies or foreign ion substitutions, domain walls are atomically coherent and charge balanced. In other words, there are no ions missing or gain inside a domain wall, but only the ordering of ions is distorted. We often call a structure containing two domains plus a domain wall in between as a "twin." Twinning provides a new functional mechanism for easy global shape deformation via the movement of domain walls.

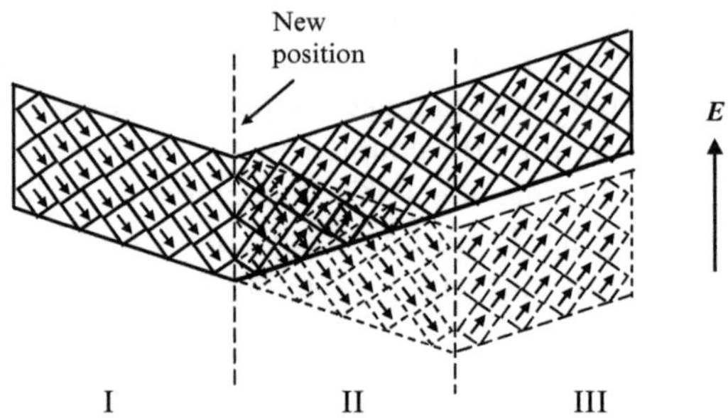

Fig. 7.11 Domain wall movement in a ferroelectric twin structure under an external electric field **E**. Regions II and III move up as the domain wall moves to the left

If the low-temperature states are polarized, domain wall movements cause the polar vector to rotate in the region swept by the moving domain wall. This situation is illustrated in Fig. 7.11 for a ferroelectric twin. Under an upward electric field, the domain wall moves to the left. At the same time, the whole regions II and III on the right-hand side of the wall move up relative to region I. The dipoles in region II are switched to more favorable positions under the applied external field, which contributes to the total dielectric response, and the response from the domain wall movements is usually called the "extrinsic" contribution. The global shape change caused by the domain wall movement could also be substantial as shown in the figure, which constitutes the extrinsic contribution to the piezoelectric effect. In other words, the switching of dipoles in region II gives an extrinsic contribution to the dielectric susceptibility, while the shape deformation caused by the wall movement contributes extrinsically to the macroscopic piezoelectric effect. Because the motion of domain walls is mechanical, it often produces mechanical losses. Hence, the extrinsic contributions to functional properties are associated with some degree of loss.

Domain walls are a special kind of defects. They create localized stress gradient and/or electric field gradient [30] that can strongly interact with other defects, such as dislocations, vacancies, and aliovalent dopants. This interesting nature of domain walls enables us to control domain patterns and domain wall densities through chemical doping of different elements.

It is a common practice to dope aliovalent ions in ferroelectric materials to create multivalence and/or vacancies in the material so that domain walls could interact with them. The charged defects created by doping can either pin the domain walls or make the walls more mobile. This doping method has proven effective to enhance the functionality of some ferroelectric materials, for example, the La- or Nb-doped PZTs have much larger piezoelectric and dielectric properties than those of the nondoped PZTs, and they were often called "soft" PZT because their coercive field is reduced.

Inhomogeneous stresses produced by localized defects may induce local phase transitions above the bulk phase transition temperature T_c, causing the material to have mixed low and high symmetry phases in certain temperature region. Such two-phase mixtures are usually very sensitive to external fields or stresses since the phase change among the mixture may become barrierless even for a first-order phase transition [31]. The so-called morphotropic phase boundary composition is where a solid solution system has a composition-driven phase transition and often contains two-phase mixture.

The formation of domain structures and the available variants in the low symmetry phase is dictated by the crystal symmetry of the high-temperature phase. However, because the existence of domain structures may produce new symmetrics at the mesoscopic scale, it is the global macroscopic symmetry, not the local crystal symmetry, which controls the functionality of the material. Therefore, at the macroscopic level, one can use complementary constituents to make composite structures of designed average symmetries to produce better functional properties. In a multidomain system, domain walls are being used to separate individual domains and to make nano- to microsize domain patterns and to produce new mesoscopic symmetries. In recent years, using domain walls to improve functional properties of ferroic materials has evolved into a new branch of material engineering, i.e., "domain engineering."

Domain engineering on ferroelectric materials in the mesoscopic level is to manipulate domain structures and their mobility in order to increase the functional properties or to create new functional properties. Aliovalent doping in a ferroelectric system can create strong localized electric forces that may facilitate or hinder the movements of domain walls, and hence, influence the extrinsic contributions to functional properties. This method has been used to improve the piezoelectric properties in soft and hard PZT systems as described above. Defects, including dislocations and point defects, can also be rearranged to accommodate the stress field generated by the formation of domain walls.

The most noteworthy examples for the success of domain engineering are the relaxor-based multidomain single crystals $(1-x)Pb(Zn_{1/3}Nb_{2/3})O_3$–$xPbTiO_3$ (PZN–PT) and $(1-x)Pb(Mg_{1/3}Nb_{2/3})O_3$–$xPbTiO_3$ (PMN–PT) solid solution systems. Their emergence has created a true excitement in the transducer and actuator communities due to the ultrahigh electromechanical coupling factor $k_{33} > 90\%$ (compared to 68% for PZT) and the very large piezoelectric coefficient ($d_{33} > 2000\,pC/N$) [32–34]. Although these single crystal materials had been discovered in 1969 [35], they did not generate enough interest because the crystal size was very small and they could not retain high remnant polarization along the threefold polar axis in the rhombohedral phase. Moreover, the piezoelectric d_{33} coefficient in the single domain state was not very impressive. The relaxor-based single crystals combined the disorder induced by the B-site aliovalent ions and the domain walls induced structural instability to create an unusual category of piezoelectric materials, whose piezoelectric coefficient is almost five times of the best traditional piezoelectric PZT ceramics. Because they are ferroelectric materials, domain pattern manipulation can be easily done through the applied electric poling field.

The crystal symmetry of these relaxor-based ferroelectric crystals is rhombohedral $3m$ with the dipoles in each unit cell pointing to the <111> (along body diagonals) of the original perovskite cubic cells. It was found that the system could sustain large polarization if the poling field is applied along the <100> family directions. After poling long [001], each unit cell has a dipole moment along one of the four upper corners of the cubic directions as shown in Fig. 7.12 The polarization projections onto the directions perpendicular to the polar axis are randomly oriented so that the global symmetry of the multidomain system (macroscopic average) is pseudo-tetragonal. A representative domain pattern of a [001] direction polarized $0.68Pb(Mg_{1/3}Nb_{2/3})O_3$–$0.32PbTiO_3$ single crystal is shown in Fig. 7.13 [36]. One can see that the interwoven domain pattern produced a macroscopic averaged 4*mm* symmetry. One can see that domain walls play an important role in the creation of such domain patterns and they are also responsible in the ultrahigh piezoelectric properties observed in these domain-engineered single crystals.

Strong elastic interactions among neighboring cells help stabilize the poled multidomain configuration. It is interesting to mention that misorientational poling produced a new type of domain pattern symmetry that is very different from the underlying crystal symmetry. If domain walls are taken into consideration, much more complex domain patterns can be created with symmetries varying from

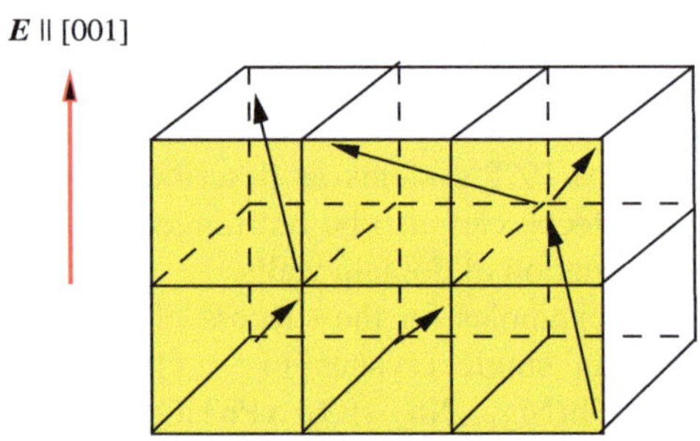

Fig. 7.12 Illustration of misorientational poling in PZN–PT single crystal system. The field is applied along [001] and the dipoles in each unit cell are pointing to the four upper corners of body diagonals.

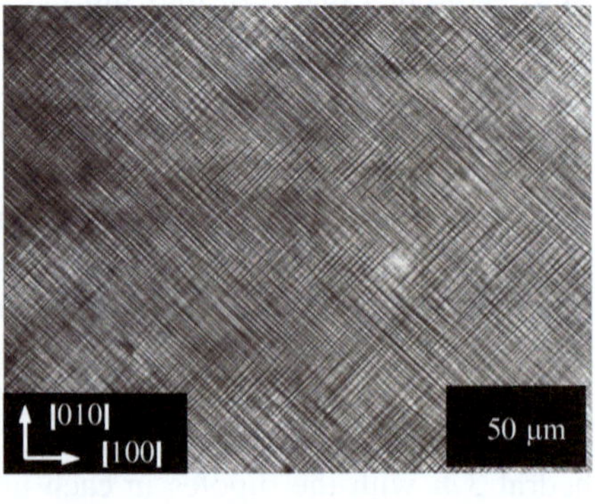

Fig. 7.13 Domain structures observed from a [001] polarized PMN-33%PT single crystal samples with a field level of 0.25 kV/cm [36]

the highest cubic all the way down to the lowest triclinic [37]. Because such multidomain states are much more responsive or unstable under external stimuli, superior functional properties are created.

7.7 Size Effects and Surface in Ferroelectric Materials

In the real world, we deal with finite size materials, including ceramics, thin films, or crystals. Surface of a finite system forms a natural defective layer, in which the periodic lattice structure terminates. For ferroelectric materials, this surface layer plays an important role in controlling the functional properties, particularly when the grain size goes down to nanometer scale and in ferroelectric thin films, for which the surface-to-volume ratio becomes significant. Due to the difference in crystal structures inside the grain and in the surface layer, properties can differ substantially. Such differences are intentionally enhanced using different doping techniques and fabrication techniques to make functional ferroelectric materials, such as temperature stable materials with very high dielectric susceptibility for MLCC applications.

One type of MLCC is the X7R series, in which $BaTiO_3$ nanoceramics are doped with different oxides, such as Y_2O_3, MgO, MnO_2, $CaCO_3$, and SiO_2, to create the so-called core–shell structure. The core is ferroelectric with high dielectric constant, while the shell is paraelectric with much lower dielectric constant [38–40]. Figure 7.14 is a typical TEM micrograph of doped nanosize $BaTiO_3$ ceramic with a core–shell structure [41]. Such core–shell structure can have very large dielectric coefficients (around $2,300 \sim 2,500\,\varepsilon_0$) near room temperature, and because the materials have a much broader dielectric peak, they have very good temperature stability for applications near room temperature.

In other cases, surface effects caused the change of domain structures, which in turn influence the peak value of the dielectric constant. If the particle size is not too small, the change of domain structures does not cause significant change of the

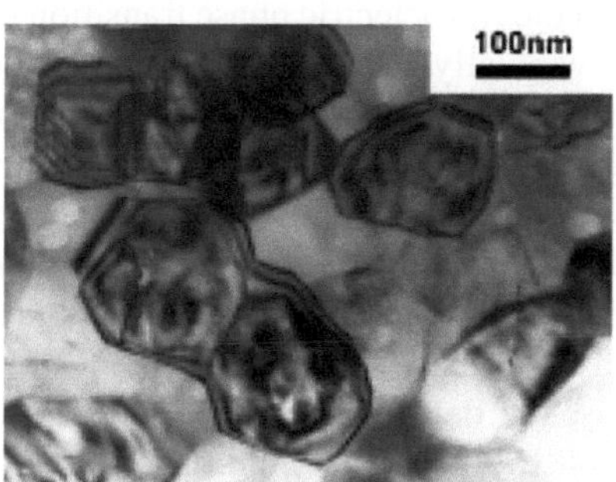

Fig. 7.14 Typical core–shell structure of nanoceramic $BaTiO_3$ [41]

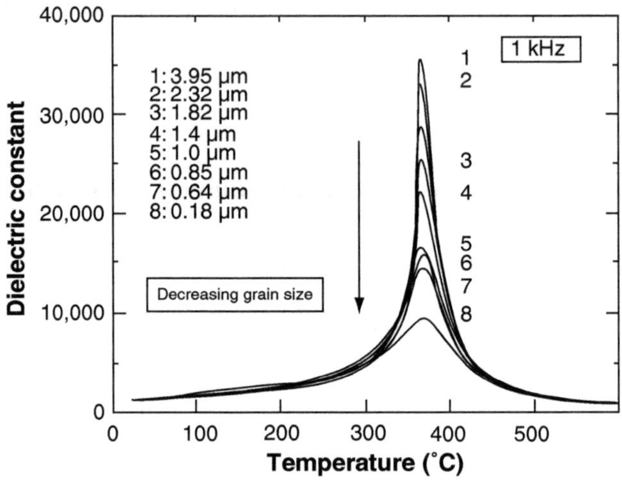

Fig. 7.15 Changes of the dielectric constant vs. temperature curve with grain size for Nb-doped PZT ceramic. The base composition of the PZT ceramic has the Ti/Zr ratio of 48/52 [42]

phase transition temperature T_c. Shown in Fig. 7.15 is the measured temperature dependence of the dielectric constant of doped fine grain PZT ceramics with different grain sizes. One can see that the Curie temperature change is very small, but the peak value of the dielectric constant is suppressed substantially with the decrease of grain size [42].

Many physical properties of ferroelectric materials depend on the domain structures. Because the grain size directly determines the type of domain pattern being formed and the size of the domains, some physical properties, such as piezoelectric and elastic properties, are also effected significantly by the reduction of grain size [43, 44].

Another important effect of surface is its contribution to the internal stress buildup in ceramic and thin film materials. This internal stress effect becomes more significant as the grain size of the ceramic goes into the nanometer range. Such internal stresses will push the ferroelectric phase transition temperature to lower and lower temperature, and eventually, totally suppress the ferroelectric phase transition at a critical size. There were numerous experimental and theoretical studies on the critical sizes on 1D nanorods, 2D film, and 3D nanograins. Some papers use the extended dimensions to represent the dimension of the systems under study. For example, the nanoparticle is called $0 - d$, nanorods $1 - d$, and thin film $2 - d$ [45]. Shown in Fig. 7.16 is a summary plot from [45]. Figure 7.16a, b is for $BaTiO_3$ and $PbTiO_3$, respectively. The experimental results in the figure are from several sources. For the $BaTiO_3$, the data marked by ▲ are from [46], • are from [47], and ■ and ○ are from [48]. For $PbTiO_3$, the data marked by ♦ are from [49], • are from [50], and ■ are from [51]. The curves are theoretical estimates based on phenomenological theories from [45]. One can see that the critical size is the

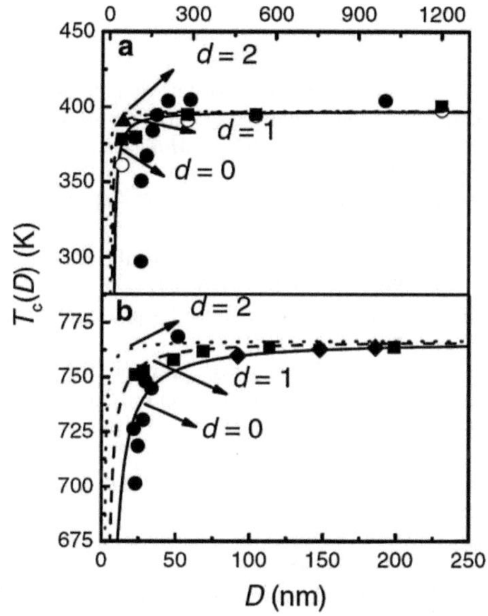

Fig. 7.16 Grain size dependence of the phase transition temperature Tc of nanograin ceramics. (**a**)$BaTiO_3$ and (**b**) $PbTiO_3$ [45]

largest for the nanoparticles, and the smallest for thin films. Below the critical size, ferroelectricity no longer exists.

7.8 Summary

In this chapter, we have described the role of defects in ferroelectric materials. Overall, there are intrinsic structural defects produced during processing, i.e., vacancies, domain walls, grain boundaries, and surfaces; substitutional defects from doping, particularly aliovalent doping defects; and interstitial defects from special processing. These defects are mostly for the benefit of functional properties of ferroelectric materials. Researchers have invented many creative ways to enhance the dielectric, piezoelectric, and thermal-electric properties of ferroelectric single crystals, thin films, and ceramics by introducing different types of defects or controlling defect amount in ferroelectric materials. Manipulating defects could be via chemical means, such as doping of aliovalent elements, or physical means, such as domain engineering. It was found that interactions between different types of defects, such as between aliovalent doping ions with domain walls, often produce large enhancement of certain functional properties. Therefore, controlling defects and utilizing defects to our advantage to gain stronger functional properties of ferroelectric materials will continuously be an important research area in the near future.

References

1. H.F. Kay, P. Vousden, Symmetry changes in barium titanate at low temperatures and their relation to its ferroelectric properties. Philos. Mag. **40**, 1019–1040 (1949)
2. A.J. Moulson, J.M. Herbert, *Electroceramics: Materials, Properties, and Applications* (Chapman-Hall, London, 1990)
3. D.E. Rase, R. Roy, Phase equilibria in the system $BaO-TiO_2$. J. Am. Ceram. Soc. **38**(3), 102–113 (1955)
4. H.M. O'bryan Jr., J. Thomson Jr., Phase equilibria in the TiO_2-rich region of the system $BaO-TiO_2$. J. Am. Ceram. Soc. **57**(12), 522–526 (1974)
5. T. Negas, R.S. Roth, H.S. Parker, D. Minor, Subsolidus phase relations in the $BaTiO_3-TiO_2$ system. J. Solid State Chem. **9**(3), 297–307 (1974)
6. S. Lee, Z.K. Liu, C.A. Randall, Modified phase diagram for the barium oxide-titanium dioxide system for the ferroelectric barium titanate. J. Am. Ceram. Soc. **90**(8), 2589–2594 (2007)
7. G.V. Lewis, C.R.A. Catlow, Computer modeling of barium titanate. Radiat. Eff. **73**(1–4), 307–314 (1983)
8. G.V. Lewis, C.R.A. Catlow, Defect studies of doped and undoped barium titanate using computer simulation techniques. J. Phys. Chem. Solids **47**(1), 89–97 (1986)
9. F.A. Kröger, H.J. Vink, Relations between the concentrations of imperfections in crystalline solids. Solid State Phys. **3**, 307–435 (1956)
10. A.M.J.H. Seuter, Defect chemistry and electrical transport properties of barium titanate. Philips Res. Rep., **3** 1–84 (1974)
11. A. Hitomi, Y. Tsur, C.A. Randall, I. Scrymgeour, Site occupancy of rare-earth cations in $BaTiO_3$. Jpn. J. Appl. Phys. **40**(1), 255–258 (2001)
12. R.D. Shannon, Synthesis of some new perovskites containing indium and thallium. Inorg. Chem. **6**, 1474–1478 (1967)
13. Y. Tsur, C.A. Randall, Charge-compensation in barium titanate. In *Proceedings of the 12th IEEE International Symposium on the Applications of Ferroelectrics* 1, 151–154 (2000)
14. H.M. Chan, M.P. Harmer, D.M. Smyth, Compensating defect in highly donor-doped $BaTiO_3$. J. Am. Ceram. Soc. **69**, 507–510 (1986)
15. W.D. Kingery, H.K. Bowen, D.R. Uhlmann, *Introduction to Ceramics* (Wiley, New York, 1976)
16. Y. Tsur, T.D. Dunbar, C.A. Randall, Crystal and defect chemistry of rare earth cations in $BaTiO_3$. J. Electroceram. **7**(1), 25–34 (2001)
17. W. Heywang, Resistivity anomoly in doped barium titanate. J. Am. Ceram. Soc., **47**(10), 484–490 (1964)
18. H.M. Al-Allak, J. Illingsworth, A.W. Brinkman, J. Woods, Permittivity-temperature behaviour of donor-doped positive temperature coefficient of resistance $BaTiO_3$ ceramics. J. Phys. D: Appl. Phys. **22**(12), 1920–1923 (1989)
19. M.J. Pan, R.J. Rayne, B.A. Bender, Dielectric properties of niobium and lanthanum doped lead barium zirconate titanate relaxor ferroelectrics. J. Electroceram. **14**(2), 139–148 (2005)
20. V.V. Mitic, Z.S. Nikolic, V.B. Pavlovic, V. Paunovic, M. Miljkovic, B. Jordovic, L. Zivkovic, Influence of rare-earth dopants on barium titanate ceramics microstructure and corresponding electrical properties. J. Am. Ceram. Soc. **93**(1), 132–137 (2010)
21. Y.X. Li, X. Yao, L.Y. Zhang, Studies of resistivity and dielectric properties of magnesium doped barium titanate sintered in pure nitrogen. J. Electroceram. **21**, 557–560 (2008)
22. J.Q. Qi, W.P. Chen, Y. Wang, H.L.W. Chan, L.T. Li, Dielectric properties of barium titanate ceramics doped by B2O3 vapor. J. Appl. Phys. **96**, 6937–6939 (2004)
23. W. Heywang, Barium titanate as a semiconductor with blocking layers. Solid-State Electron. **3**(l), 51–58 (1961)
24. P. Gerthsen, K.H. Haerdtl, A method for direct observation of conductivity inhomogeneities at grain boundaries. Z. Naturforsch. A; Astrophys., Phy. Phys. Chem. **18**, 423–424 (1963)
25. H. Rehme, Electron microscope investigation of the mechanism of barium titanate PTC Ceramics. Phys. Status Solidi **26**, Kl–K3 (1968)

26. H.B. Haanstra, H. Ihrg, Voltage contrast imaging of PTC-Type $BaTiO_3$ ceramics having low and high titanium excess. Phys. Status Sofdi **39**, K7–K10 (1977)
27. H. Ihrg, M. Klerk, Visualization of the grain boundary potential barriers of PTC-type $BaTiO_3$ ceramics by cathodoluminescence in an electron-probe microanalyzer. Appl. Phys. Lett. **35**(4), 307–309 (1979)
28. Da Yu Wang, Kazumasa Umeya, Electrical properties of PTCR barium titanate. J. Am Ceram. Soc. **73**(3), 669–677 (1990)
29. N.W. Ashcroft, N.D. Mermin, in *Solid State Physics* (Saunders College, Philadelphia, 1976)
30. W. Cao, L.E. Cross, Theory of tetragonal twin structures in ferroelectric Perovskites with a first-order phase transition. Phys. Rev. B **44**, 5–12 (1991)
31. W. Cao, J.A. Krumhansl, R. Gooding, Defect-induced heterogeneous transformations and thermal growth in athermal martensite. Phys. Rev. B **41**, 11319–11327 (1990)
32. S.E. Park, T. Shrout, Relaxor based ferroelectric single crystals for electro-mechanical actuators. J. Mat. Res. Innov. **1**, 20–25 (1997)
33. J. Yin, B. Jiang, W. Cao, Elastic, piezoelectric and dielectric properties of $0.955Pb(Zn_{1/3}Nb_{2/3})O_3$–$0.045PbTiO_3$ single crystal with designed multi-domains. IEEE Trans. Ultrson. Ferroelectric. Freq. Contr. **47**(1), 285–291 (2000)
34. R. Zhang, B. Jiang, W. Cao, Elastic, piezoelectric and dielectric properties of multi-domain $0.67Pb(Mg_{1/3}Nb_{2/3})O_3$–$0.33PbTiO_3$ single crystal. J. Appl. Phys. **90**, 3471–3475 (2001)
35. S. Noemura, T. Takahashi, Y. Yokomizo, Ferroelectric properties in the system $Pb(Zn_{1/3}Nb_{2/3})O_3 - PbTiO_3$. J. Phys. Soc. Jpn **27**, 262 (1969)
36. Jiaping Han, Wenwu Cao, Interweaving domain configurations in [001] poled rhombohedral phase $0.68Pb(Mg_{1/3}Nb_{2/3})O_3$–$0.32PbTiO_3$ single crystals. Appl. Phys. Lett. **83**, 2040–2042 (2003)
37. J. Erhart, W. Cao, Permissible symmetries of multi-domain configurations in Perovskite ferroelectric crystals. J. Appl. Phys. **94**(5), 3436–3445 (2003)
38. D. Hennings, G. Rosenstein, Temperature-stable dielectrics based on chemically inhomogeneous $BaTiO_3$. J. Am. Ceram. Soc. **67**(4), 249–254 (1984)
39. T.R. Armstrong, L.E. Morgens, A.K. Maurice, R.C. Buchanan, Effects of zirconia on microstructure and dielectric properties of barium titanate ceramics. J. Am. Ceram. Soc. **72**(4), 605–611 (1989)
40. Y. Park, H.G. Kim, Dielectric temperature characteristics of cerium-modified barium titanate based ceramics with core–shell grain structure. J. Am. Ceram. Soc. **80**(1), 106–112 (1997)
41. Zhibin Tian, Xiaohui Wang, Yichi Zhang, Jian Fang, TaeHo Song, Kang Heon Hur, Seungju Lee, Longtu Li, Formation of core-shell structure in ultrafine-grained $BaTiO_3$-based ceramics through nanodopant method. **93**(1), 171–175 (2010)
42. C.A. Randall, N. Kim, J.P. Kucera, W. Cao, T.R. Shrout, Intrinsic and extrinsic effects in fine-grained morphotropic-phase-boundary lead zirconate titanate ceramics. J. Am. Ceram. Soc. **81**(3), 677–688 (1998)
43. G. Arlt, The influence of microstructure on the properties of ferroelectric ceramics. Ferroelectrics **104**, 217–227 (1990)
44. W. Cao, C.A. Randall, Grain size and domain size relations in bulk ceramic ferroelectric materials. J. Phys. Chem. Solids **57**(10), 1499–1505 (1996)
45. X.Y. Lang, Q. Jiang, Size and interface effects on Curie temperature of perovskite ferroelectric nanosolids. J. Nanop. Res. **9**, 595–603 (2007)
46. M.T. Buscaglia, V. Buscaglia, M. Viviani, J. Petzelt, M. Savinov, L. Mitoseriu, A. Testino, P. Nanni, C. Harnagea, Z. Zhao M. Nygren, Ferroelectric properties of dense nanocrystalline $BaTiO_3$ ceramics. Nanotechnology **15**, 1113–1117 (2004)
47. Q. Jiang, X.F. Cui, M. Zhao, Size effects on Curie temperature of ferroelectric particles. Appl. Phys. **A78**, 703–704 (2004)
48. Z.V. Buscaglia, M. Viviani, M.T. Buscaglia, L. Mitoseriu, A. Testino, M. Nygren, M. Johnsson, P. Nanni, Grain-size effects on the ferroelectric behavior of dense nanocrystalline BaTiO3 ceramics. Phys. Rev. B **70**, 024107 (2004)

49. S. Chattopadhyay, P. Ayyub, V.R. Palkar, M. Multani, Size-induced diffuse phase transition in the nanocrystalline ferroelectric $PbTiO_3$. Phys. Rev. B **52**, 13177–13183 (1995)
50. K. Ishikawa, K. Yoshikawa, N. Okada, Size effect on the ferroelectric phase transition in $PbTiO_3$ ultrathin particles. Phys. Rev. B **37**, 5852–5855 (1988)
51. W.L. Zhong, B. Jiang, P.L. Zhang, J.M. Ma, H.M. Cheng, Z.H. Yang, L.X. Li, Phase transition in $PbTiO_3$ ultrafine particles of different sizes. J. Phys.: Condens. Matter **5**, 2619–2624 (1993)

Chapter 8
High-Resolution Visualization Techniques: Structural Aspects

D. Schryvers and S. Van Aert

Abstract This chapter discusses a number of conventional and advanced techniques in transmission electron microscopy used for the visualization of structural aspects of disorder and strain-induced complexity in a selection of real materials. Most examples relate to shape memory materials such as Ni–Al and Ni–Ti(–X) and some to plasticity in bulk and thin films. The techniques are chosen in view of existing or potential quantitative output such as Geometric Phase Imaging based on atomic resolution images, statistical parameter estimation, tomography, and conical dark-field imaging. Clearly, this overview does not provide a complete list of present day methods for high-resolution imaging, but it should give the reader a flavour of the possibilities and potentials of transmission electron microscopy for the quantitative study of complex materials.

The study of materials can be conducted on many length scales and by many different techniques and methods. For visualization techniques, despite efforts on multi-scale exercises, often the scale of the details aimed for relates closely to the dimensions of the device in mind or at most one order of magnitude smaller. A typical example of macroscopic imaging techniques is automated camera-assisted strain measurements using surface labelling techniques. Correlations between macroscopic properties and much smaller dimensions, e.g., at the nano-level, often still suffer from serious gaps in connecting results from different length scales. For functional materials, however, with properties sensitive to a change in the environment such as temperature, pressure, electric field, magnetic field, and chemical interactions, the working dimensions often immediately fall within the micro- or nano-scale so that no or little scale differences exist between the properties and the high-resolution imaging techniques. Moreover, the continuing evolution towards miniaturization of devices from functional materials even further calls for special imaging techniques with very high spatial resolution.

D. Schryvers (✉) · S. Van Aert
EMAT, University of Antwerp, Groenenborgerlaan 171, 2020 Antwerp, Belgium
e-mail: nick.schryvers@ua.ac.be; sandra.vanaert@ua.ac.be

T. Kakeshita et al. (eds.), *Disorder and Strain-Induced Complexity in Functional Materials*, Springer Series in Materials Science 148,
DOI 10.1007/978-3-642-20943-7_8, © Springer-Verlag Berlin Heidelberg 2012

In this chapter, the focus is on atomic or high-resolution transmission electron microscopy (HRTEM) used to collect data on a variety of real materials and problems, with the emphasis on shape memory materials. Some examples also include spectroscopic data from energy-dispersive X-ray analysis (EDX) or electron energy loss spectroscopy (EELS) and novel TEM techniques.

8.1 Earlier Results on Tweed Patterns in Ni–Al

Tweed and other features precursing displacive phase transformations have been a topic of intensive research over the years, both experimentally and theoretically. Due to the sensitive nature of the many parameters playing a role in these phenomena, obtaining accurate quantitative data on the atomic level has not been obvious. At most, some average numbers such as the position of an intensity peak on the $<110>_{B2}$ diffuse intensity streak in electron diffraction patterns or the shift of the dip in the TA_2 phonon branch with composition or temperature are available. Moreover, when interpreting conventional electron microscopy (EM) images such as the original tweed patterns in, e.g., Ni–Al, care needs to be taken about the actual imaging conditions. Indeed, the observed length scale is not only related to the underlying strain or disorder pattern, but also to the actual orientation of the sample in the used two-beam bright- or dark-field procedure. Since the strain in the matrix is expected to degrade away from any stress or disorder locus, a feature resulting in a tail of diffuse intensity around affected diffraction spots in reciprocal space (e.g., electron diffraction), the chosen orientation of observation selects one particular magnitude for the imaged modulation periodicity [1]. Ultimately, atomic resolution images need to be used to reveal directly the local strains underlying the tweed pattern.

In Fig. 8.1a, a $Ni_{62.5}Al_{37.5}$B2 austenite matrix splat-cooled with a cooling rate of approx. 9×10^5 K/s is shown in high-resolution transmission electron microscopy (HRTEM) revealing the atomic lattice strained in the precursor condition occurring in a temperature region above the bcc (B2) to fct ($L1_0$) martensitic transformation [2]. Although the overall symmetry of the corresponding selected area diffraction pattern (SAED) is perfectly square, the image clearly does not reveal a homogeneous microstructure, but instead pockets of around 2 nm across each with a slightly different image resolution can be recognized. Rows of strong dots aligned along the indicated <100> direction correspond to one of the sublattices of the ordered B2 structure (e.g., at A), while rows of dots of equal intensity along one of the <110> directions reveal the entire bcc lattice (e.g., at B). Since large steps in sample thickness are improbable at such a small scale, such local image differences are seen as an imaging signature of local distortions. A clearer signature of lattice deformations is seen in the regions indicated by arrows where the lattice is tilted or strained, resulting in line contrast instead of atom column imaging. Moreover, when looking along a grazing incidence, all resolved lines are seen to be distorted yielding a wave-like pattern indicating local strains. These strains result in the diffuse streaks

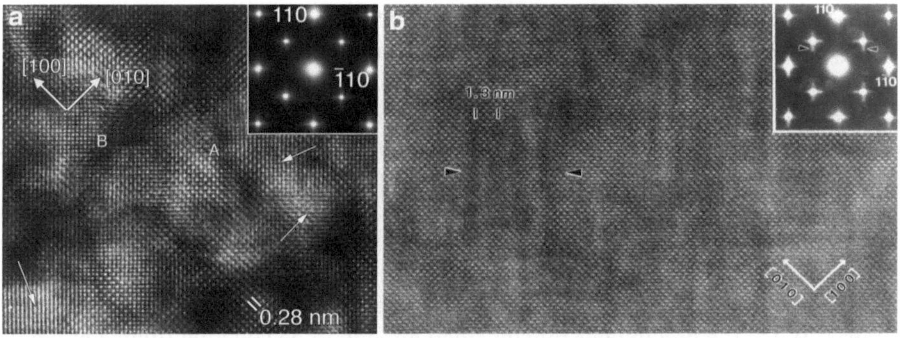

Fig. 8.1 HRTEM images of tweed in (**a**) splat-cooled and (**b**) water quenched $Ni_{62.5}Al_{37.5}$ revealing, respectively, pockets of homogeneously strained domains and of domains with a modulated strain pattern

in the <110> directions observed in the inset SAED pattern. In Fig. 8.1b, an HRTEM image of the same system but water quenched in a conventional way reveals extra modulations with an average periodicity of $6.5\{110\}_{B2}$ lattice planes. This periodicity perfectly fits with the position of the intensity peak observed in corresponding electron diffraction patterns, as shown by arrowheads in the inset, as well as in elastic phonon scattering [3, 4]. The observed contrast can moreover be explained by assuming transverse $\{110\}_{B2}<1{-}10>_{B2}$ displacements of the atoms [3]. It is assumed that these lattice modulations are induced by atomic scale defects such as vacancies or anti-site atoms in the centre of the modulated domain and inducing a strain field that couples with the anisotropy of the lattice, yielding so-called embryos for the ensuing martensitic nucleation [5–9]. However, no direct observations of such defects have been performed as yet, but models including defects as disorder in the austenite confirm such a coupling [9, 10]. Although the observed lattice displacements are of the same nature as those occurring during the ensuing displacive martensitic transformation, their maximum amplitude only reaches about 10% of the latter, fading out farther away from the central defect [3]. In other words, no full-grown product phase nuclei are observed in the tweed structure precursing the first-order transformation, as has been suggested in some earlier theoretical models [11].

8.2 Matrix Deformation and Depletion from Precipitation in Ni–Ti

The Ni_4Ti_3 structure observed in lenticular precipitates in Ni-rich Ni–Ti material annealed at moderate temperatures (around 400–500°C) has a rhombic unit cell with space group R-3. This space group was first suggested by Tadaki et al. [12] and the atom positions were later refined using the Multi-Slice Least Squares

(MSLS) technique optimizing atomic and experimental parameters to fit dynamic electron diffraction intensities [13]. Due to this lower symmetry with respect to the cubic B2 austenite matrix four orientation variants, each having a different $\{111\}_{B2}$ plane as a habit plane or interface with the cubic B2 matrix, can exist. Moreover, each orientation variant further consists of two ordering variants, but since this has no effect on the precipitate morphology or the surrounding matrix, this will be disregarded in the following. The different crystal structures induce a small lattice mismatch between the cubic matrix and the rhombic precipitate along the interface. Essentially, the Ni_4Ti_3 structure is compressed by about 3% in the $<111>_{B2}$ direction perpendicular to the central plane when compared to the B2 matrix. When the precipitates remain small enough to be coherent (diameter below approx. 300 nm) the surrounding matrix will be strained to accommodate this lattice mismatch [14]. Moreover, due to the higher Ni-content of the precipitates w.r.t. the B2 matrix, which is close to the stoichiometric 50:50 composition, a Ni-depletion region surrounding each precipitate exists [15, 16].

Quantifying a nano-scale strain field around a precipitate can be done by measuring atomic positions from HRTEM images directly captured on CCD or via a scanning procedure from conventional photographic plates. In this case, such images were obtained with a LaB_6 top-entry JEOL 4000EX microscope avoiding serious image delocalization effects. In the first step, the strain field was characterized for a $<110>_{B2}$ viewing orientation in which the precipitate–matrix interface is viewed edge-on. As a result, no overlapping between both structures exists and displacement measurements can be obtained with a variety of quantification techniques, mostly including some form of Fourier transform. The results show that the nano-scale precipitates induce a nano-scale deformation region confirming the expected lattice extension close to the large interface plane and lattice compression at the precipitate edge area. Since in this case the observation is limited to a single orientation, however, only two-dimensional (2D) information could be obtained [14].

Combining two different orientations with independent lattice directions, we were able to obtain three-dimensional (3D) data on the strain field in the matrix [17]. In this case the, $[10{-}1]_{B2}$ and $[1{-}11]_{B2}$ directions were used, with the remark that for the latter the precipitate–matrix interface is inclined over an angle of 19°, leading to a small overlapping region (a schematic of both viewing orientations w.r.t. the position of the precipitate in the matrix shown in Fig. 8.2a). Consequently, these results concern a small region about 10 nm inside the matrix (white squares in Fig. 8.2b) in a direction along the central normal to the basal plane of the precipitate, i.e. not including the overlapping region. The high-resolution images were treated with the Geometric Phase Analysis (GPA) method [18], yielding a precision of 0.6% on the atomic displacements. In Fig. 8.2b, the colour maps of both GPA analyses showing plots of ε_{xx} are presented using the undeformed matrix lattice as reference. From these (and similar datasets for ε_{yy} and ε_{zz}), the principal strain components of the chosen deformation region can be calculated, yielding the E values in Table 8.1. Also, the precipitates are seen to retain their fixed lattice parameter, confirming an

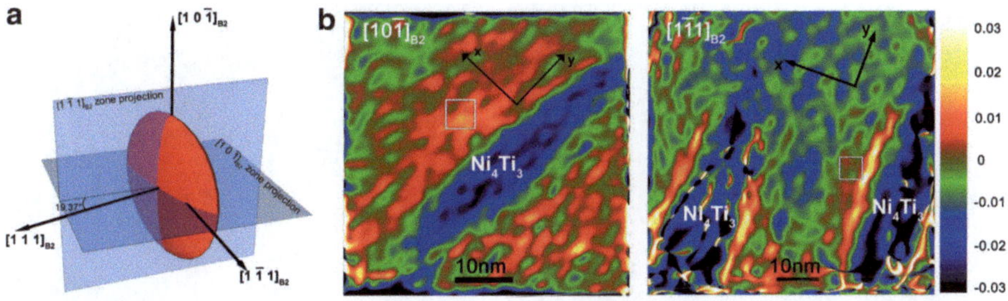

Fig. 8.2 (**a**) Schematic representation of the precipitate w.r.t. the matrix orientations and HRTEM observation directions $[10\text{–}1]_{B2}$ and $[1\text{–}11]_{B2}$ and (**b**) the ε_{xx} strain component determined from HRTEM measurements with the undeformed precipitate used as reference

Table 8.1 Principal strains and directions as measured from the combined HRTEM lattice images of the strained matrix and those calculated for the B2 → R transformation (single variant)

	Principal strains	Principal directions
E	$E = \begin{pmatrix} -0.0056 & 0 & 0 \\ 0 & -0.0033 & 0 \\ 0 & 0 & 0.0110 \end{pmatrix}$	$\begin{pmatrix} -0.2618 \\ -0.4903 \\ 0.8313 \end{pmatrix} \begin{pmatrix} -0.7553 \\ 0.6403 \\ 0.1398 \end{pmatrix} \begin{pmatrix} 0.6008 \\ 0.5913 \\ 0.5380 \end{pmatrix}$
R1	$E = \begin{pmatrix} -0.0059 & 0 & 0 \\ 0 & -0.0059 & 0 \\ 0 & 0 & 0.0121 \end{pmatrix}$	$\begin{pmatrix} -0.2991 \\ -0.5084 \\ 0.8075 \end{pmatrix} \begin{pmatrix} -0.7598 \\ 0.6389 \\ 0.1209 \end{pmatrix} \begin{pmatrix} 0.5774 \\ 0.5774 \\ 0.5574 \end{pmatrix}$

earlier observation by electron energy low-loss spectroscopy revealing larger elastic moduli for the precipitate than for the matrix [16].

Table 8.1 also shows the transformation strain values for a single variant of the R-phase, a structure which is often seen to appear before the martensitic transformation and in the vicinity of the Ni_4Ti_3 precipitates [19]. The correspondence between both sets of values is apparent, with the same signs and order of magnitude for the strains, indicating proper correspondence between lattice compression and tension and only a difference of maximum 2.5° between the respective principal directions. In reality, however, the nucleating R-phase will appear with at least two variants, which is explained by the competition of the above correspondence with the need for energy minimization and the fit at the habit plane with the matrix. Moreover, it is believed that the scale of the distance between adjacent precipitates can further inhibit the martensitic transformation.

The Ni-depletion zone surrounding the precipitates can be measured by different local spectroscopy techniques in a TEM such as EDX or EELS [15, 16]. Using mapping techniques, the result can directly be visualized, such as in the example of Fig. 8.3 including traces crossing single precipitates from which a depletion zone with approximately the same width as the precipitate can be recognized. Quantification by EELSMODEL [20] confirms the 4:3 composition ratio of the precipitates and reveals a dip by about 4% from which a good matching between the concentration in the precipitate and the depletion zone can be concluded.

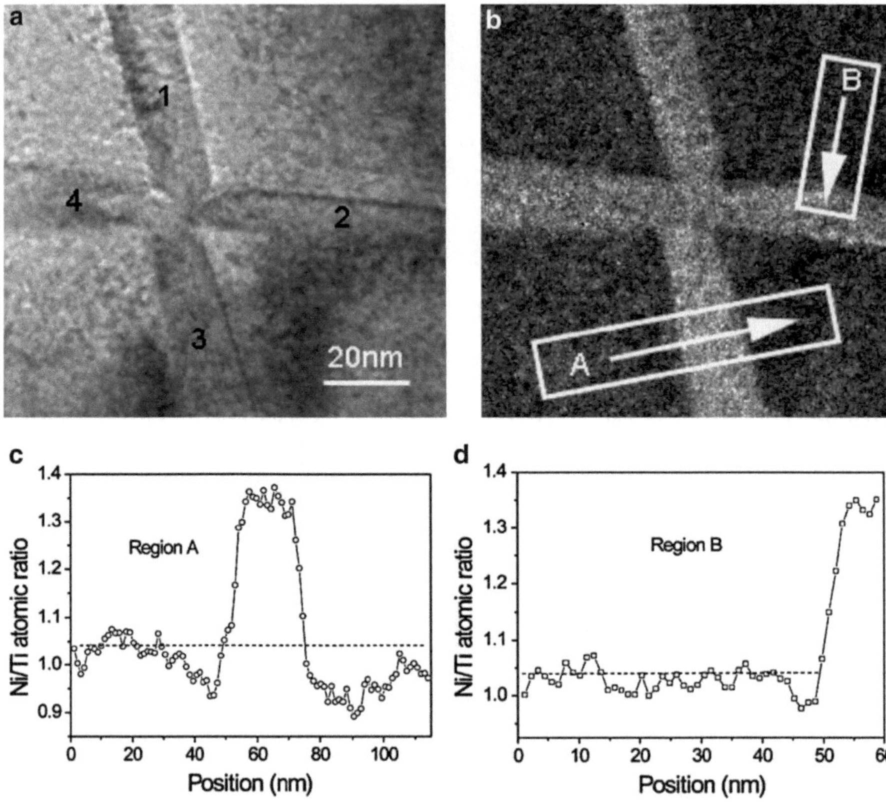

Fig. 8.3 (**a**) Zero-loss image of several large Ni_4Ti_3 precipitates. (**b**) EFTEM image revealing the excess of Ni in the precipitates. (**c**) and (**d**) line traces along A and B indicated in (**b**) revealing Ni depletion close to the precipitate

8.3 Minimal Strain at Austenite – Martensite Interface

Local strains at the austenite–martensite interface or habit plane are often compensated for by introducing lamellar twinning in the product phase. The twin width ratio between both twin variants depends on the ratios between the parent and product lattice parameters, while the actual twin width is a function of a competition between surface and volume energies. Generally, however, using twinning (or the alternative of slip) a perfect atomic match at the habit plane is never reached, which is believed to be related to energy dissipation mechanisms revealed as hysteresis and the formation of interface trailing dislocation when the transformation proceeds. However, it was recently shown that under certain particular conditions of the lattice parameters of the parent and product phases such a perfect match can exist and no twinning is needed to accommodate any lattice mismatch [21, 22]. This is the case when the middle eigenvector λ_2 of the transformation matrix equals 1 (with $\lambda_1 \leq 1$ and $\lambda_3 \geq 1$), which leads to a twin width ratio of 0 implying the disappearance of one of the twin variants. Also, the hysteresis decreases to very low values for the condition $\lambda_2 = 1$ [22]. The mosaic CTEM image in Fig. 8.4 is an example of a very large martensite plate (grey area) with only a few twin variants (black platelets) as found in a $Ti_{50}Ni_{40.5}Pd_{9.5}$ sample with a λ_2 value around 0.999 [23–25]. Most of the small platelets are seen to originate at Ti_2Pd precipitates; i.e. their existence is related to local inhomogeneities in strain and/or composition, sufficient to deviate the local lattice parameters from the ideal $\lambda_2 = 1$ condition. Moreover, as seen in Fig. 8.5, obtained in a $Ti_{50}Ni_{39}Pd_{11}$ sample with $\lambda_2 = 1.0001$, at atomic resolution the habit plane indeed does not reveal any remaining strains or dislocations confirming again the perfect fit between both structures [23, 24]. Also, the observed orientation of the habit plane fits with the predicted $(75–5)_{B2}$ indices, while groups of untwinned martensite plates form self-accommodating structures following the symmetry relations deducted earlier by Watanabe et al. [26].

Fig. 8.4 Mosaic of bright-field TEM pictures stitched together to show the full extent of a very large martensite plate in $Ti_{50}Ni_{40.5}Pd_{9.5}$ containing single fine martensite platelets but no twin lamellae. The *arrows* point at Ti_2Pd precipitate acting as nucleation sites for many of the platelets

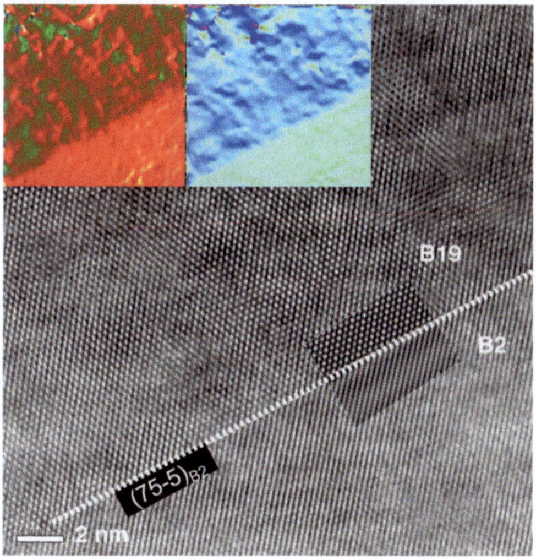

Fig. 8.5 HRTEM of the austenite/martensite habit plane in $Ti_{50}Ni_{39}Pd_{11}(\lambda_2 = 1.0001)$ showing perfect fit without the need for dislocations or twins. The location of the habit plane is best recognized from the *top two insets* revealing the changes in d-spacing (*left*) and lattice plane orientation (*right*)

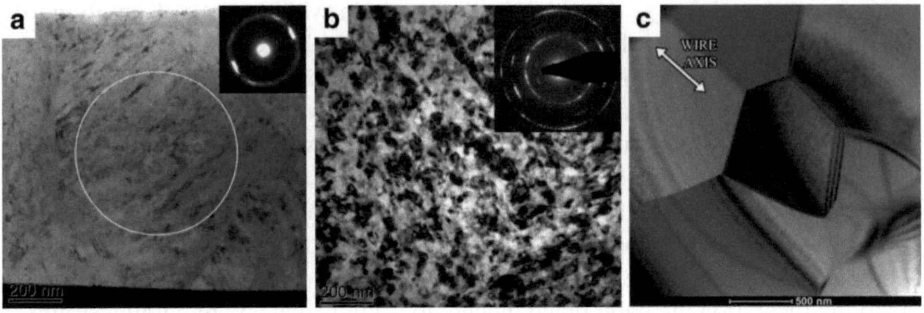

Fig. 8.6 TEM of the microstructure of as-received cold-drawn micro-wires (**a**) before and (**b**) and (**c**) after pulse-heating showing a combination of polygonized/recrystallized nano-grains at 12 ms (**b**) and fully recrystallized grains at 18 ms (**c**)

8.4 Internal Strain Control in Ni–Ti Micro-Wires

When materials with nano-scale dimensions are intentionally produced, their responses to external parameters can become substantially different from those operating at bulk level. However, the good functional properties of analogue bulk materials such as superelasticity or shape memory in Ni–Ti-based systems should not be compromised. In the example of cold-drawn Ni–Ti micro-wires with diameters below 100 μm, it is shown that electric pulse-heating with 125 W for 12 ms results in a partially polygonized/recrystallized nano-sized microstructure with grain size in the 25–50 nm range [27,28]. This microstructure allows the stress-induced martensitic transformation to fully develop, while the wire still exhibits high strength and excellent stability in cyclic mechanical loading due to the very small grain size. An example of the observed nanostructure before and after the pulse-heating is shown in Fig. 8.6a, b, respectively. A pulse-heating of 18 ms is included in Fig. 8.6c revealing fully crystallized grains of around 500 nm in diameter. After ten cycles in the superelastic regime, the latter have developed large numbers of dislocations, while the partially polygonized/recrystallized nanostructure of the 12 ms wire does not reveal any clear changes [27, 28]. As a result, the stress-induced martensitic transformation in the latter is not hindered by any particular microstructural deformations and no amnesia occurs.

8.5 Strain Effects in Metallic Nano-beams

Nanocrystalline metallic thin films often suffer from a lack of ductility due to a poor strain hardening capacity (see, e.g., [29]). This low ductility impedes the use of these structures in a wide range of applications such as flexible electronics, MEMS devices, and thin functional coatings, in which the ability of the materials to deform, stretch, or permanently change shape without cracking must be controlled and optimized to improve manufacturability and reliability. However, using proper preparation conditions very high strain hardening capacities can be found in particular metallic nanocrystalline thin films such as Al and Pd, sometimes strongly depending on specimen size and film thickness [30].

The present example refers to nanocrystalline Pd films that have been deposited by an electron-gun vacuum system on (111) Si with a 1 μm thick SiO_2 intermediate layer and a 5 nm Cr adhesion layer. The thickness of the films ranges from 80 to 310 nm and they have been reshaped by lithography to form parallel beams of 1–6 μm width. The mechanical tests have been performed using a novel concept of on-chip nanomechanical testing by which internal stresses, present in an actuating beam deposited and shaped during the same procedure (here a 100 nm thick Si_3N_4 layer), deform the attached material (here Pd) by removing the underneath sacrificial layer separating the two materials from the substrate (a detailed description of this procedure can be found elsewhere [31]). Using a high-temperature deposition process, large internal stresses up to about 1 GPa can be reached. A single point in the stress–strain curve of the deformed material is provided by the measurement of one given displacement after the release step, coupled with additional experiments used to determine the elastic properties of both the actuator and the Pd thin film. A complete stress-strain curve is then obtained by varying the length ratio between the actuator and the Pd specimen [32, 33].

Transmission electron microscopy of cross-section focussed-ion-beam (FIB) samples of the Pd beams reveals a columnar growth of nano-scale grains without clear texture. The lateral diameter of the grains appears independent of the thickness of the films, whereas the height of the columns increases with increasing film thickness. Roughly speaking, the grain height is slightly larger than half the film thickness. Irrespective of film thickness and despite the nano-scale size of the grains, a number of coherent growth twins are observed throughout the sample. An example of such a growth twin observed in unreleased material is shown in Fig. 8.7a. Such coherent growth twins offer multiple barriers to dislocation motion as well as sources for dislocation storage and multiplication. The coherency of the twin boundaries is seen to decrease after deformation with the accumulation of sessile

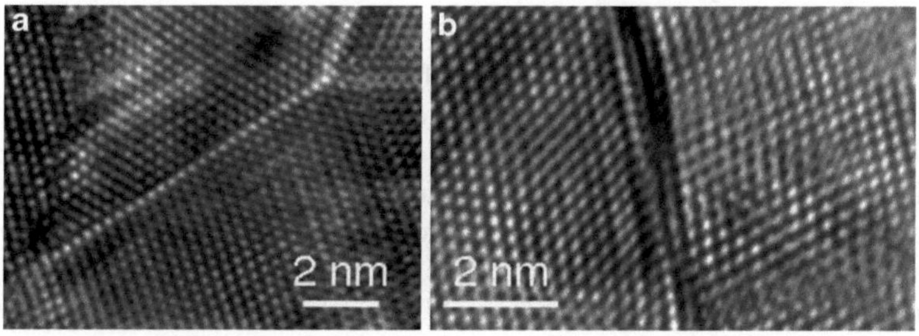

Fig. 8.7 HRTEM images of twin interfaces in sputter-deposited Pd thin films (**a**) before and (**b**) after application of mechanical stress, clearly showing the difference between coherent twin planes in the former and a strained twin interface and lattice in the latter

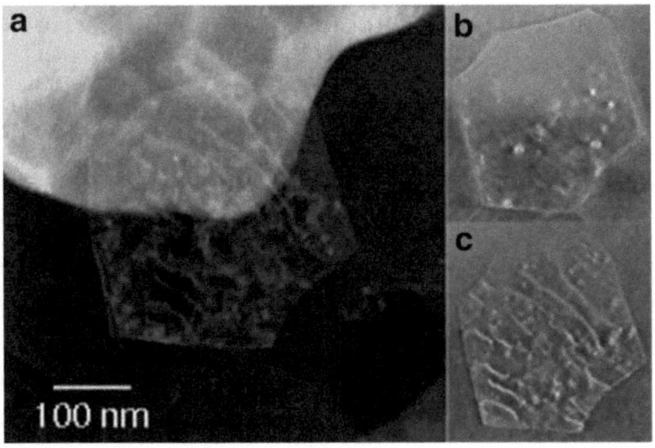

Fig. 8.8 Deformation and radiation induced dislocations observed in a nano-grain in Al micron-scale beams by means of 3D two-beam imaging electron tomography. (**a**) Snapshot from the tomography tilt series, (**b**) slice through the grain interior showing dislocation loops, and (**c**) slice close to the grain surface showing migrated dislocations

dislocations, as shown in Fig. 8.7b. The dislocation/twin boundary interactions are thought to be responsible for the remarkable mechanical properties of the Pd thin films with a long elastoplastic transition [34].

Other thin films produced by the same technique such as Al do not reveal the above-mentioned growth twins. Dislocations induced by the deformation are, however, observed and their 3D configuration can be visualized by novel electron tomography techniques. Due to the specific imaging conditions needed to properly observe dislocation contrast, a special purpose rotation-tilt holder such as the HATA holder [35] needs to be used. An example of a snapshot from such a 3D stack is given in Fig. 8.8a where the long dislocations have migrated to the sample surface as seen in Fig. 8.8c, while the dislocation loops, seen as small white dots or loops in the sample interior (Fig. 8.8b), are believed to be due to the TEM sample preparation using a Ga^+ ion beam.

8.6 Future Prospects

Quantification of HRTEM images, whether in wide-field or scanning mode (HRSTEM), as well as of spectroscopic data becomes more and more reliable due to standardized measurement methods and digital responses of cameras and other recording media. Moreover, with increasing resolution also improved precisions can be obtained (typical values for precision are approximately a factor of 100 below the instrumental resolution). Cs-corrected microscopes minimize delocalization problems typical for field emission gun (FEG) instruments, while focus variation methods provide tools for separating amplitude and phase information in a TEM image. Still, in order to properly interpret all data contained in an atomic resolution image, special techniques such as statistical parameter estimation need to be included.

In general, the aim of statistical parameter estimation theory is to determine, or more correctly, to estimate, unknown physical quantities or parameters on the basis of experimental observations. Examples of such observations in the field of electron microscopy are HRTEM images, EEL spectra, or reconstructed tomograms. Usually, these observations are not the quantities to be measured themselves but are related to certain quantities of interest. This relation is often known in the form of a mathematical function, which might be derived from physical laws. The quantities to be determined are the parameters of this function. Parameter estimation then is the computation of numerical values for the parameters from the available observations. For example, if electron microscopy images are recorded of a specific object, this function describes the electron–object interaction, the transfer of the electrons through the microscope, and the image detection. The parameters are the atom positions and atom types. The parameter estimation problem then becomes computing the atom positions and atom types from the observations. Therefore, the parameterized mathematical function is fitted to the observations using a criterion of goodness of fit, which quantifies the similarity between this function and the observations.

As is well known, observations contain noise. As a result, the above estimates computed on the basis of the observations will vary if the experiment is repeated under the same conditions. This unavoidable presence of noise thus limits the precision of the estimates. Generally, the precision will depend on the criterion of goodness of fit used in the estimation procedure. Often, use of the likelihood function as optimality criterion enhances the precision. This criterion takes the statistical nature of the observations properly into account. Under normality conditions, this criterion reduces to the well-known (weighted) least squares sum. A general overview of statistical parameter estimation theory was given by Van den Bos [36], while den Dekker [37] has provided a summary of this theory with a focus on electron microscopy.

Despite the fundamental limitation of noise in the observations, the precision of the position estimates of projected atom columns that can be obtained from HRTEM images is orders of magnitude better than the resolution of the electron microscope. Consider, for example, a small part of the phase of an experimentally reconstructed exit wave of a $Bi_4Mn_{1/3}W_{2/3}O_8Cl$ compound shown in Fig. 8.9a. Although the phase of the exit wave is often considered as the final result, it is here used as a starting point for quantitative refinement of the atom column positions using statistical parameter estimation. Nowadays, the physics behind the electron–object interaction is sufficiently well understood to have a parameterized mathematical function describing the phase of an electron exit wave. The parameters of this function have been estimated in the least squares sense. Figure 8.9b shows this function evaluated at the estimated parameters. In a sense, this figure can be regarded as an optimal reconstruction of the phase of the exit wave. In Fig. 8.9c, an overlay indicates the estimated positions of different atom column types. Interatomic distances have been computed from the estimated atom column positions. Next, mean interatomic distances and their corresponding standard deviations have been computed from sets of equivalent distances. The standard deviation, being a

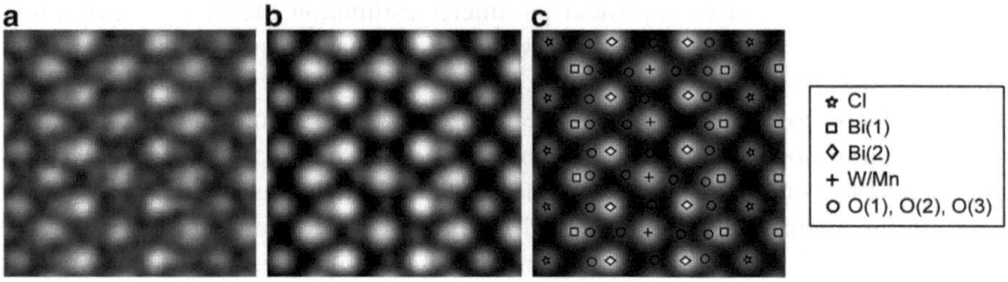

Fig. 8.9 (**a**) The experimental phase of the exit wave. (**b**) Parameterized mathematical function for the phase evaluated at the estimated parameters. (**c**) An overlay indicating the estimated positions of the different atom column types

measure of the precision, ranges from 3 to 10 pm. This is one to two orders of magnitude better than the resolution of the microscope itself (110 pm). Furthermore, a good agreement has been found when comparing these results with X-ray powder diffraction data. More details on this study can be found elsewhere [38].

Not only structure analyses but also chemical analyses may benefit from statistical parameter estimation. Recently, progress has been made in the quantitative evaluation of high-angle annular dark-field scanning transmission electron microscopy (HAADF STEM) images [39]. It is generally known that these images show Z-contrast meaning that the intensity scales with the atomic number Z. One of the advantages is therefore the possibility to visually distinguish between chemically different atomic column types. However, if the difference in atomic number of distinct atomic column types is small or if the signal-to-noise ratio is poor, direct interpretation of HAADF STEM images remains inadequate. In order to extract quantitative chemical information on a local scale, the total intensity of the scattered electrons for the individual atomic columns can be quantified using statistical parameter estimation. The thus estimated intensities can then be used as a performance criterion to identify unknown column types. As such, differences in averaged atomic number of only 3 can clearly be distinguished in an experimental image, a result which is impossible to obtain by means of visual interpretation only. This is an important advantage when studying, e.g., interfaces.

Computational optimization tools already provide automatic recognition software for kinematic diffraction experiments such as the long-standing X-ray diffraction and the more recent electron backscattered diffraction (EBSD) in an SEM. In a TEM, dynamic diffraction features lead to more complex diffraction intensities (depending, e.g., on the thickness and orientation of the sample), but these can be minimized by the novel spinning methods averaging out diffraction intensities or they can be incorporated into the optimization software leading to an increase in parameter space as done in the method of MSLS [40, 41]. Alternative applications of these novel diffraction techniques can result in more detailed nanostructural information using so-called fluctuation and conical dark-field (CDF) imaging procedures [42, 43]. In Fig. 8.10, an example of the use of CDF imaging, in which a series of dark-field images is produced by tilting the incident electron beam around

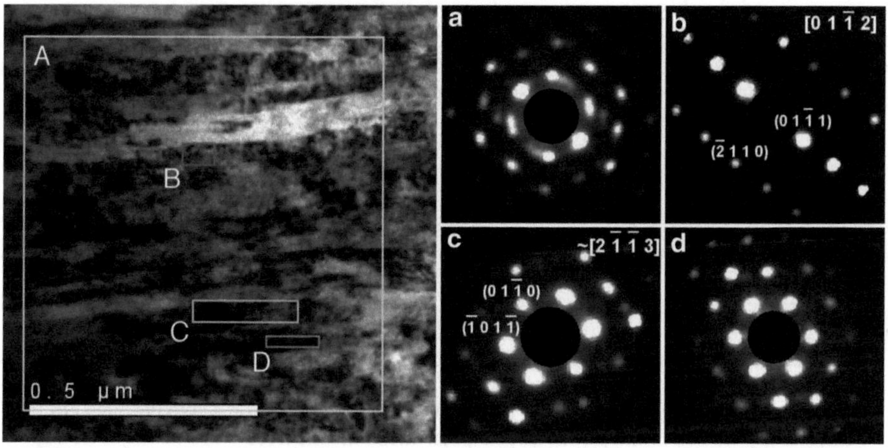

Fig. 8.10 CDF image and selected pixel diffraction patterns from an adiabatic shear region in Ti6Al4 V. The entire dataset is 256 × 256 × 4,076 pixels large

a conical surface in such a way that the ring of diffracted intensity, due to a region with nano-scale grains, passes through the objective aperture [43]. This yields a 3D data box from which the orientation of each single nano-grain can be retrieved. The present example shows an adiabatic shear band (ASB) in hcp αTi6Al4V of around 1 μm wide. The different diffraction patterns are compiled from the large CDF dataset and indicate local orientation changes in elongated sub-grains. The entire region labelled as A does not show a single zone diffraction pattern, whereas in the smaller elongated regions of B and C two different zone axes can be recognized. Region D appears to have the same zone as region C, but with a different in-plane orientation. Conventional selected area electron diffraction or even micro-diffraction cannot obtain the needed lateral resolution for these types of problems. Ultimately diffraction information from a single pixel can be obtained, which in the present example corresponds with an area of $36 \times 36\,\text{nm}^2$.

The evolution towards real 3D TEM is another promising path for future materials characterization. Different tomography techniques including dedicated reconstruction procedures are these days available or are under development. Typical applications are again direct Z-contrast imaging in HAADF STEM or conventional DF imaging. The combination with Cs-corrected machines will ultimately lead to atomic scale tomography yielding 3D chemical as well as lattice strain information.

Acknowledgments The authors thank S. Bals, W. Tirry, H. Idrissi, B. Wang, and Z.Q. Yang for support with the TEM observations. Part of this work was performed in the framework of a European FP6 project "Multi-scale modeling and characterization for phase transformations in advanced materials" (MRTN-CT-2004–505226) and an IAP program of the Belgian State Federal Office for Scientific, Technical and Cultural Affairs (Belspo), under Contract No. P6/24. Support was also provided by FWO projects G.0465.05 "The functional properties of SMA: a fundamental approach", G.0576.09 "3D characterization of precipitates in Ni–Ti SMA by slice-and-view in a FIB-SEM dual-beam microscope", G.0188.08N "Optimal experimental design for quantitative electron microscopy", G.0064.10N "Quantitative electron microscopy: from experimental measurements to precise numbers" and G.0180.08 "Optimization of Focused Ion Beam (FIB) sample preparation for transmission electron microscopy of alloys".

References

1. I.M. Robertson, C.M. Wayman, Tweed microstructures I. Characterization in β-NiAl. Phil. Mag. A **48**, 421 (1983)
2. D. Schryvers, D. Holland-Moritz, Austenite and martensite microstructures in splat-cooled Ni-Al. Intermetallics **6**, 427 (1998)
3. D. Schryvers, L.E. Tanner, On the interpretation of high-resolution electron microscopy images of premartensitic microstuctures in the Ni-Al B2 phase. Ultramicroscopy **32**, 241 (1990)
4. S. Shapiro, B. Yang, Y. Noda, L. Tanner, D. Schryvers, Neutron-scattering and electron-microscopy studies of the premartensitic phenomena in NixAl100-x alloys. Phys. Rev. B **44**, 9301 (1991)
5. B.I. Halperin, C.M. Varma, Defects and the central peak near structural phase transitions. Phys. Rev. B **14**, 4030 (1976)
6. Y. Yamada, Y. Noda, M. Takimoto, 'Modulated lattice relaxation' in β-based premartensitic phase. Sol. St. Comm. **55**, 1003 (1985)
7. R. Gooding, J. Krumhansl, Symmetry-restricted anharmonicities and the CsCl-to-7R martensitic structural phase transformation of the NixAl1-x system, Phys. Rev. B **39**, 1535 (1989)
8. W. Cao, J. Krumhansl, R. Gooding, Defect-induced heterogeneous transformations and thermal growth in athermal martensite. Phys. Rev. B **41**, 11319 (1990)
9. W. Zhang, Y.M. Jin, A.G. Khachaturyan, Phase field microelasticity modeling of heterogeneous nucleation and growth in martensitic alloys. Acta. Mat. **55**, 565 (2007)
10. P. Lloveras, T. Castán, M. Porta, A. Planes, A. Saxena, Influence of elastic anisotropy on structural nanoscale textures. Phys. Rev. Lett. **100**, 165707 (2008)
11. Y. Wang, A.G. Khachaturyan, Three-dimensional field model and computer modeling of martensitic transformations. Acta. Mat. **45**, 759 (1997)
12. T. Tadaki, Y. Nakata, K.I. Shimizu, K. Otsuka, Crystal structure, composition and morphology of a precipitate in an aged Ti-51 at.% Ni shape memory alloy. Trans. JIM. **27**, 731 (1986)
13. W. Tirry, D. Schryvers, K. Jorissen, D. Lamoen, Electron-diffraction structure refinement of Ni4Ti3 precipitates in Ni52Ti48. Acta. Cryst. B **62**, 966 (2006)
14. W. Tirry, D. Schryvers, Quantitative determination of strain fields around Ni4Ti3 precipitates in NiTi. Acta. Mat. **53**, 1041 (2005)
15. Z. Yang, W. Tirry, D. Schryvers, Analytical TEM investigations on concentration gradients surrounding Ni4Ti3 precipitates in Ni-Ti shape memory material. Scripta. Mat. **52**, 1129 (2005)
16. Z. Yang, W. Tirry, D. Lamoen, S. Kulkova, D. Schryvers, Electron energy-loss spectroscopy and first-principles calculation studies on a Ni-Ti shape memory alloy. Acta. Mat. **56**, 395 (2008)
17. W. Tirry, D. Schryvers, Linking a completely three-dimensional nanostrain to a structural transformation eigenstrain. Nat. Mater. **8**, 752 (2009)
18. M. Hÿtch, E. Snoeck, R. Kilaas, Quantitative measurement of displacement and strain fields from HRTEM micrographs. Ultramicroscopy **57**, 131 (1998)
19. L. Bataillard, J.E. Bidaux, R. Gotthardt, Interaction between microstructure and multiple-step transformation in binary NiTi alloys using in-situ transmission electron microscopy observations. Phil. Mag. A **78**, 327 (1998)
20. J. Verbeeck, G. Bertoni, Model-based quantification of EELS spectra: Treating the effect of correlated noise, Ultramicroscopy **108**, 74 (2008)
21. J. Cui, Y.S. Chu, O.O. Famodu, Y. Furuya, J. Hattrick-Simpers, R.D. James, A. Ludwig, S. Thienhaus, M. Wuttig, Z. Zhang, I. Takeuchi, Combinatorial search of thermoelastic shape-memory alloys with extremely small hysteresis width. Nat. Mater. **5**, 286 (2006)

22. Z. Zhang, R.D. James, S. Müller, Energy barriers and hysteresis in martensitic phase transformations. Acta. Mat. **57**, 4332 (2009)
23. R. Delville, D. Schryvers, Z. Zhang, R.D. James, Transmission electron microscopy investigation of microstructures in low-hysteresis alloys with special lattice. Scripta. Mat. **60**, 293 (2009)
24. R. Delville, S. Kasinathan, Z. Zhang, J. Van Humbeeck, R.D. James, D. Schryvers, Transmission electron microscopy study of phase compatibility in low hysteresis shape memory alloys. Phil. Mag. **90**, 177 (2010)
25. R. Delville, From functional properties to micro/nano-structures: a TEM study of TiNi(X) shape memory alloys (Antwerp, 2010)
26. Y. Watanabe, T. Saburi, Y. Nakagawa, S. Nenno, Jpn. Inst. Met. **54**, 861 (1990)
27. R. Delville, B. Malard, J. Pilch, P. Sittner, D. Schryvers, Microstructure changes during non-conventional heat treatment of thin Ni-Ti wires by pulsed electric current studied by TEM. Acta. Mat. **58**, 4503 (2010)
28. R. Delville, B. Malard, J. Pilch, P. Sittner, D. Schryvers, Transmission electron microscopy investigation of dislocation slip during superelastic cycling of Ni-Ti wires. Int. J. Plast. **27**, 282 (2011)
29. M.A. Haque, M.T.A. Saif, Strain gradient effect in nanoscale thin films. Acta. Mat. **51**, 3053 (2003)
30. M. Coulombier, A. Boe, C. Brugger, J.P. Raskin, T. Pardoen, Imperfection sensitive ductility of aluminium thin films. Scripta. Mat. **62**, 742 (2010)
31. S. Gravier, M. Coulombier, A. Safi, N. André, A. Boé, J.P. Raskin, T. Pardoen, J. Microelectromech. Syst. **18**, 555 (2009)
32. A. Boé, A. Safi, M. Coulombier, D. Fabrègue, T. Pardoen, J.-P. Raskin, MEMS-based microstructures for nanomechanical characterization of thin films. Smart. Mat. Struct. **18**, 115018 (2009)
33. A. Boé, A. Safi, M. Coulombier, T. Pardoen, J.P. Raskin, Internal stress relaxation based method for elastic stiffness characterization of very thin films. Thin Solid Films **518**, 260 (2009)
34. H. Idrissi, B. Wang, M.S. Colla, J.P. Raskin, D. Schryvers, T. Pardoen, Ultrahigh Strain Hardening in Thin Palladium Films with Nanoscale Twins. Advanced Materials **23**, 2119 (2011)
35. http://www.melbuild.com/
36. A. Van den Bos, Parameter estimation for scientists and engineers (Wiley-Interscience, 2007)
37. A.J. den Dekker, S. Van Aert, A. van den Bos, D. Van Dyck, Maximum likelihood estimation of structure parameters from high resolution electron microscopy images. Part I: a theoretical framework. Ultramicroscopy **104**, 83 (2005)
38. S. Bals, S. Van Aert, G. Van Tendeloo, D. Avila-Brande, Statistical estimation of atomic positions from exit wave reconstruction with a precision in the picometer range. Phys. Rev. Lett. **96**, 096106 (2006)
39. S. Van Aert, J. Verbeeck, R. Erni, S. Bals, M. Luysberg, D. Van Dyck, G. Van Tendeloo, Quantitative atomic resolution mapping using high-angle annular dark field scanning transmission electron microscopy. Ultramicroscopy **109**, 1236 (2009)
40. J. Jansen, D. Tang, H.W. Zandbergen, H. Schenk, MSLS, a least-squares procedure for accurate crystal structure refinement from dynamical electron diffraction patterns. Acta. Cryst. A **54**, 91 (1998)
41. J. Jansen, H.W. Zandbergen, Determination of absolute configurations of crystal structures using electron diffraction patterns by means of least-squares refinement. Ultramicroscopy **90**, 291 (2002)
42. P.M. Voyles, J.M. Gibson, M.M. Treacy, Fluctuation microscopy: a probe of atomic correlations in disordered materials. J. Elect. Microsc. **49**, 259 (2000)
43. G. Wu, S. Zaefferer, Advances in TEM orientation microscopy by combination of dark-field conical scanning and improved image matching. Ultramicroscopy **109**, 1317 (2009)

Chapter 9
High-Resolution Visualizing Techniques: Magnetic Aspects

Yasukazu Murakami

Abstract Magnetic imaging plays an important role in clarifying the phase transformation mechanisms and/or the correlation between nanostructures and materials functions in magnetic compounds. Methods based on transmission electron microscopy (Lorentz microscopy and electron holography) are powerful tools for analyzing the aforementioned aspects, because these methods allow for nanometer-scale resolution and can be effectively combined with peripheral techniques such as electron diffraction and high-resolution transmission electron microscopy (lattice imaging). This chapter explains the essence of magnetic imaging and discusses recent topical studies on magnetic functional materials, to which these techniques have been applied to obtain useful information that helps understand the mechanisms underlying the extraordinary material properties. In particular, this chapter focuses on colossal magnetoresistive manganites and ferromagnetic shape-memory alloys, both of which are closely related to the issues of the disorder and strain-induced complexity.

9.1 Introduction

Since the early 1990s, researchers have developed fascinating magnetic compounds that exhibit gigantic responses to stimuli such as a magnetic field, electric field, stress, or temperature change. Many of these compounds exhibit a severe competition between antipathetic crystalline phases and/or strong correlation between spin and lattice degrees of freedom. Colossal magnetoresistive manganites [1, 2] and ferromagnetic shape-memory alloys (FMSMAs) [3] are probably the most representative cases belonging to this category. Applying a magnetic field to

Y. Murakami (✉)
Institute of Multidisciplinary Research for Advanced Materials, Tohoku University, Sendai, Japan
e-mail: murakami@tagen.tohoku.ac.jp

T. Kakeshita et al. (eds.), *Disorder and Strain-Induced Complexity in Functional Materials*, Springer Series in Materials Science 148,
DOI 10.1007/978-3-642-20943-7_9,

hole-doped manganites that exhibit colossal magnetoresistance (CMR) stabilizes the metallic ferromagnetic (FM) phase that competes with the charge-ordered (CO) insulating phase [4]; this effect produces macroscopic conduction paths that reduce the resistivity by a thousand fold [5]. In FMSMAs, applying a magnetic field causes rearrangement of both the magnetic and crystallographic domains due to the significant magnetoelastic interplay, which gives rise to a giant magnetostriction of the order of 10^{-2} [3,6]. To understand the microscopic mechanisms that underlie these interesting phenomena, it is important to consider issues relevant to complex magnetic microstructures, including the configuration of the nanoscale phase separation, the correspondence between magnetic and crystallographic domains, and the nucleation and growth processes of the FM phase. In this context, magnetic imaging is vitally important in the research and development of these functional materials. This chapter presents topics related to the magnetic imaging with transmission electron microscopy (TEM).

Several magnetic imaging techniques have been developed including the Bitter method that employs a magnetic colloid as an indicator [7], magneto-optical methods (such as Kerr microscopy) [8], magnetic force microscopy (MFM) [9], scanning superconducting quantum interference device (SQUID) [10], spin-polarized scanning electron microscopy [11], and methods based on TEM. Among these techniques, TEM observations provide unique information about the magnetic flux distribution and/or the magnetic domain wall positions with sufficiently high resolutions [12–14]. In TEM observations, magnetic information is acquired not only from the specimen surface but also from the interior region of the specimen, although specimens must be thin films with thicknesses of typically 100 nm. The magnetic imaging can be effectively combined with other functions of TEM, such as electron diffraction, dark-field imaging (which is useful for observing structural defects), high-resolution electron microscopy for observing lattice images, energy-dispersive X-ray spectroscopy, and electron energy-loss spectroscopy. Furthermore, as a result of recent advancements in peripheral techniques (e.g., tools for controlling the magnetic field around the specimen), it is possible to perform various sorts of in situ observations with TEM. These aspects make TEM a powerful tool for studying phase transformation mechanisms and the domain structures in magnetic materials.

This chapter first explains the principle of Lorentz microscopy [13] and electron holography [12, 14], which are magnetic imaging methods based on TEM. In the subsequent sections, we introduce recent studies on CMR manganites and FMSMAs for which TEM observations have revealed essential information that provides a deeper understanding of the microscopic mechanisms underlying their extraordinary characteristics.

9.2 Magnetic Imaging by TEM

This section describes the fundamentals of Lorentz microscopy and electron holography. Both these techniques skillfully use magnetic deflection and/or interference of electrons to acquire the magnetic information from a thin-foil specimen, although they employ different experimental/analytical processes.

9.2.1 Lorentz Microscopy

Lorentz microscopy is a useful method that visualizes the magnetic domain walls in a thin film. The image contrast is due to the deflection of incident electrons by the Lorentz force exerted by the magnetic specimen. Let us consider a simple configuration of 180° magnetic domains in a thin-foil specimen; i.e., magnetic domains A and B are separated by a 180° domain wall, as shown in Fig. 9.1a. The magnetization in domain A deflects the trajectory of electrons to the left via the Lorentz force, whereas the electrons are deflected to the right by the adjacent domain B. The divergent mode of deflection reduces the beam intensity beneath the domain wall W1. In contrast, the beam is intensified beneath the neighboring domain wall W2, due to the convergent mode of deflection. Thus, an image obtained under defocus conditions (i.e., overfocus conditions, as illustrated in Fig. 9.1a), can reveal the positions of magnetic domain walls – W1 and W2 appear as the deficient intensity (dark) line and excess intensity (bright) line, respectively, as shown in Fig. 9.1b. Note that the image contrast of the walls should be converted with underfocus conditions since these lines behave as Fresnel fringes; thus, this observation mode is referred to as the *Fresnel mode*. Although Fig. 9.1 illustrates domain wall observation, this method can also be used to identify a tiny single domain produced in a nonmagnetic matrix (this is described in more detail in the Sect. 9.3.1). All of the Lorentz microscope images shown in this chapter were obtained in the Fresnel mode.

The other observation mode is the *Foucault mode* [13, 14]. The advantage of the Foucault mode is that it can image the entire area with a specific orientation of magnetic domains, rather than imaging the magnetic domain walls. Its principle is similar to that of the conventional dark-field method [15]. For example, in the case of Fig. 9.1a, the magnetic deflections produce a diffraction spot that is split into two. Using an aperture in the microscope to select one of these two spots excites only the domain that contributes to the selected spot. This method is particularly useful for

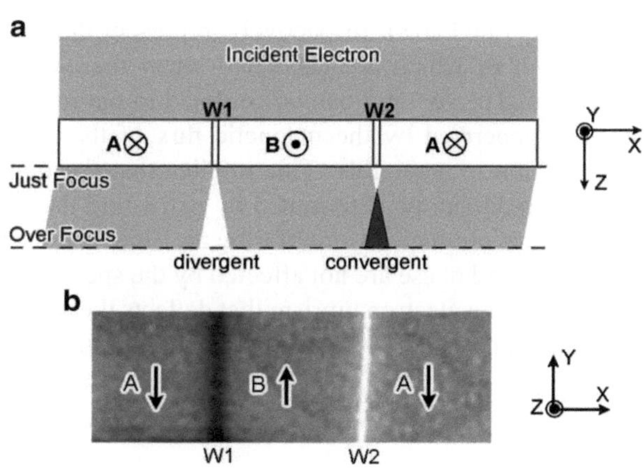

Fig. 9.1 Principle of contrast formation in Lorentz microscopy. (**a**) Schematic illustration of the deflection of electrons by the Lorentz force (cross-sectional view of a thin-foil specimen with 180° domains). (**b**) Example of a Lorentz microscope image. W1 and W2 indicate the positions of 180° magnetic domain walls that separate domains A and B

revealing the shape of magnetic domains when the split of the diffraction spot is sufficiently large. However, the magnetic deflection is approximately two orders of magnitude smaller than typical Bragg reflection angles. For example, in a Nd–Fe–B film with thickness of 100 nm, which has a significant magnetic flux density of 1.6 T [16], the Bragg angle of the 002 reflection is 2.1×10^{-3} rad for an acceleration voltage of 200 kV, whereas the magnetic deflection is only 9.7×10^{-5} rad under these conditions.

It is not easy to simply define the spatial resolution of Lorentz microscopy since it depends on several factors in both the microscope and specimen. However, computer simulations conducted by De Graef [13] provided useful information about the resolution. His study showed that the magnetic component of the contrast in Co nanoparticles (i.e., the asymmetric pattern observed in the Lorentz micrograph, which is representative of the state of single magnetic domain) can be detected down to a radius of 10–20 nm. Refer to [13] for further details.

9.2.2 Electron Holography

Electron holography is a method that retrieves phase information of the electron wave, which can be used to determine the magnetic flux distribution in a specimen. The process of electron holography consists of two steps. In the first step, a hologram (a type of interference pattern) is formed with a biprism, whereby an object wave passing through the thin-foil specimen interferes with a reference wave passing through a vacuum (Fig. 9.2a). In the second step, phase information is extracted from the digitized hologram using the Fourier transform (Fig. 9.2b-d).

It is useful to explain the phase retrieval process using wave functions. We express the object wave in the form of wave function

$$q(\vec{r}) = a(\vec{r}) \exp(i \o(\vec{r})), \tag{9.1}$$

where $a(\vec{r})$ and $\phi(\vec{r})$, respectively, represent the amplitude change and the phase shift, both of which are generated when the electron wave traverses a thin-foil specimen. The vector $\vec{r}$ can be confined to the foil plane. Note that the phase shift $\phi(\vec{r})$ is generated by the magnetic flux of the specimen when the electric field contribution is negligible. This implies that the magnetic flux distribution in the viewing field can be determined by extracting the phase information. An electron hologram is produced by interfering the object wave with the reference wave whose amplitude and phase are not affected by the specimen. An electron hologram can be obtained by an electron biprism that deflects the object wave and the reference wave by $-\alpha_h/2$ and $\alpha_h/2$, respectively. The hologram intensity is expressed by

$$I_h(\vec{r}) = 1 + a^2(\vec{r}) + 2a(\vec{r}) \cos(2\pi \frac{\alpha_h}{\lambda} x - \phi(\vec{r})). \tag{9.2}$$

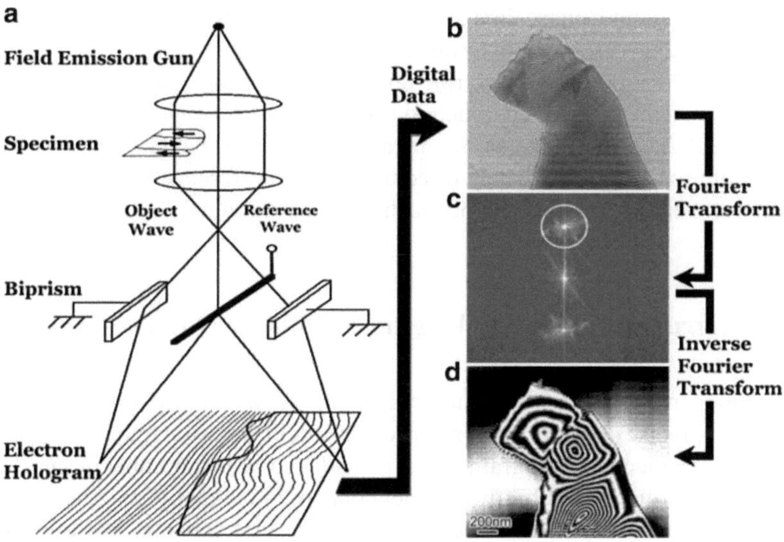

Fig. 9.2 Principle of electron holography. (**a**) Geometric configuration for forming an electron hologram. (**b**)–(**d**) Analytical process to extract phase information

This equation indicates that the period λ/α_h of the interference fringes is modulated by the phase shift $\phi(\vec{r})$: in other words, the information of $\phi(\vec{r})$ is stored in the hologram (Fig. 9.2b) as the modulation of interference fringes. It is convenient to take the Fourier transform ($\mathscr{F}$) to analyze the modulation of the periodic pattern. The Fourier transform of the hologram is given by

$$\mathscr{F}\left[I_h(\vec{r})\right] = \delta\left(\vec{u}\right) + \mathscr{F}\left[a^2(\vec{r})\right] + \mathscr{F}\left[a\left(\vec{r}\right)\exp\left(i\phi\left(\vec{r}\right)\right)\right] * \delta\left(\vec{u} + \frac{\alpha_h}{\lambda}\right)$$
$$+\mathscr{F}\left[a\left(\vec{r}\right)\exp\left(-i\phi\left(\vec{r}\right)\right)\right] * \delta\left(\vec{u} - \frac{\alpha_h}{\lambda}\right), \tag{9.3}$$

where $*$ represents the convolution operation. Note that the phase information $\phi(\vec{r})$ is reserved in the third and fourth terms on the right-hand side. By selecting the third term, shifting it by α_h/λ, and performing the inverse Fourier transform ($\mathscr{F}^{-1}$) on the third term (Fig. 9.2c), we obtain

$$\mathscr{F}^{-1}\left[\mathscr{F}\left[a(\vec{r})\exp\left(i\phi\left(\vec{r}\right)\right)\right] * \delta\left(\vec{u}\right)\right] = a\left(\vec{r}\right)\exp\left(i\phi\left(\vec{r}\right)\right). \tag{9.4}$$

Thus, both the amplitude change $a(\vec{r})$ and the phase shift $\phi(\vec{r})$ can be retrieved. The result of phase shift can be displayed in the form of a contour map, as shown in Fig. 9.2d. In this reconstructed phase image, the contour lines indicate the lines of magnetic flux (in-plane component of the magnetic flux) when the electric field contribution is negligible. In most of the experimental results presented in this chapter, reconstructed phase images are given in terms of $\cos\phi(\vec{r})$, except for Fig. 9.10a, which is displayed in terms of $\phi(\vec{r})$.

The spatial resolution of electron holography depends on the radius of the mask [17], which is inserted when performing the inverse Fourier transform (see Fig. 9.2c). The mask radius is typically $1/3(\alpha_h/\lambda)$, which indicates that the spatial resolution is approximately three times the interference fringe spacing. The resolution in our experiments can be reached at approximately 7 nm [18].

9.2.3 Instrumentation for Magnetic Domain Observations

The methods of magnetic imaging based on TEM require coherent incident electrons for precise observations. In particular, a field emission gun is indispensable for obtaining an electron hologram which provides sufficient number of interference fringes and/or sufficient visibility. Another concern in magnetic imaging by TEM is the significant magnetic field in the objective lens. In a conventional transmission electron microscope, the specimen is placed in the gap of the pole piece and it is subjected to a strong magnetic field of approximately 2 T. Most specimens are immediately magnetized on being inserted in the electron microscope. Thus, we need some modifications of the microscope, in order to observe the original magnetic domain structure. For example, our electron microscope (JEM-3000F) is equipped with a specially designed pole piece, which is sometimes called a Lorentz lens (see Fig. 9.3) [18]. With the aid of the magnetic shield, the magnetic field at the specimen position can be reduced to 0.04 mT, which is comparable to the geomagnetic field.

In situ observations of the magnetization process and/or magnetic phase transformations are especially important when we conduct studies of magnetic compounds. It is possible to apply a magnetic field to the specimen, which is placed in the Lorentz lens, by using a special specimen holder equipped with an electromagnet

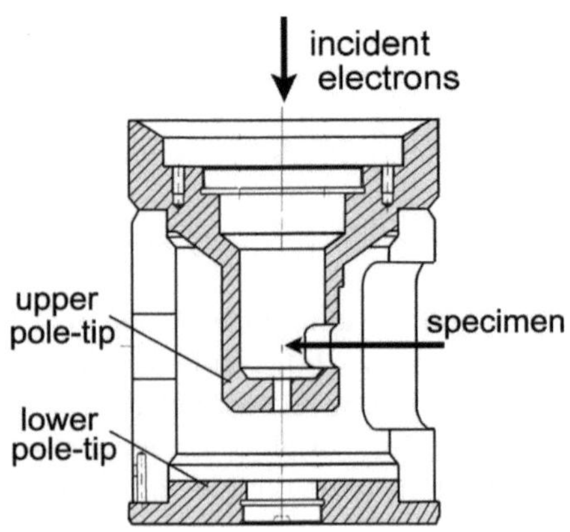

Fig. 9.3 Schematic diagram (cross section) of a Lorentz lens installed in a 300 kV transmission electron microscope (JEM-3000 F)

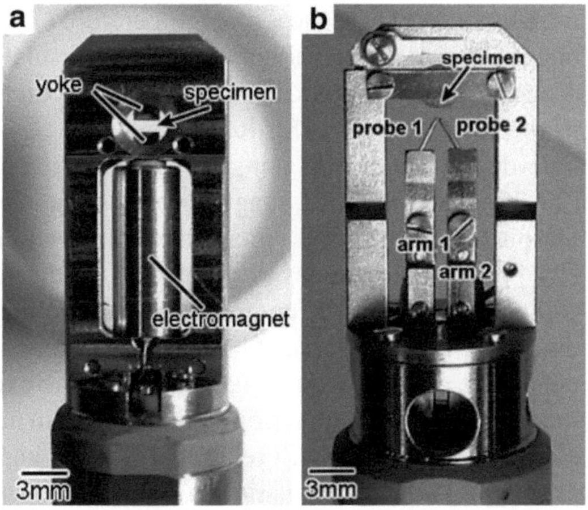

Fig. 9.4 View of TEM specimen holders that can be used in in situ observations. (**a**) Magnetizing holder. (**b**) Double-probe piezodriving holder

(see Fig. 9.4a). The magnetic field is efficiently guided to the specimen position by a magnetic circuit made of pure iron embedded in the holder frame [19]. This technique has been successfully applied to several in situ observations, including the formation of a magnetic flux channel connecting FM islands in a hole-doped manganite [20] and domain wall pinning by structural imperfections in an FMSMA [21]. Figure 9.4b depicts another tool that is useful for simultaneously examining the conductivity, magnetism, and crystal structure [22]. Its two arms can be manipulated in the limited space of the Lorentz lens by using piezoelectric elements. The $I - V$ (current vs. voltage) characteristics of a local area can be easily evaluated by bringing the probe tips into contact with the region of interest (e.g., interface, nanometer-sized area in a mixed-phase state, and individual particles). Lorentz microscopy and electron holography can be used to obtain magnetic information from the same area. This method is also useful in studies of multiferroics and spintronics where the correlation between nonconjugate parameters plays a crucial role in their extraordinary functions.

9.3 Study of Magnetic Microstructure in Colossal Magnetoresistive Manganite

Since its discovery in the early 1990s, the phenomenon of CMR (a dramatic reduction in resistivity when a magnetic field is applied [1, 2, 23]) in hole-doped manganites has attracted considerable attention because of its potential applicability to advanced technologies related to magnetic data storage and spintronics. The key process underlying the CMR mechanism appears to be microscopic phase separation caused by competition between crystal phases with different structural orders. Many manganites exhibit severe competition between the FM metal phase and the CO insulator phase [4, 5, 24–26]. Applying an external field (magnetic field or temperature) near the Curie temperature (T_C) effectively stabilizes the FM metal phase relative to the nonmagnetic CO insulator phase. This effect can produce a macroscopic network of FM domains, which results in a remarkable reduction in the resistivity [27, 28]. Researchers have studied magnetic phase separation in hole-doped manganites using several magnetic imaging techniques, including MFM [29, 30], TEM [31–33], magneto-optical methods [34], and such other techniques. Despite these extensive microscopy observations, it is still under debate how the FM phase predominates in the crystal in the face of severe competition with the CO phase. Neither the complex structure in the mixed-phase state nor the domain dynamics under external fields is fully understood. Furthermore, it is difficult to determine the magnetic parameters (e.g., the exchange stiffness constant A and the magnetocrystalline anisotropy constant K_u) in nanoscale magnetic phase separation. Hence, we performed Lorentz microscopy and electron holography studies on a prototypical CMR manganite $La_{0.25}Pr_{0.375}Ca_{0.375}MnO_3$ [35].

9.3.1 Ferromagnetic Domain Nucleation and Growth

The nucleation and growth process of the FM phase in $La_{0.25}Pr_{0.375}Ca_{0.375}MnO_3$ was examined by cryogenic Lorentz microscopy. As shown in the phase diagram in Fig. 9.5, $La_{0.25}Pr_{0.375}Ca_{0.375}MnO_3$ is a paramagnetic (PM) insulator above T_{CO} (the onset temperature of charge ordering, which is characterized by a regular array of Mn^{3+} and Mn^{4+}) [36]. A different phase is stable in the temperature window between T_{CO} and T_C – we call this phase the *mother phase* since it produces another phase on further cooling (see below). Superlattice spots in diffraction patterns and

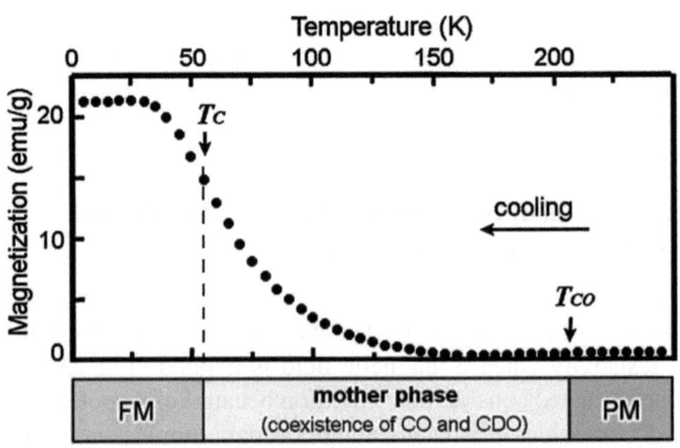

Fig. 9.5 Phase transitions undergone by $La_{0.25}Pr_{0.375}Ca_{0.375}MnO_3$: (*upper panel*) Magnetization vs. temperature curve measured on cooling in a magnetic field of 1 kOe; (*lower panel*) Phase diagram that represents the state(s) dominant in each temperature region. Reprinted with permission from [35]. See text for details

a gradual increase in magnetization on cooling imply the coexistence of CO and charge-disordered (CDO) states (presumably with a short-range order) in the mother phase, although the complex internal structure is not fully understood yet [27]. Further cooling below T_C yields a distinct metal phase in which long-range FM order prevails. A large temperature hysteresis in the thermomagnetization curve [36, 37] indicates that the transformation from the mother phase to the FM phase is a first-order transformation. We focused on the FM domain formation process near T_C. In fact, the CMR effect is most pronounced near T_C.

The Lorentz micrograph in Fig. 9.6a shows a tiny FM region produced in the nonmagnetic mother phase on zero-field cooling (ZFC). The asymmetric image contrast in the central portion (i.e., a pair of the bright and dark dots) reveals that the FM phase initially forms in a *single domain*. This FM phase is converted into a form of *double domain* with the volume increase on cooling to reduce the demagnetization energy; this is shown in Fig. 9.6b, where the phase boundary (outer frame) is bright and the internal magnetic domain wall is dark. The domain shape became more ellipsoidal at lower temperatures, as shown in Fig. 9.6c, presumably due to elastic and magnetic anisotropies. The definite nucleation and growth process is consistent with the characteristic feature of the first-order phase transformation, which is predicted by the thermomagnetization measurements in this compound.

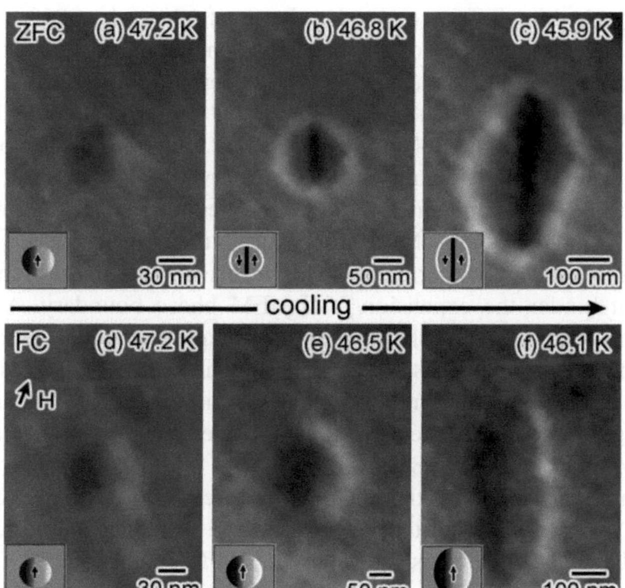

Fig. 9.6 Nucleation and growth of the FM phase in $La_{0.25}Pr_{0.375}Ca_{0.375}MnO_3$ during cooling. (**a**)–(**c**) Lorentz microscope images of the FM phase obtained during ZFC. (**d**)–(**f**) Lorentz microscope images of the FM phase during FC (120 Oe). The *inset* figures schematize the image contrast of a single domain (**a**, **d–f**) and of double domains (**b**, **c**). *Small arrows* in the *insets* represent magnetization vectors in each domain. Reprinted with permission from [35]

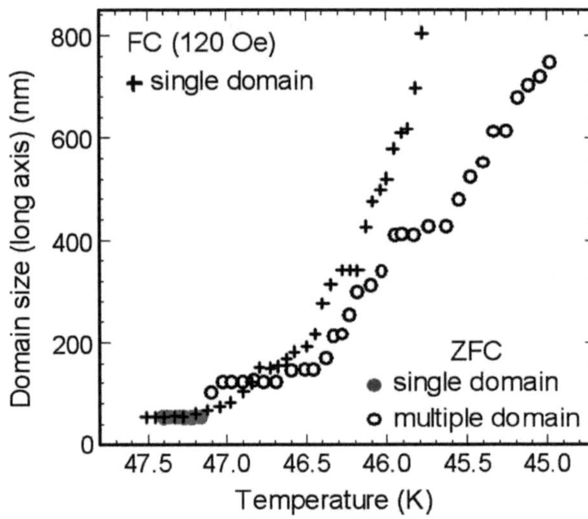

Fig. 9.7 Size of the FM domains in $La_{0.25}Pr_{0.375}Ca_{0.375}MnO_3$ plotted as a function of temperature. Reprinted with permission from [35]

Figure 9.6d-f shows another set of Lorentz micrographs of an FM phase that was produced in a magnetic field of 120 Oe (i.e., field cooling (FC)). When a magnetic field is applied in the direction indicated by the arrow labeled H, the FM phase remained a single domain form even when it grew to the order of 10^2 nm.

Figure 9.7 shows the FM domain diameter (the long axis in the case of the ellipsoidal domain) plotted as a function of temperature. The growth rate appears to be slightly enhanced by the applied field: ZFC required supercooling to 2.3 K to attain a domain size of 700 nm, whereas FC required supercooling to only 1.5 K to achieve the same domain size. This observation is consistent with several characteristics of CMR manganites, including the reduction in the thermal hysteresis on applying a magnetic field. More importantly, the growth curves are stepwise in both the ZFC and FC experiments. This result implies the presence of robust obstacles that hinder the expansion of the FM phase boundary. Another result supporting this scenario is the shape of the FM phase boundary – it has many dips protruding into the mother phase (see Fig. 9.8). The average dip sizes, which may be related to the scale of the pinning obstacles, were 70 nm for ZFC and 79 nm for ZC. This length scale is consistent with the average step sizes in the growth curves in Fig. 9.7 (69 nm for ZFC and 82 nm for FC).

A reasonable explanation is that the pinning force is due to the structural antipathy between the CO state and the FM phase. The former is an antiferromagnetic (AFM) insulator state, whereas the latter is an FM metal [5, 27]. The CO state and the FM phase have different lattice parameters [38]. These results indicate the presence of a significant potential barrier during the phase transformation, which requires further supercooling to generate a strong driving force that can produce motion of the phase boundary. As mentioned above, the presence of the CO regions is evidenced by superlattice reflections observed in the temperature range $T_C < T < T_{CO}$ [35]. We anticipate that the scales of the dips and the steps are

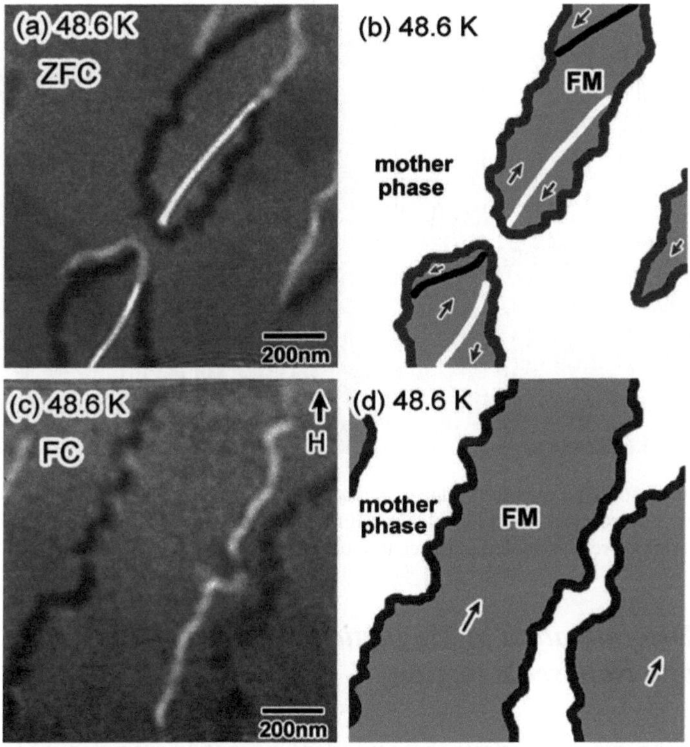

Fig. 9.8 Boundary structure of the long-range ordered FM phase in $La_{0.25}Pr_{0.375}Ca_{0.375}MnO_3$. (**a**) Lorentz microscope image of multiple domains formed by ZFC. (**b**) A schematic illustration of the boundary structure. (**c**) Lorentz microscope image of single domains formed by FC. (**d**) A schematic illustration of the boundary structure. Reprinted with permission from [35]

related to the correlation length of the charge ordering in the mother phase, although imaging of the CO regions in $La_{0.25}Pr_{0.375}Ca_{0.375}MnO_3$ has not yet been completed.

With respect to the microstructure and/or evolution process of CO regions on cooling, a recent dark-field image observation [39] appears to offer useful information, although the TEM observation was carried out by using a different system of layered manganite $La_{0.5}Sr_{1.5}MnO_4$. Figure 9.9 shows the evolution of the CO regions in $La_{0.5}Sr_{1.5}MnO_4$ during cooling. The appearance of superlattice reflections (as indicated by the arrows in Fig. 9.9a) demonstrates that charge ordering starts to occur at approximately 220 K. At 183 K, nanometer-sized dots (i.e., CO regions) are clearly observable in the dark-field image (Fig. 9.9b). The size of the bright dots increases on cooling, as shown in Fig. 9.9c–f): the observation is consistent with the increase in the intensity of superlattice reflections by cooling. Interestingly, the size of the bright dots at 113 K (approximately 200 nm) was comparable to the roughness of the CO domain interface (i.e., dip size in the CO domain interface) observed in this compound: refer to [39] for more details.

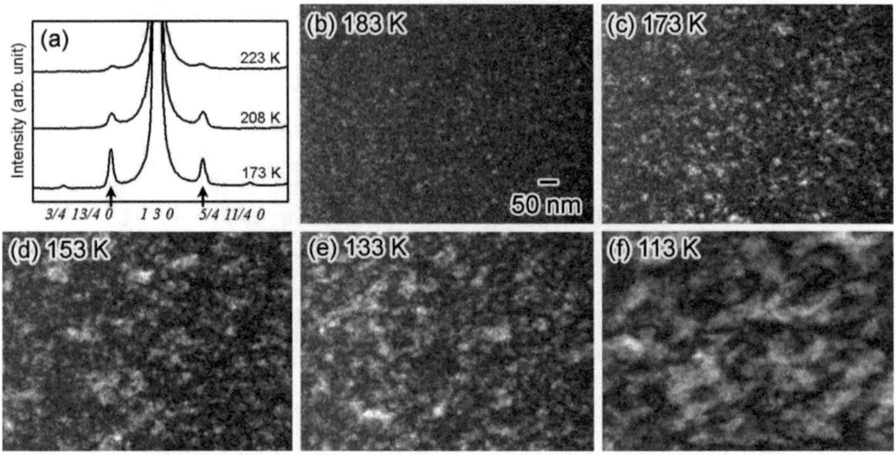

Fig. 9.9 Charge-ordered regions produced in $La_{0.5}Sr_{1.5}MnO_4$ during cooling. (**a**) Intensity profiles of superlattice reflections; results of electron diffraction. (**b**)–(**f**) Dark-field images obtained by using a superlattice reflection due to the charge ordering. Reprinted with permission from [39]

9.3.2 Determination of Magnetic Parameters of a Nanoscale Region

In this section, we demonstrate that important magnetic parameters of $La_{0.25}Pr_{0.375}Ca_{0.375}MnO_3$ (such as A and K_u) can be determined from only electron microscopy data [35]. Using TEM, we can measure the magnetic domain wall energy per unit area γ_d and the magnetic domain wall width (180° wall) W; these parameters are respectively represented by the following equations [40], where both A and K_u are unknown parameters:

$$\gamma_d = 4\sqrt{AK_u}, \tag{9.5}$$

$$W = \pi\delta = \pi\sqrt{A/K_u}, \tag{9.6}$$

where δ represents the magnetic exchange length. Thus, we have two equations with two unknown parameters. Solving these equations yields A and K_u.

We start by evaluating γ_d. In situ Lorentz microscopy observations (such as shown in Fig. 9.6a–c) indicate that the critical radius (r_c; half the domain size) at which a single domain changes into a double domain is approximately 39 nm. We assume that the domain is spherical and, at the point r_c, the domain wall formation energy ($\gamma_d \pi r_c^2$) equals to the reduction in the demagnetization energy by assuming the double domain form. The demagnetization energy reduction is approximately $I_s^2 \pi r_c^3/9\mu_0$ (i.e., half the original demagnetization energy), where I_s is the saturation magnetization and μ_0 is the permeability of vacuum. These results lead to an estimation of γ_d to be $1.8 \times 10^{-3}\,J/m^2$.

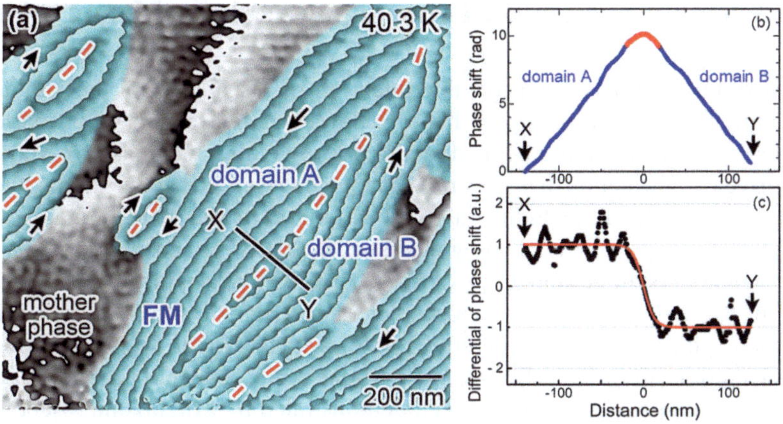

Fig. 9.10 Determination of the width of the 180° wall in $La_{0.25}Pr_{0.375}Ca_{0.375}MnO_3$. (**a**) Magnetic flux distribution determined by electron holography. *Arrows* indicate the direction of the lines of magnetic flux. (**b**) Phase shift of the electron wave plotted along the line XY in (**a**). (**c**) Differential of the phase shift shown in (**b**). The *red line* in (**c**) represents the curve fitting result. Reprinted with permission from [35]

The parameter W can be directly determined from electron holography observations. Figure 9.10a shows a reconstructed phase image obtained from $La_{0.25}Pr_{0.375}Ca_{0.375}MnO_3$ at 40.3 K. The contour lines represent the in-plane component of the magnetic flux. We focus on the 180° wall (indicated by the dotted line) that separates the large FM domains A and B. Figure 9.10b is a plot of the phase shift $\phi(x)$ of the electron wave measured along the line XY across the 180° domain wall (x indicates the position along the line XY). The sign of $\phi(x)$ changes from positive to negative at the 180° domain wall (refer to the transient area in the plot of Fig. 9.10b). Using a classical approximation of spin twisting near the domain wall [40, 41], the distribution of the in-plane magnetic flux component can be approximated by $c \cdot \tanh\{x/(-\delta)\}$, where c is a constant. Since the differential of $\phi(x)$ is related to the in-plane magnetic flux component, the parameter δ can be determined by curve fitting $\mathrm{d}\phi(x)/\mathrm{d}x$ (Fig. 9.10c). Thus, the domain wall width $\pi\delta$ is evaluated to be 39 nm.

By substituting the values of γ_d and W into (9.5) and (9.6), the unknown parameters are found to be $K_u = 3.6 \times 10^4\,\mathrm{J/m^3}$ and $A = 5.6 \times 10^{-12}\,\mathrm{J/m}$. These results are in excellent agreement with K_u determined from a magnetic torque measurement of a single crystalline film ($3.6 \times 10^4\,\mathrm{J/m^3}$, $La_{0.7}Ca_{0.3}MnO_3$ at 77 K [41]) and A deduced from a magnon measurement by neutron scattering ($3.3 \times 10^{-12}\,\mathrm{J/m}$, $La_{0.67}Ca_{0.33}MnO_3$ at 50 K [42]). We emphasize that our method can determine the magnetic parameters of a nanoscale region and that it does not require either bulk magnetization measurements or neutron scattering data. Thus, this technique can be used in studies of nanodevices and/or nanomaterials, for which it is difficult to determine K_u and A by conventional methods.

9.4 Magnetic Imaging of Ferromagnetic Shape-Memory Alloys

FMSMAs have attracted considerable attention of researchers in the last decade due to the giant magnetostriction that is achieved by rearrangement of martensite variants (namely, crystallographic domains with twinning relationship). Magnetic-field driven deformation has several advantages over the conventional thermal shape-memory effect, including a rapid response to the field and noncontact operation. A key point in explaining magnetic-field driven shape deformation is the high magnetocrystalline anisotropy of martensites, which is typically of the order of $10^5\ \mathrm{J/m^3}$ [3, 43–49]. This leads to a characteristic magnetic domain structure in the martensitic phase (the low-temperature phase), which exhibits a definite correlation with the martensite variants. In fact, a one-to-one correspondence between magnetic and crystallographic domains has been demonstrated by Lorentz microscopy [50–53], MFM [45], electron holography [51, 54, 55], and other techniques.

An interesting aspect of magnetic domain observations of the parent phase (the high-temperature phase) is the effect of structural defects (such as antiphase boundaries (APBs)) on the magnetization distribution. Actually, many FMSMAs (Ni_2MnGa [3], Ni_2MnAl [56], Ni_2MnIn [57, 58], Ni_2MnSn [57, 59, 60], Ni_2FeGa [61], etc.) have an $L2_1$-ordered (Heusler-type) structure in the parent phase. Therefore, these alloys contain many APBs that are formed by chemical ordering from the B2-disordered state to the $L2_1$-ordered state (Fig. 9.11). APBs produced by thermal diffusion have a finite width, in which the $L2_1$ order appears to be depressed [62, 63]. Consequently, the magnetization at APB positions can be reduced when the magnetic structure strongly depends on the degree of $L2_1$ order. Due to this local modulation of magnetism, APBs are regarded as an important factor in controlling the magnetic properties of ordered alloys. This section presents the results of intensive TEM studies on the impact of APBs, observed in the $L2_1$-type FMSMA Ni_2Mn (Al,Ga) [21]. Regarding the correlation between APBs and magnetic domain walls, also refer to the reports by Venkateswaran et al. [64] who examined the domain structure of the Ni–Mn–Ga alloy system. Furthermore, in the last subsection, we briefly mention the peculiar phenomenon observed in a Ni–Fe–Ga alloy, in which a definite magnetic pattern intimating the martensitic domain structure was observed even in the parent phase [54]; this phenomenon was observed

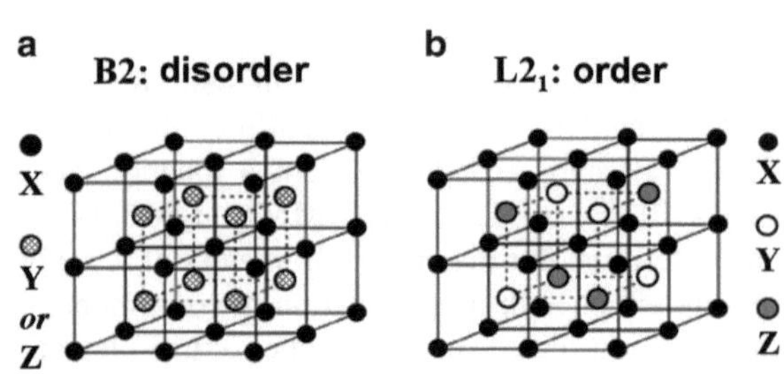

Fig. 9.11 Schematic illustrations of (**a**) B2-type and (**b**) $L2_1$-type structures

in an alloy in which the magnetic modulation by APBs had been considerably weakened by heat treatment.

9.4.1 Impact of APBs on the Local Magnetization Distribution

APBs in the parent phase can be readily observed by using a diffraction contrast, i.e., a dark-field method that uses a superlattice reflection (e.g., the 111 reflection) related to the $L2_1$ ordering. A typical contrast of APBs in a $Ni_{50}Mn_{25}Al_{12.5}Ga_{12.5}$ alloy is shown in Fig. 9.12a: refer to the gray meandering lines [65]. Figure 9.12b shows a Lorentz micrograph of the same viewing field in which the magnetic domain walls appear as bright or dark lines. These observations reveal that the magnetic domain walls trace the positions of the APBs: note the perfect coincidence in the positions, such as those indicated by the arrows labeled 1–6. In other words, the magnetic domain energy is minimized at the position of APBs. Figure 9.13a shows another Lorentz micrograph that visualizes the magnetic domain walls tracing APBs [65]. Lines of magnetic flux in the same area were observed by electron holography (see Fig. 9.13b). A remarkable feature is that there are many magnetic flux vortices. The superimposed image in Fig. 9.13c indicates that these magnetic vortices are formed in the closed and/or highly curved APBs.

The observations in Figs. 9.12 and 9.13 suggest that APBs provide significant pinning sites for the motion of magnetic domain walls during the magnetization process. This prediction is supported by in situ Lorentz microscopy observations. Figure 9.14a shows a Lorentz microscope image obtained in a negligible magnetic field [66]. It reveals two types of chirality in the magnetic vortices formed in the

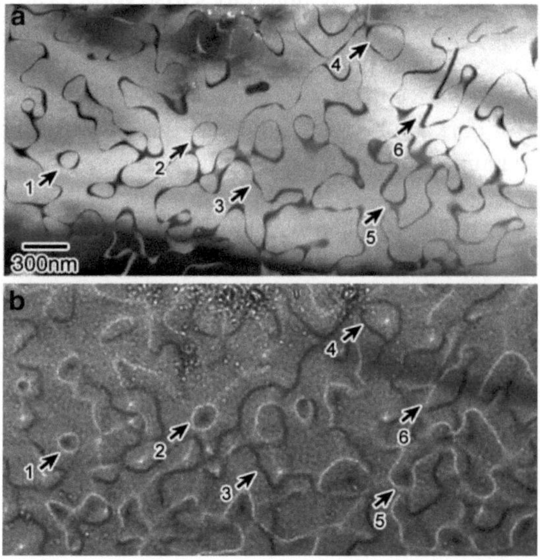

Fig. 9.12 Correspondence between APBs and magnetic domain walls in $Ni_{50}Mn_{25}Al_{12.5}Ga_{12.5}$. (**a**) Dark-field image and (**b**) Lorentz microscope image of the parent phase ($L2_1$ phase). Reprinted with permission from [65]

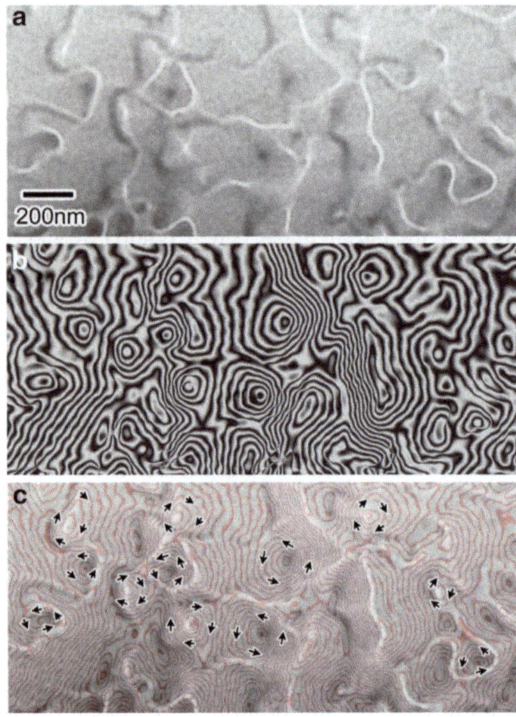

Fig. 9.13 Magnetic vortices formed in $Ni_{50}Mn_{25}Al_{12.5}Ga_{12.5}$. (**a**) Lorentz microscope image and (**b**) reconstructed phase image representing the lines of magnetic flux in the parent phase. (**c**) Superimposed images of (**a**) and (**b**). The *small arrows* indicate the direction of the lines of magnetic flux. Reprinted with permission from [65]

closed APBs. The arrows labeled V1, V3, and V5 indicate clockwise vortices, which have a central bright spot (vortex core) and an outer dark frame (domain wall). The other type of vortices is counterclockwise, which is indicated by the arrows labeled V2 and V4: these counterclockwise vortices are characterized by a central dark spot and an outer bright frame. Refer to the upper panels in Fig. 9.14c to distinguish these vortices patterns. When a magnetic field (H) of 60 Oe was applied along the white arrow, the vortices cores moved to the left (for bright spots) and to the right (for dark spots), depending on the chirality. This phenomenon can be reasonably explained by the increase in the magnetic component (within the closed APBs) that is aligned to the applied magnetic field, as illustrated in the lower panels in Fig. 9.14c. Despite this change in the inner magnetization distribution, the magnetic domain walls remained trapped by the APBs due to the significant pinning.

We discuss here the width of magnetic domain walls that are located at the positions of APBs. The result is also useful for obtaining a deeper understanding of the nature of the significant pinning effect by APBs. The wall width was determined by using Lorentz microscopy observations [21]: we focused on a dark line (i.e., deficient intensity line representing a magnetic domain wall) tracing the APB position (Fig. 9.15a). The wall was identified to be a 180° wall, based on the electron holography observation acquired from the same region: refer to the red contour lines superposed to the Lorentz microscope image in Fig. 9.15a. The full width at

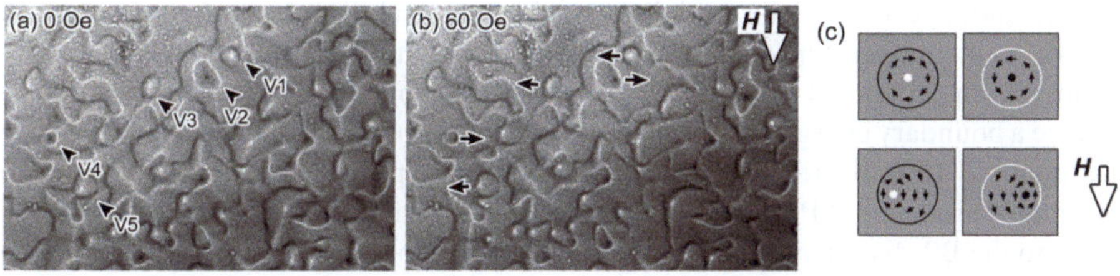

Fig. 9.14 Pinning of magnetic domain walls by APBs in $Ni_{50}Mn_{25}Al_{12.5}Ga_{12.5}$. Lorentz micrographs obtained (**a**) before applying a magnetic field and (**b**) when a magnetic field of 60 Oe is applied. (**c**) Schematic representation of the change in the vortex-like patterns caused by applying a magnetic field. Reprinted with permission from [66]

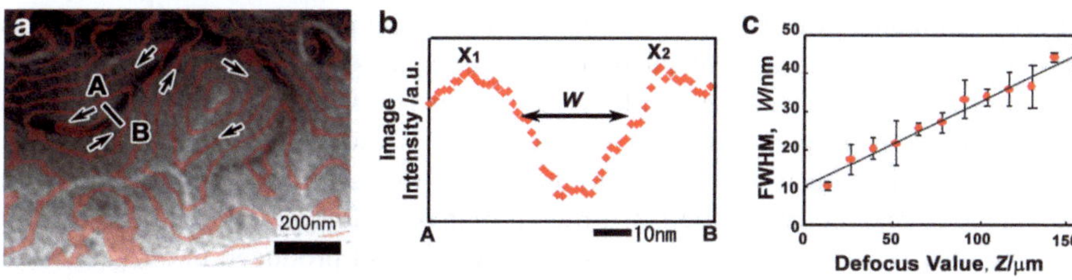

Fig. 9.15 Determination of the width of the 180° wall located at the position of APB in $Ni_{50}Mn_{25}Al_{12.5}Ga_{12.5}$. (**a**) Lorentz microscope image showing a 180° wall (*dark line*). The result of electron holography (*red contour lines*) is superposed to the Lorentz microscope image. (**b**) Intensity profile measured along the line AB in (**a**). (**c**) Full width at half maximum (FWHM) of the intensity profile plotted as a function of the defocus value Z. Reprinted with permission from [21]

half maximum W in the intensity profile (result of Lorentz microscopy), which was obtained along the line AB crossing the 180° wall, can be a measure of the wall width, as shown in Fig. 9.15b. The background was defined as the straight line that intersects the intensity maxima at X_1 and X_2. Note that W increases with increasing defocus value Z; the relationship can be approximated by a linear function when Z is small. This means that a good approximation of the wall width can be obtained by plotting W as a function of Z and extrapolating it to zero defocus [67]. If the plane of APBs and magnetic domain walls is tilted off that of the incident electrons, this plot overestimates the width. To minimize this overestimation, we selected a region in which the observed APB width is narrow and does not change greatly with the specimen thickness. Least-squares fitting of the plots yields a wall width of 10 nm (Fig. 9.15c). Interestingly, this wall width is considerably smaller than that of 180° walls in other cubic systems. We recently determined that the width of a 180° wall in the $Ni_2Mn(Al,Ga)$ alloy is about 48 nm when it is produced in the $L2_1$-ordered matrix region (namely, a magnetic domain wall that is free from trapping by APB) [68]. Thus, the result of Fig. 9.15c indicates that the width of domain walls at APB positions (10 nm) is dominated by a distinct mechanism from that of conventional 180° walls.

The above result may be explained by the chemical disorder that must occur in APBs. As mentioned earlier, APBs are a type of structural defect, which is thermally formed in the chemical ordering from the B2 state to the $L2_1$ state. A condition to reduce a boundary energy requires a gradual change in the chemical order parameter (as opposed to a discrete change) at the boundary position [62]. Eventually, the thermally produced APBs have a finite width, in which the atomic order may be close to the B2-disordered state, rather than the $L2_1$-ordered state. It is important to be aware of the peculiar magnetism of this alloy. The $Ni_2Mn(Al,Ga)$ alloy is FM in the $L2_1$-ordered state, whereas it is AFM in the disordered B2 state [69]. This means that the FM order should be considerably depressed at the positions of the APBs, as schematically shown in Fig. 9.16b. In other words, the effective magnetic parameters may be different from those in the $L2_1$-ordered matrix regions. Consequently, the magnetic domain wall energy can be minimized at the positions of the APBs; this appears to explain the significant pinning force against the domain wall motion. It is probable that the narrow wall width (10 nm) is also due to the depression of ferromagnetism near APBs. At the positions of the APBs in $Ni_2Mn(Al,Ga)$, the magnetic domain wall is probably no longer expressed simply by (9.6) with the magnetic parameters of the $L2_1$-ordered matrix region: i.e., the feature of spin twisting may be different from the case of a Bloch wall produced in a region with uniform magnetization (Fig. 9.16a). We anticipate that for a magnetic domain wall trapped by an APB the observable wall width is related to the APB thickness. In fact, recent electron microscopy observations reveal that the width of an APB formed in a $Ni_2Mn(Al,Ga)$ alloy is approximately 5 nm [68]. The result does not deviate greatly from the magnetic domain wall width of 10 nm.

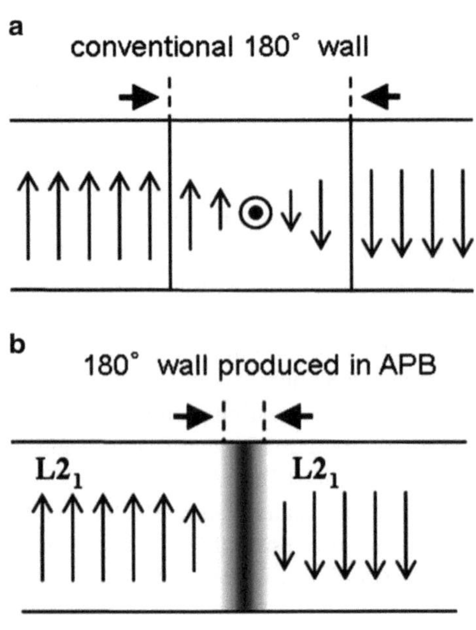

Fig. 9.16 Schematic illustrations of the structure of 180° walls. (**a**) Typical case of a Bloch wall formed in a ferromagnetic matrix. (**b**) Magnetic domain wall formed at the position of APB (*gray portion*) in $Ni_{50}Mn_{25}Al_{12.5}Ga_{12.5}$

9.4.2 Magnetic Pattern Formation Triggered by Premartensitic Lattice Anomaly

This section presents another topic of magnetic imaging with an FMSMA [54, 70]. The experiments were performed using a $Ni_{51}Fe_{22}Ga_{27}$ alloy, which undergoes a cubic ($L2_1$) $\rightarrow$ monoclinic (14 M or 10 M) martensitic transformation at 141 K. Like many other shape-memory alloys (both magnetic and nonmagnetic alloys), the parent phase in $Ni_{51}Fe_{22}Ga_{27}$ shows lattice modulation, which is reminiscent of the structural unit of martensite. Such lattice modulation is sometimes referred to as the precursor effect of martensitic transformations, and it has attracted considerable attention of researchers for many years [71–75]. We here report that the magnetic microstructure in the parent phase offers information that deepens our understanding of the premartensitic lattice modulation.

The $Ni_{51}Fe_{22}Ga_{27}$ alloy is FM in both the $L2_1$-ordered and B2-disordered states, although its magnetic moment is reduced by the chemical disorder in the B2 state [61]. Furthermore, it is possible to reduce the size of antiphase domain (APD) to approximately 20 nm or smaller by applying an appropriate heat treatment. As a result, the magnetic perturbation caused by the APBs can be weakened relative to that in a $Ni_2Mn(Al,Ga)$ alloy. Using a $Ni_{51}Fe_{22}Ga_{27}$ alloy, we examined the temperature dependence of the magnetic domain structure (in the parent phase), which can be an indicator of the premartensitic lattice anomaly. In fact, the magnetization distribution is sensitive to the lattice modulation which occurs prior to the onset of a martensitic transformation.

The left column of Fig. 9.17 shows the variations in Lorentz micrographs on cooling. Fig. 9.17a–c) shows images of the parent phase, whereas Fig. 9.17d shows an image obtained from the martensitic phase. At 295 K, there are only two large magnetic domains (of the order of 10^3 nm); they are separated by a straight domain wall, as indicated by the straight dark line. When the temperature is reduced, low-contrast speckles (of the order of 10 nm) become visible in Lorentz micrographs due to local fluctuations caused by small APDs. More importantly, the original microscale domain appears to divide into domains of the order of 10^2 nm (the arrowheads in Fig. 9.17b, c indicate the formation of domain walls). The subdivided magnetic domains have comparable sizes and shapes as the martensite phase. This is shown in Fig. 9.17d, in which there are two martensite variants (V1 and V2) containing several magnetic domains of the order of 10^2 nm.

In this case, electron holography provides more insight into the variation of the magnetic microstructure. The right column in Fig. 9.17 shows magnetic flux maps obtained from the same area as that shown in the left column. At 295 K (Fig. 9.17e), the magnetic flux lines are linear, although their directions change sharply at the magnetic domain wall; this is a typical magnetic domain structure in a cubic system. However, when the parent phase is cooled to a temperature near M_s (i.e., the martensitic transformation start temperature), a distinct macroscopic pattern formation is obtained (see Fig. 9.17f, g). The most interesting point is that the wavy magnetic flux pattern, which is formed in the parent phase on cooling,

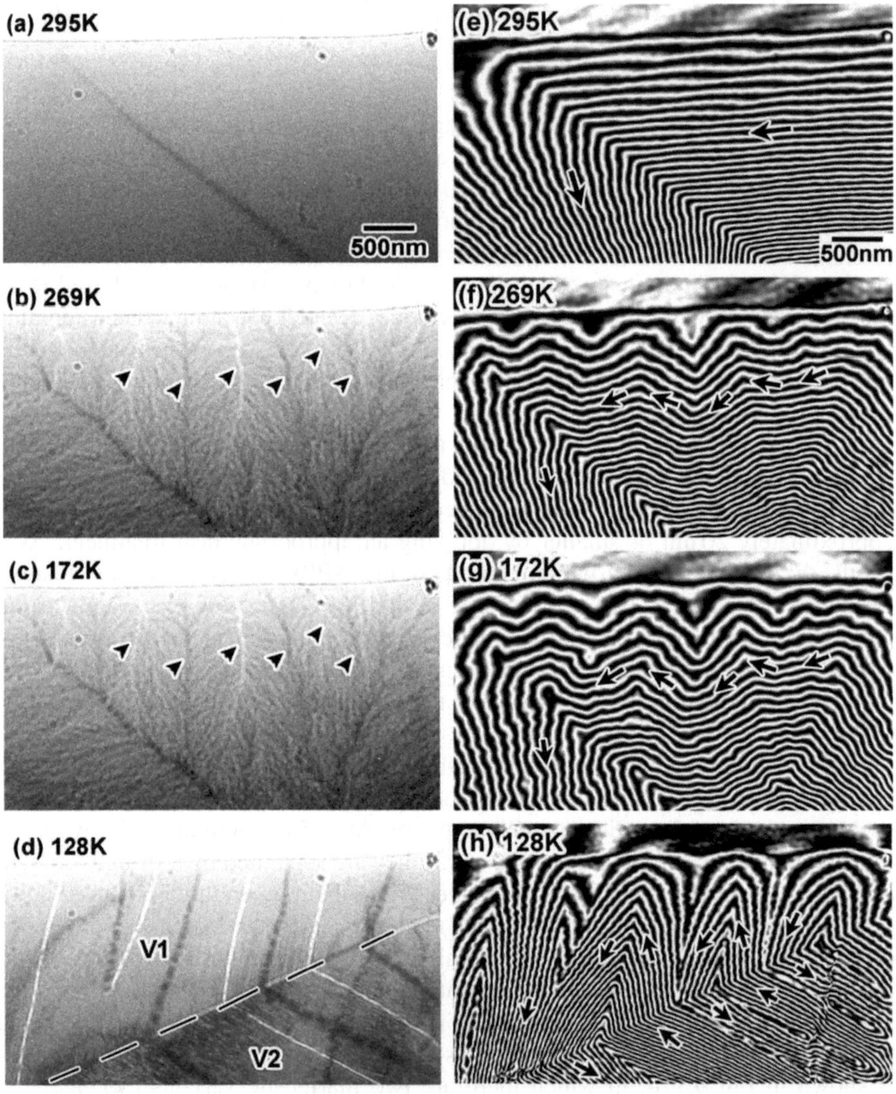

Fig. 9.17 Macroscopic pattern formation observed in $Ni_{51}Fe_{22}Ga_{27}$ before the onset of the martensitic transformation. (**a**)–(**d**) Change in Lorentz micrographs with cooling. The *broken line* in (**d**) represents the interface of martensite variants V1 and V2. (**e**)–(**h**) Change in reconstructed phase images (magnetic flux maps determined by electron holography) observed in the same field of view. Reprinted with permission from [54]

appears to be inherited by the martensite phase. There is a remarkable similarity between the two flux maps shown in Fig. 9.17g, h.

Electron diffraction studies provide important information clues about the origin of this macroscopic pattern formation. Like other martensitic alloys, this alloy does

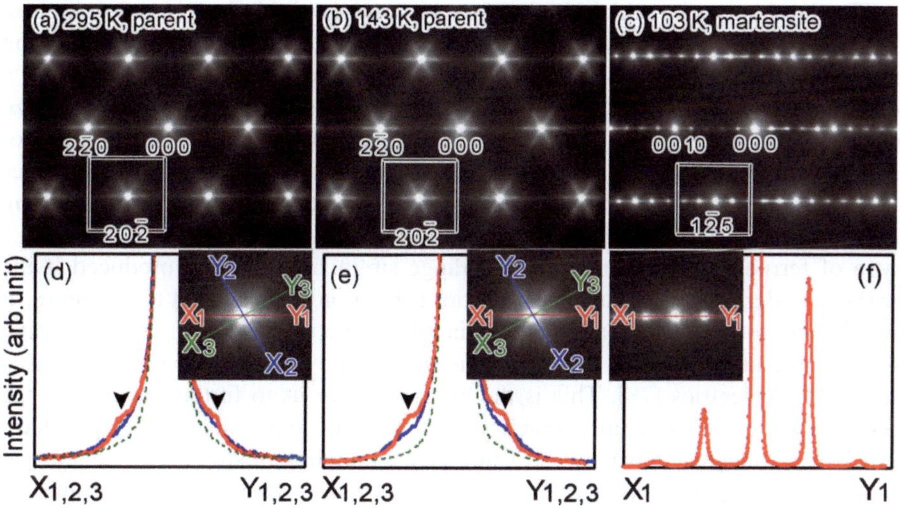

Fig. 9.18 Diffraction patterns showing the lattice anomaly in the parent phase of $Ni_{51}Fe_{22}Ga_{27}$. (**a**)–(**c**) Change in the electron diffraction patterns with cooling. (**d**)–(**f**) Intensity profiles observed in the closed areas in (**a**), (**b**), and (**c**), respectively. Reprinted with permission from [70]

undergo a transverse lattice displacement in the {110} planes of the parent phase due to the anomaly in shear modulus $(c_{11}-c_{12})/2$ [70, 76]. This effect is confirmed by the observed rod-like diffuse scattering (see Fig. 9.18a). The diffuse scattering intensity increases when the parent phase is cooled to 143 K (Fig. 9.18b) due to the pronounced lattice modulation in the parent phase. This observation indicates a significant correlation between the magnetic pattern formation (such as that shown in Fig. 9.17) and the premartensitic lattice modulation, which develops as M_s is approached. Note that, in the diffuse scattering, there is an intensity maximum (diffuse spot) near $0.18 \cdot \boldsymbol{q}_{\mathbf{hh0}}$. This is close to $0.20 \cdot \boldsymbol{q}_{\mathbf{hh0}}$ representing the structural unit of the martensite phase. At 295 K, these diffuse spots are approximately symmetric about the fundamental reflection (see Fig. 9.18d) and the intensity profile along X_1–Y_1 is comparable to that along X_2–Y_2. However, at 143 K, the spot intensity is asymmetric (Fig. 9.18e); i.e., the diffuse spots along X_1–Y_1 are more intense than those along X_2–Y_2. This observation implies that an asymmetric strain field (due to the premartensitic lattice modulation) might have developed in this viewing field. This long-range strain field may induce a favored martensite variant in the same area. In fact, further cooling produced a martensite variant that gives superlattice reflections along the same lines as X_1–Y_1 (Fig. 9.18c, f). Asymmetry in the diffuse scattering intensity has also been reported for a Ni_2MnGa alloy by Tsuchiya et al. [77]. We concluded that the macroscopic pattern formation was triggered by the long-range strain field that developed in the parent phase; the change in the magnetic microstructure is a consequence of the magnetoelastic interaction.

The mechanism by which the long-range strain field develops remains unclear. Diffraction studies will presumably be the key to understanding the underlying mechanism. The diffuse spots indicate that the parent lattice is subjected to a distortion that is reminiscent of the martensite structure. We estimated the size of these distorted regions (scattered in the parent phase) to be of the order of nanometers from the broadness of the diffuse spots. Since there are several choices for a propagation vector of the lattice displacement in the cubic parent phase (along $\boldsymbol{q}_{110}$, $\boldsymbol{q}_{101}$, $\boldsymbol{q}_{011}$, etc.), the distorted regions can be classified into variants (i.e., a type of ferroelastic domains). A long-range strain field may be produced if the variants undergo an alignment via elastic interaction at a reduced temperature. It will be interesting to compare the peculiar nanostructure formed in the martensitic alloy with that observed in other systems such as the polar nanoregions formed in relaxor ferroelectrics [78]. This is the motivation for us to further investigate the premartensitic nanostructures, which we are currently doing. Saxena et al. [79] have proposed a "magnetoelastic tweed", in which a coupling of strain with magnetism produces a magnetic modulation in the parent phase. Our observations may be related to the mechanism that gives rise to this magnetoelastic tweed.

9.5 Concluding Remarks

It appears that structural disorder is an important concept in the research and development of magnetic functional materials. For example, it is widely accepted that the CMR effect (in manganites) is influenced by quenched-in disorder. In the hole-doped manganites, the structural disorder affects the itinerancy of 3d electrons, which are responsible for both conduction and magnetism. Attainable magnetoresistance also depends on other structural factors, such as the grain boundary structure that is responsible for the spin-orientation dependence of the electronic transportation. In this context, simultaneous observation of the crystallographic and magnetic microstructures must be the key to revealing the mechanism of extraordinary phenomenon observed in magnetic functional materials. Due to recent technical advances (such as aberration correction and the development of coherent sources of electron beams, both of which have significantly improved the microscope resolution), methods based on TEM will play a crucial role in characterizing those compounds. We anticipate that it will be applied to other recent topics on magnetism, such as multiferroics, spin textures, spintronics, and magnetic nanoparticles; these all require precise magnetic imaging with a nanoscale resolution.

Acknowledgments The experimental results presented in this chapter were acquired in collaborations with researchers in Tohoku University, Okinawa Institute of Science and Technology (OIST), Osaka Prefecture University, and JEOL Co. The author expresses his sincere gratitude to Professors D. Shindo, R. Kainuma, T. Arima, K. Oikawa, K. Ishida, Dr. T. Yano, Mr. S. Konno (Tohoku), Dr. A. Tonomura, Mr. H. Kasai, Dr. J.J. Kim, Mr. S. Mamishin (OIST), Prof. S. Mori (Osaka), and Mr. T. Suzuki (JEOL) for the very helpful discussions regarding the topics presented in this chapter.

References

1. S. Jin, T.H. Tiefel, M. McCormack, R.A. Fastnacht, R. Ramesh, L.H. Chen, Thousand-fold change in resistivity in magnetoresistive La-Ca-Mn-O films. Science **264**, 413 (1994)
2. Y. Tokura, A. Urushibara, Y. Moritomo, T. Arima, A. Asamitsu, G. Kido, N. Furukawa, Giant magnetotransport phenomena in filling-controlled Kondo lattice system: $La_{1-x}Sr_xMnO_3$, J. Phys. Soc. Jpn. **63**, 3931 (1994)
3. K. Ullakko, J.K. Huang, C. Kantner, R.C. O'Handley, V.V. Kokorin, Large magnetic-field-induced strains in Ni_2MnGa single crystals. Appl. Phys. Lett. **69**, 1966 (1996)
4. N. Mathur, P. Littlewood, Mesoscopic texture in manganites. Phys. Today **56**, 25 (2003)
5. E. Dagotto, *Nanoscale Phase Separation and Colossal Magnetoresistance* (Springer-Verlag, Berlin, 2003).
6. T. Kakeshita, T. Takeuchi, T. Fukuda, M. Tsujiguchi, T. Saburi, R. Oshima, S. Muto: Giant magnetostriction in an ordered Fe_3Pt single crystal exhibiting a martensitic transformation, *Appl. Phys. Lett.*, **77**, 1502 (2000)
7. F. Bitter, Experiments on the nature of magnetism. Phys. Rev. **41**, 507 (1932)
8. H.J. Williams, F.G. Foster, E.A. Wood, Observation of magnetic domains by the Kerr effect. Phy. Rev. **82**, 119 (1951)
9. Y. Martin, H.K. Wickramasinghe, Magnetic imaging by "force microscopy" with 1000 Åresolution. Appl. Phys. Lett. **50**, 1455 (1987)
10. J.R. Kirtley, M.B. Ketchen, K.G. Stawisaz, J.Z. Sun, W.J. Gallagher, S.H. Blanton, S.J. Wind, High-resolution scanning SQUID. Appl. Phys. Lett. **66**, 1138 (1995)
11. K. Koike, K. Hayakawa, Scanning electron microscope observation of magnetic domains using spin-polarized secondary electrons. Jpn. J. Appl. Phys. **23**, L187 (1984)
12. A. Tonomura, *Electron Holography* (Springer, Berlin, 1999)
13. M. De Graef, Y. Zhu, *Magnetic Imaging and its Applications to Materials* (Academic, San Diego, 2001)
14. D. Shindo, T. Oikawa, *Analytical Electron Microscopy for Materials Science* (Springer, Tokyo, 2002)
15. P. Hirsch, A. Howie, R. Nicholson, D.P. Pashley, M.J. Whelan, *Electron Microscopy of Thin Crystals* (Krieger, Malabar, Florida, 1977)
16. M. Sagawa, S. Fujiwara, N. Togawa, H. Yamamoto, Y. Matsuura, New material for permanent magnets on base of Nd and Fe. J. Appl. Phys. **55**, 2083 (1994)
17. H. Lichte, M. Lehmann, Electron holography–basics and applications. Rep. Prog. Phys. **71**, 1 (2008)
18. D. Shindo, Y. Murakami, Electron holography of magnetic materials. J. Phys. D: Appl. Phys. **41**, 183002 (2008)
19. M. Inoue, T. Tomita, M. Naruse, Z. Akase, Y. Murakami, D. Shindo, Development of a magnetizing stage for *in situ* observations with electron holography and Lorentz microscopy. J. Electron Microsc. **54**, 509 (2005)
20. J.H. Yoo, Y. Murakami, D. Shindo, T. Atou, M. Kikuchi, Interaction of separated ferromagnetic domains in a hole-doped manganite achieved by a magnetic field, Phys. Rev. Lett. **93**, 047204 (2004)
21. T. Yano, Y. Murakami, R. Kainuma, D. Shindo, Interaction between magnetic domain walls and antiphase boundaries in $Ni_2Mn(Al,Ga)$ studied by electron holography and Lorentz microscopy. Mater. Trans. **48**, 2636 (2007)
22. Y. Murakami, N. Kawamoto, D. Shindo, I. Ishikawa, S. Deguchi, K. Yamazaki, Y, Kondo, K. Suganuma: Simultaneous measurements of conductivity and magnetism by using microprobes and electron holography, *Appl. Phys. Lett.* **88**, 223103 (2006)
23. A. Millis, Lattice effects in magnetoresistive manganese perovskites. Nature **392**, 147 (1998)
24. G.C. Milward, M.J. Calderón, P.B. Littlewood, Electronically soft phases in manganites. Nature **433**, 607 (2005)
25. Y. Tokura: *Colossal Magnetoresistive Oxides* (Gordon and Breach Science, Amsterdam, 2000)

26. A. Moreo, S. Yunoki, E. Dagotto, Phase separation scenario for manganese oxides and related materials. Science **283**, 2034 (1999)
27. M. Uehara, S. Mori, C.H. Chen, S.-W. Cheong, Percolative phase separation underlies colossal magnetoresistance in mixed-valent manganites. Nature **399**, 560 (1999)
28. M. Mayr, A. Moreo, J.A. Verges, J. Arispe, A. Feiguin, E. Dagotto, Resistivity of mixed-phase manganites. Phys. Rev. Lett. **86**, 135 (2001)
29. L. Zhang, C. Israel, A. Biswas, R.L. Greene, A. de Lozanne, Direct observation of percolation in a manganite thin film. Science **298**, 805 (2002)
30. W. Wu, C. Israel, N.J. Hur, S.Y. Park, S.-W. Cheong, A. de Lozanne, Magnetic imaging of a supercooling glass transition in a weakly disordered ferromagnet. Nature Mater. **5**, 881 (2006)
31. T. Asaka, Y. Anan, T. Nagai, S. Tsutsumi, H. Kuwahara, K. Kimoto, Y. Tokura, Y. Matsui, Phys. Rev. Lett. **89**, 207203 (2002)
32. Y. Murakami, J.H. Yoo, D. Shindo, T. Atou, M. Kikuchi, Magnetization distribution in the mixed-phase state of hole-doped manganites, Nature **423**, 965 (2003)
33. J.C. Loudon, M.D. Mathur, P.A. Midgley, Charge-ordered ferromagnetic phase in $La_{0.5}Ca_{0.5}MnO_3$, Nature **420**, 797 (2002)
34. M. Tokunaga, Y. Tokunaga, T. Tamegai, Imaging of percolative conduction paths and their breakdown in phase-separated $(La_{1-y}Pr_y)_{0.7}Ca_{0.3}MnO_3$ with $y = 0.3$. Phys. Rev. Lett. **93**, 037203 (2004)
35. Y. Murakami, H. Kasai, J.J. Kim, S. Mamishin, D. Shindo, S. Mori, A. Tonomura, Ferromagnetic domain nucleation and growth in colossal magnetoresistive manganite. Nature Nanotech. **5**, 37 (2010)
36. H.H. Kim, M. Uehara, C. Hess, P.A. Sharma, S.-W. Cheong, Thermal and electronic transport properties and two-phase mixtures in $La_{5/8-x}Pr_xCa_{3/8}MnO_3$. Phys. Rev. Lett. **84**, 2961 (2000)
37. L. Ghivelder, F. Parisi, Dynamic phase separation in $La_{5/8-y}Pr_yCa_{3/8}MnO_3$. Phys. Rev. B **71**, 184425 (2005)
38. V.Yu Pomjakushin, D.V. Sheptyakov, K. Conder, E.V. Pomjakushina, A.M. Balagurov, Effect of oxygen substitution and crystal microstructure on magnetic ordering and phase separation in $(La_{1-y}Pr_y)_{0.7}Ca_{0.3}MnO_3$. Phys. Rev. B **75**, 054410 (2007)
39. Y. Murakami, S. Konno, D. Shindo, T. Arima, T. Suzuki, Electric-field-induced domain switching in the charge-orbital-ordered state of manganite $La_{0.5}Sr_{1.5}MnO_4$. Phys. Rev. B **81**, 140102(R) (2010)
40. A. Hubert, R. Schäfer, *Magnetic Domains* (Springer-Verlag, Berlin, 2000)
41. N.D. Mathur, M.-H Jo, J.E. Evetts, M.G. Blamire, Magnetic anisotropy of thin film $La_{0.7}Ca_{0.3}MnO_3$ on untwinned paramagnetic $NdGaO_3$ (001). J. Appl. Phys. **89**, 3388 (2001)
42. J.W. Lynn, R.W. Erwin, J.A. Borchers, Q. Huang, A. Santoro, J-L. Peng, Z.Y. Li: Unconventional ferromagnetic transition in $La_{1-x}Ca_xMnO_3$. Phys. Rev. Lett. **76**, 4046 (1996)
43. T. Kakeshita, T. Fukuda, T. Takeuchi, Magneto-mechanical evaluation for twinning plane movement driven by magnetic field in ferromagnetic shape memory alloys. Mater. Sci. Eng. A **438–440**, 12 (2006)
44. R.C. O'Handley, Model for strain and magnetization in magnetic shape-memory alloys. J. Appl. Phys. **83**, 3263 (1998)
45. Q. Pan, R.D. James, Micromagnetic study of Ni_2MnGa under applied field. J. Appl. Phys. **87**, 4702 (2000)
46. T. Fukuda, T. Sakamoto, T. Kakeshita, T. Takeuchi, K. Kishio, Rearrangement of martensite variants in iron-based ferromagnetic shape memory alloys under magnetic field. Mater. Trans. **45**, 188 (2004)
47. M. Wuttig, J. Li, C. Craciunescu, A new ferromagnetic shape memory alloy system. Scr. Mater. **44**, 2393 (2001)
48. R.D. James, M. Wuttig, Magnetostriction of martensites. Phil. Mag. A **77**, 1273 (1998)
49. V.A. Chernenko, V.A. L'vov, S.P. Zagorodnyuk, T. Takagi, Ferromagnetism of thermoelastic martensites: Theory and experiment. Phys. Rev. B **67**, 064407 (2003)
50. M. De Graef, M.A. Willard, M.E. McHenry, Y. Zhu, *In-situ* Lorentz TEM cooling study of magnetic domain configurations in Ni_2MnGa. IEEE Trans. Magn. **37**, 2663 (2001)

51. Y. Murakami. D. Shindo, K. Oikawa, R. Kainuma, K. Ishida, Magnetic domain structures in Co-Ni-Al shape memory alloys studied by Lorentz microscopy and electron holography. Acta Mater. **50**, 2173 (2002)
52. S.P. Venkateswaran, N.T. Nuhfer, M. De Graef: Magnetic domain memory in multiferroic Ni_2MnGa, *Acta Mater.* **55**, 5419 (2007)
53. B. Bartova, N. Wiese, D. Schryvers, J.N. Chapman, S. Ignacova, Acta Mater. **56**, 4470 (2008)
54. Y. Murakami, D. Shindo, K. Oikawa, R. Kainuma, K. Ishida, Microstructural change near the martensitic transformation in a ferromagnetic shape memory alloy $Ni_{51}Fe_{22}Ga_{27}$ studied by electron holography. Appl. Phys. Lett. **85**, 6170 (2004)
55. V.C. Solomon, M.R. McMaryney, D.J. Smith, Y.J. Tang, A.E. Berkowitz, R.C. O'Handley: Magnetic domain configurations in spark-eroded ferromagnetic shape memory Ni-Mn-Ga particles, *Appl. Phys. Lett.* **86**, 192503 (2005)
56. A. Fujita, K. Fukamichi, F. Gejima, R. Kainuma, K. Ishida, Magnetic properties and large magnetic-field-induced strains in off-stoichiometric Ni-Mn-Al Heusler alloys. Appl. Phys. Lett. **77**, 3054 (2000)
57. Y. Sutou, Y. Imano, N. Koeda, T. Omori, R. Kainuma, K. Ishida, K. Oikawa, Magnetic and martensitic transformations of NiMnX ($X = $ In, Sn, Sb) ferromagnetic shape memory alloys. Appl. Phys. Lett. **85**, 4358 (2004)
58. R. Kainuma, Y. Imano, W. Ito, Y. Sutou, H. Morito, S. Okamoto, O. Kitakami, K. Oikawa, A. Fujita, T. Kanomata, K. Ishida, Magnetic-field-induced shape recovery by reverse phase transformation. Nature **439**, 957 (2006)
59. T. Krenke, E. Duman, M. Acet, E.F. Wassermann, X. Moya, L. Manosa, A. Planes, Inverse magnetocaloric effect in ferromagnetic Ni-Mn-Sn alloys. Nature Mater. **4**, 450 (2005)
60. R. Kainuma, Y. Imano, W. Ito, H. Morito, Y. Sutou, K. Oikawa, A. Fujita, K. Ishida, Metamagnetic shape memory effect in a Heusler-type $Ni_{43}Co_7Mn_{39}Sn_{11}$ polycrystalline alloy. Appl. Phy. Lett. **88**, 192513 (2006)
61. K. Oikawa, T. Ota, T. Ohmori, Y. Tanaka, H. Morito, A. Fujita, R. Kainuma, K. Fukamichi, K. Ishida, Magnetic and martensitic phase transitions in ferromagnetic Ni-Ga-Fe shape memory alloys. Appl. Phys. Lett. **81**, 5201 (2002)
62. M. Ohno, T. Mohri, Relaxation kinetics of the long-range order parameter in a non-uniform system studied by the phase field method using the free energy obtained by the cluster variation method. Philos. Mag. **83**, 315 (2003)
63. Ch. Ricolleau, A. Loiseau, F. Ducastelle, R. Caudron, Logarithmic divergence of the antiphase boundary width in Cu-Pd (17%). Phys. Rev. Lett. **68**, 3591 (1992)
64. S.P. Venkateswaran, N.T. Nuhfer, M. De Graef, Anti-phase boundaries and magnetic domain structures in Ni_2MnGa-type Heusler alloys. Acta Mater. **55**, 2621 (2007)
65. Y. Murakami, T. Yano, D. Shindo, R. Kainuma, T. Arima, Transmission electron microscopy on magnetic phase transformations in functional materials. Metall. Mater. Trans. A **38**, 815 (2007)
66. Y. Murakami, D. Shindo, K. Kobayashi, K. Oikawa, R. Kainuma, K. Ishida, TEM studies of crystallographic and magnetic microstructures in Ni-based ferromagnetic shape memory alloys. Mater. Sci. Eng. A **438–440**, 1050 (2006)
67. V.V. Volkov, Y. Zhu, Magnetic structure and microstructure of die-upset hard magnets $RE_{13.75}Fe_{80.25}B_6$ ($RE = $ Nd, Pr): A possible origin of high coercivity. J. Appl. Phys. **85**, 3254 (1999)
68. Y. Murakami, T. Yano, R.Y. Umetsu, R. Kainuma, D. Shindo, Suppression of ferromagnetism within antiphase boundaries in $Ni_{50}Mn_{25}Al_{12.5}Ga_{12.5}$, *Scripta Mater.*, in press.
69. H. Ishikawa, R.Y. Umetsu, K. Kobayashi, A. Fujita, R. Kainuma, K. Ishida, Atomic ordering and magnetic properties in $Ni_2Mn(Ga_xAl_{1-x})$ Heusler alloys. Acta Mater. **56**, 4789 (2008)
70. Y. Murakami, D. Shindo, K. Oikawa, R. Kainuma, K. Ishida, Macroscopic pattern formation preceding martensitic transformation in a ferromagnetic shape memory alloy $Ni_{51}Fe_{22}Ga_{27}$. Appl. Phys. Lett. **92**, 102512 (2008)
71. I.M. Robertson, C.M. Wayman, Tweed microstructures: I. Characterization in $\beta -$ NiAl. Phil. Mag. A **48**, 421 (1983)

72. D. Schryvers, L.E. Tanner, On the interpretation of high resolution electron microscopy images of premartensitic microstuctures in the Ni-Al β_2 phase. Ultramicroscopy **32**, 241 (1990)
73. S.M. Shapiro, Y. Noda, Y. Fujii, Y. Yamada, X-ray investigation of the premartensitic phase in $Ni_{46.8}Ti_{50}Fe_{3.2}$. Phys. Rev. B **30**, 4314 (1984)
74. R. Oshima, M. Sugiyama, F.E. Fujita, Tweed structures associated with Fcc-Fct transformations in Fe-Pd alloys. Metall. Trans. A **19**, 803 (1988)
75. Y. Noda, M. Takimoto, T. Nakagawa, Y. Yamada, X-ray study of the premartensitic phenomena in AuCd. Metall. Trans. A **19**, 265 (1988)
76. J.I. Pérez-Landazábal, V. Recarte, V. Sánchez-Alarcos, J.A. Rodríguez-Velamazán, M. Jiménez-Ruiz, P. Link, E. Cesari, Y.I. Chumlyakov, Lattice dynamics and external magnetic-field effects in Ni-Fe-Ga alloys, *Phys. Rev. B* **80**, 144301 (2009)
77. K. Tsuchiya, A. Tsutsumi, H. Nakayama, S. Ishida, M. Umemoto, Displacive phase transformations and magnetic properties in Ni-Mn-Ga ferromagnetic shape memory alloys. J. Phys. IV **112**, 907 (2003)
78. A.A. Bokov, Z.-G. Ye, Recent progress in relaxor ferroelectrics with perovskite structure. J. Mater. Sci. **41**, 31 (2006)
79. A. Saxena, T. Castán, A. Planes, M. Porta, Y. Kishi, T.A. Lograsso, D. Viehland, M. Wuttig, M. De Graef, Origin of magnetic and magnetoelastic tweedlike precursor modulations in ferroic materials. Phys. Rev. Lett. **92**, 197203 (2004)

Chapter 10
Understanding Glassy Phenomena in Materials

David Sherrington[1]

Abstract A basis for understanding and modelling glassy behaviour in martensitic alloys and relaxor ferroelectrics is discussed from the perspective of spin glasses.

10.1 Introduction

There has been much activity in the last few decades in understanding, analysing and applying knowledge of the character and properties of many-body systems with complex behaviour arising cooperatively through the combination of competitive interactions and quenched disorder, even where the individual entities, their interactions and any global constraints are simple. Example areas cover condensed matter physics, hard optimization and computer science, information science, biology and economics. They have been conceptually and technically studied and related through statistical physics, which has itself undergone major stimulation and development in the process [1].

Much progress has been made, both experimentally and theoretically, within the area of magnetic alloys exemplified by spin glasses and simple models devised

[1] Caveat: The author is not a materials scientist, but a theoretical statistical physicist concerned with modelling and understanding complex cooperative behaviour in disordered and frustrated many-body systems in idealized contexts in a number of application areas. He makes no claim to expertise in the literature of the materials systems discussed in this article, but hopes that his complementary perspective can be stimulating.

D. Sherrington (✉)
Santa Fe Institute, 1399 Hyde Park Road, Santa Fe, NM 87501, USA

Rudolf Peierls Centre for Theoretical Physics, University of Oxford, 1 Keble Rd., Oxford OX1 3NP, UK
e-mail: d.sherrington1@physics.ox.ac.uk

T. Kakeshita et al. (eds.), *Disorder and Strain-Induced Complexity in Functional Materials*, Springer Series in Materials Science 148,
DOI 10.1007/978-3-642-20943-7_10,

to capture their essence [2, 3]. The mathematical techniques and concepts that have been developed in spin glass theory have led to several valuable applications in the other areas outside of conventional condensed matter physics mentioned above [4–6], as well as in probability theory [7, 8]. This chapter is concerned with condensed matter, but in systems where the interest is in structural rather than magnetic behaviour. It uses phenomenological arguments to employ knowledge of the magnetic systems and their models to gain insight into, explain and anticipate behaviour in martensitic alloys and relaxor ferroelectrics, as well as to consider how the structurally deformable systems can provide "laboratories" to examine novel issues less accessible to real magnetic systems and suggest new problems for study in statistical physics.[2]

10.2 Spin Glasses: A Brief Review

Experimental spin glasses [9, 10] are alloys of magnetic and non-magnetic ions, exhibiting frozen magnetic behaviour without periodic order, preparation dependence, rejuvenation[3], memory[4] and aging [11], features generically described as "glassy". They can be metallic (e.g. $Au_{1x}Fe_x$) or insulating/semiconducting (e.g. $Eu_xSr_{1x}S$), where x gives the concentration of magnetic atoms and the spin glass features occurring for x less than (system-dependent) critical values. They are commonly of substitutional character, i.e. with periodic lattice structure but random-site occupation, but this is not essential. The characteristic ingredients believed to lead to their unusual cooperative behaviour are competition (or "frustration") between different microscopic spin interactions (some separations favouring ferromagnetic pairing and others anti-ferromagnetic pairing) and spatial (atomic) disorder, quenched on relevant timescales.

Their glassy cooperative behaviour is a consequence of the existence of many metastable macroscopic states without periodic order and with significant barriers to moving from one such state to another, with hierarchical organization and preferences changing as control parameters, such as applied fields, are varied, and with the configurational entropy of the metastable states increasing as the temperature is reduced. These features appear to be ubiquitous, given the ingredients above.

---

[2]The style will be tutorial/expository rather than attempting to give all historical originality credits.

[3]Rejuvenation refers to a situation in which, after a perturbation, a system starts a process anew as though previous events had not occurred. In spin glasses, it is observable in χ'', which decays with time, where a sudden reduction in the temperature after decay at the higher temperature causes it to return quickly to a higher value (closer to the original) and then start to decay again. See [15] and also E. Vincent in [11].

[4]Memory refers to a system storing knowledge of its history. For example, in the previous footnote, a further sudden resumption of the earlier higher temperature makes χ'' jump to the value it had just before the temperature was reduced. See [15] and also Vincent in [11].

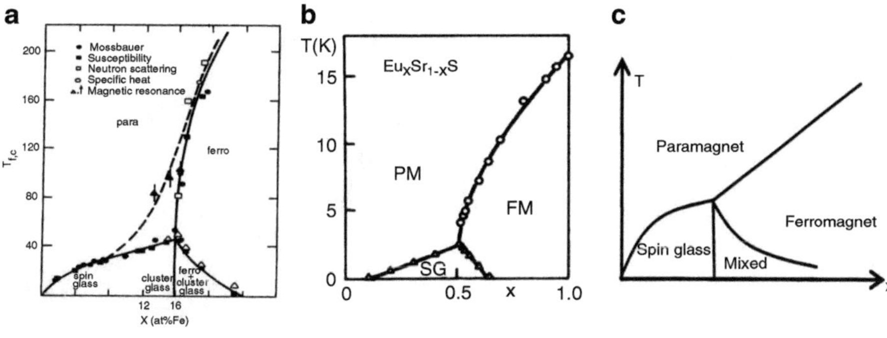

Fig. 10.1 Phase diagrams of (**a**) metallic spin glass system $Au_{1-x}Fe_x$(reprinted with permission from [12]; http://www.informaworld.com), (**b**) semiconducting spin glass system $Eu_xSr_{1-x}S$ (reprinted with permission from [13]; © (1979) American Physical Society; http://link.aps.org/abstract/PRL/v42/p108) and (**c**) mean-field theory for the SK spin glass model [14] with random bonds of mean and variance both scaling as x

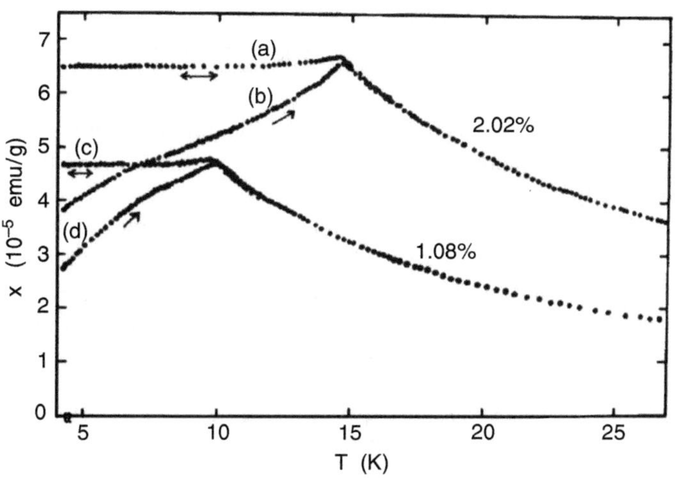

Fig. 10.2 Susceptibilities of $Cu_{1-x}Mn_x$ as measured under field cooling (**a** and **c**) and zero-field cooling (**b** and **d**); reprinted with permission with from [16], © (1979) American Physical Society; http://link.aps.org/abstract/PRB/v19/p1633

Figures 10.1–10.3 illustrate typical properties of spin glass alloys; Fig. 10.1 shows two phase diagrams of temperature against concentration of magnetic atoms, showing that the cooperative order which appears as the temperature is reduced from the paramagnetic state is periodic for large x but spin glass for smaller x; Fig. 10.2 shows results of a typical experiment demonstrating preparation dependence and non-equilibration, and implying the metastability discussed above – it shows the differences in the susceptibility (magnetization/field) measured by applying a field only after cooling (ZFC, zero-field cooled) and that obtained by cooling in the field (FC, field cooled); Fig. 10.3 shows rejuvenation and memory in an experiment in which the out-of-phase susceptibility is measured as a function of time during a protocol in which the temperature is stepped down and up again.

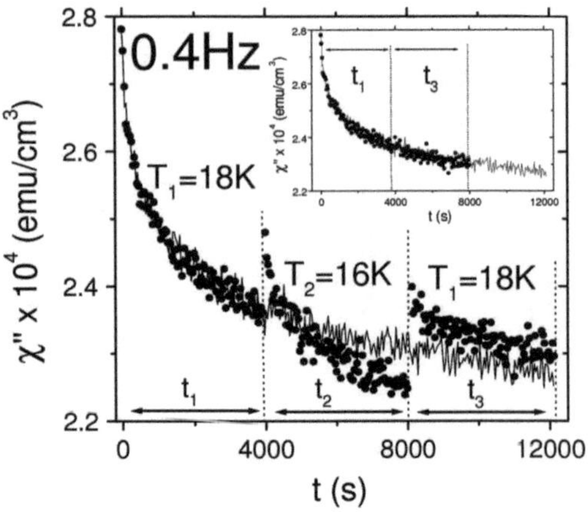

Fig. 10.3 Out-of-phase susceptibility of $Fe_{0.5}Mn_{0.5}TiO_3$ measured during field cycling as indicated, showing rejuvenation and memory; reprinted with permission from [15], © (2001) American Physical Society; http://link.aps.org/abstract/PRB/v64/p174204

As noted, the principal qualitative behaviour of such spin glass systems is rather universal. It is captured by the simple Hamiltonian

$$H = -\sum\nolimits_{\mathrm{magnetic}(ij)} J(\mathbf{R}_i - \mathbf{R}_j)\mathbf{S}_i.\mathbf{S}_j, \qquad (10.1)$$

where the $\mathbf{S}_i$ label the spins and the $\mathbf{R}_i$ their locations, $J(\mathbf{R})$ is the exchange interaction and is frustrated (competitive at different ranges), and the sum is only over the sites occupied by magnetic atoms.

Theoretical and computer simulational studies have played a major role in understanding spin glasses, but have almost exclusively concentrated on random-bond (rather than random-site) quenched disorder, in the belief that the key ingredients are frustration and disorder and since the introduction of the model of Edwards and Anderson (EA) [17] characterized by

$$H = -\sum\nolimits_{\mathrm{all\ sites}} J_{ij}\mathbf{S}_i.\mathbf{S}_j \qquad (10.2)$$

with spins on every site, but the J_{ij} drawn randomly and independently from distributions $P_{\mathrm{sep}}(J)$.[5] The EA model with only nearest-neighbour interactions, uniformly distributed around $J = 0$ and with Ising spins, has received much simulational study, verifying that it has features similar to those of experimental spin glasses (in the spin glass phase[6]) and exposing a multiplicity of chaotically evolving metastable macrostates[7], as well as many further important aspects.

[5] In general, the distribution depends on the separation of the relevant sites and hence on the subscript *sep*.

[6] Note that, in accord with a common practice, we use the expression 'spin glass' to describe both a material exhibiting a spin glass phase and the phase itself.

[7] The rejuvenation and memory effects observed in spin glasses are explainable in terms of the hierarchical yet evolving metastable state structure, with the free energy acquiring more and more nested metastability as the temperature is lowered but melting as it is raised again.

The finite-range EA model is not analytically soluble, but a modification to include interactions of any range with identical distance-independent probability distributions, the Sherrington–Kirkpatrick (SK) [14] model, is soluble, albeit that its solution is very subtle [4–8], and has been proved to exhibit a non-trivially evolving hierarchy of metastable macrostates, non-ergodic and aging dynamics, and the breakdown of the normal fluctuation–dissipation relation (FDR) and its replacement by a modified relation. Extensions of the SK model, together with the conceptual and mathematical tools its examination has engendered, have led to a broad conceptual and technical application [1, 4–6] and demonstrated further subtleties, believed of relevance to understanding conventional structural glasses and several other systems.

Let us now turn to materials systems in which the interesting effects are structural rather than magnetic.

10.3 Martensites

Martensitic materials [18] exhibit structural phase transitions from higher temperature phases of higher symmetry to lower temperature phases of lower symmetry, through first-order transitions. One such example, on which we shall concentrate for illustration, is from high-temperature cubic austenite to a lower temperature phase of alternating planes of complementary tetragonal character, alias twins. We shall consider these systems at a phenomenological level [19].

The macroscopic behaviour of pure martensites is often considered in terms of continuum elasticity theory [18]. We also shall consider martensitic materials, including alloys[8], as being driven by elastic considerations but shall analyse them via pseudo-spin mappings and analogies, employing experience from spin glasses.[9]

Our discussion will be at the level of a type of mean-field/Landau–Ginzburg free-energy theory [20, 21] and will not consider critical fluctuations. We shall mostly treat temperature simply as a means of varying effective parameters in an energy-minimization exercise. The spatial scale of the 'microscopic' variables of our modelling is coarse on the atomic scale but much smaller than macroscopic material scales.

Our starting point is to model phenomenologically the existence locally of transitions from cubic austenite to different tetragonal variants as the temperature T is reduced. For further conceptual simplicity, we shall initially idealize further by considering a two-dimensional analogue in which the transition is from a locally

[8] In fact, our main interest for glassiness is in alloys.

[9] The first recognition that there should be a spin glass analogue in martensitic alloys was by Kartha et al. [20], looking for an explanation of 'tweed', with similarities of ideas to those discussed here, but without the direct mappings and specificity reported in this chapter, which the present author believes provide conceptual and quantitative underpinning.

square structure to two orthogonal rectangular structures.[10] Denoting the local deviatoric strain over coarse-grained regions i by $\phi_i = \varepsilon_i^{xx} - \varepsilon_i^{yy}$, this can be emulated by a local free energy

$$F_L = \sum_i \{a_i\phi_i^2 - b_i\phi_i^4 + c_i\phi_i^6\} \qquad (10.3)$$

with the $\{b, c\}$ all positive (and of qualitatively unimportant variation) but with the $\{a_i\}$ reducing importantly as T is reduced, in such a way that the local minimum changes discontinuously from $\phi_i = 0$ for large T to $\phi_i = \pm\widetilde{\phi}$, where $\widetilde{\phi}$ is finite, as T is reduced through a local value T_i^{L}. This can be further simplified by discretizing (and rescaling) to a description in terms of scalar pseudo-spin variables $S_i = 0, \pm 1$ [19,22], with $S_i = 0$ corresponding to austenite and $S_i = \pm 1$ to the two martensitic variants, and correspondingly considering an effective local "Hamiltonian"

$$H_{\mathrm{L}} = -\sum_i D_i S_i^2, \qquad (10.4)$$

to be minimized. Lowering the temperature T of the real system is emulated by reducing the D. For D_i positive, the local i-minimum is at $S_i = 0$, while for D_i negative, there are equivalent minima at $S_i = \pm 1$.

Next we need to include effective pseudo-spin interactions between different local regions,

$$H_{\mathrm{I}} = -\sum_{(ij)} J(\mathbf{R}_i - \mathbf{R}_j) S_i S_j; \, J(\mathbf{R}) = J_{\mathrm{SR}}(\mathbf{R}) + J_{\mathrm{LR}}(\mathbf{R}). \qquad (10.5)$$

There are two types of contribution to $J(\mathbf{R})$, a short-ranged "ferromagnetic" term $J_{\mathrm{SR}}(\mathbf{R})$ representing the inclination to follow neighbours[11] and an effective long-range interaction $J_{\mathrm{LR}}(\mathbf{R})$ arising through integrating out the non-ordering strains while taking account of the St Venant elasticity compatibility constraints [21, 23]; this scales as $\mathbf{R}^{-d}$ in d dimensions and varies from ferromagnetic to anti-ferromagnetic depending on the angle subtended by $\mathbf{R}_i$ relative to an austenitic cell edge. In $d = 2$, $J_{\mathrm{LR}}(\mathbf{R})$ scales with distance as R^{-2} with a multiplicative angular factor that yields an anti-ferromagnetic interaction at angles $\theta = (2n+1)\pi/4$ and a ferromagnetic interaction at angles $\theta = n\pi/2$, where θ is the angle subtended by $\mathbf{R}$ relative to an austenite cell edge [23], as $\cos(4\theta_{ij})$ where θ_{ij} is the polar angle of $\mathbf{R}_{ij}$.

We now consider the behaviour resulting from minimizing the total $H = H_{\mathrm{L}} + H_{\mathrm{I}}$, with temperature reduction emulated by reducing the $\{D\}$, and using experience of magnetic systems to make deductions about the martensites.

---

[10] We shall briefly discuss extension to three dimensions later, but note at this time that since we are employing mean-field theory considerations of critical dimensions caused by fluctuations are irrelevant.

[11] This is the usual Ginzburg $(\nabla\phi)^2$ term in a spatially continuous formulation.

Let us first consider a pure system in which all the $\{D_i\}$ are the same. In this case, there is a first-order transition as D is lowered, from a state with all $S_i = 0$ to one with $\{S_i = \sigma_i\}$, where the $\{\sigma_i\}$ are the ground state solutions of the Ising Hamiltonian

$$H_\sigma = -\sum\nolimits_{(ij)} J(\mathbf{R}_{ij})\sigma_i\sigma_j; \mathbf{R}_{ij} = \mathbf{R}_i - \mathbf{R}_j; \sigma = \pm 1. \tag{10.6}$$

The transition value of D is positive and given by balancing the energetic increase on taking $S = \pm 1$ in H_L and the corresponding energetic reduction in H_I. In the lower D region, this ground state consists of alternating stripes of $S = +1$ and $S = -1$ at angles $\pi/4$ or $3\pi/4$.[12] These are the twins of pure martensite.

Now let us turn to alloys, emulated by a random distribution of the $\{D_i\}$ over the lattice. Again the criterion of whether any S_i is 0 or ± 1 is given by the balance of H_L and H_I, determined self-consistently across the whole system.

For conceptual orientation, it is useful to consider first a scenario where the D are distributed randomly and independently across the i, at each site taking one of two values; with probability $(1 - x)$ a large D_0, such that sites with this D always have $S_i = 0$; and with probability x a smaller D_1 that can be varied across a phase transition (emulating reduction in temperature). For large enough D_1, the ground state is austenitic (all $\{S_i = 0\}$). Lowering D_1 further, a transition will occur into a phase with $\{S_i = \sigma_i\}$ on the sites having $D_i = D_1$ when there is first a solution of

$$\sum\nolimits_i c_i D_1 - \sum\nolimits_{(ij)} c_i c_j J(\mathbf{R}_{ij})\sigma_i\sigma_j = 0; \sigma = \pm 1, \tag{10.7}$$

where $c_i = 1, 0$ for $D_i = D_1, D_0$. By comparison with (10.1), the second term of (10.7) is recognized as the Hamiltonian of a site-disordered Ising, $\{\sigma = \pm 1\}$, spin glass system with exchange $J(\mathbf{R})$ and whose magnetic sites correspond to those of the original $\{S = 0, \pm 1\}$ system that have $D_i = D_1$ The nature of the ordered phase depends on the character of the ground state of $H_\mathrm{eff} = -\sum_{ij} c_i c_j J(\mathbf{R}_{ij})\sigma_i\sigma_j$ and can be either ferromagnetic or spin glass.

Currently, we have no precise calculations for the ground state energies of H_eff with the specific interaction of (10.2).[13] However, the (T, x) phase diagrams of conventional spin glasses give an indication of what to expect, since the transition temperatures at magnetic concentrations x provide estimates of the corresponding ground state energies. In particular, there is a critical x_c (depending

[12] There has been much interest recently in stripe ordering in systems with a combination of short-range ferromagnetic and long-range power-law-decreasing anti-ferromagnetic interactions and it has been proved that the preferred order is of stripes for $d < p \leq d + 1$, where d is the spatial dimensionality and $(-p)$ is the power of the long-range decay [52]. Stripe widths are determined by the relative strengths of the two types of interaction. Here, $p = d = 2$ and the system is marginal with relevant boundary size L and it has been shown that the average twin stripe width depends on L (as the square root); see [53].

[13] Indeed, even with specified interactions, its evaluation is surely NP-hard [54].

on system details) separating a high x periodically order regime from a lower x spin glass phase, with the ordering temperature (and correspondingly the ground state energy) growing with x in both spin glass and periodic phases, with a discontinuous increase in $\mathrm{d}T_\mathrm{c}/\mathrm{d}x$ (and correspondingly the binding energy per non-zero spin) at x_c. Consequently, for the martensitic alloys we expect transitions as D is reduced; (1) from austenite to twinned martensite (the analogue of the ferromagnet in the magnetic examples shown) for $x < x_\mathrm{c}$, with x_c dependent on details of H_1, (2) from austenite to a pseudo-spin glass frozen amorphous martensitic state for $x > x_\mathrm{c}$, with the critical D for the transition increasing monotonically with x and with a positive discontinuity in $\mathrm{d}D_\mathrm{c}/\mathrm{d}x$ at x_c. The pseudo-spin glass state would be expected to exhibit non-ergodic behaviour analogous to that found in spin glasses, such as differences between FC and ZFC uniaxial compressibilities. This behaviour was observed recently [24, 25] and the pseudo-spin glass state named "strain glass".[14]

In reality, one might expect a more quasi-continuous range of local D-values, particularly allowing for the coarse-graining implicit in our effective site description. Hence, it is reasonable to consider the case in which the D_i are chosen independently from a distribution $P(D)$ of mean D_0 and standard deviation Δ and study the behaviour as D_0 is reduced, emulating reduction in temperature of the real materials. This will lead to different local penalties for $S \neq 0$ at different sites and hence different amounts of bootstrapped interaction energy needed to convert locally to favourable $S = \pm 1$. Specifically, any site i will convert from $S_i = 0$ to $S_i = \pm 1$ at a critical D_i given by $D_i = -\delta H_\mathrm{I}^i(c)$, where $\delta H_\mathrm{I}^i(c) < 0$ is the resultant change in the value of H_I, with already a fraction c of sites converted. The actual sign choice of S_i will depend upon the specific instance of the $\{D\}$ and the states of the other $\{S_j\}$, but the magnitude of $\delta H_\mathrm{I}^i(c)$ is expected to be dominantly self-averaging and again it can be estimated from the (T, c) phase diagram of the corresponding spin glass system[15] or, in its absence, qualitatively from those of known spin glasses. Thus, we expect the transition from austenite as D_0 is reduced to martensitic twins for $\Delta < \Delta_\mathrm{c}$ and to pseudo-spin (strain) glass for $\Delta > \Delta_\mathrm{c}$. Again, this is in accord with observation, noting that Δ is expected to be a monotonically increasing function of the defect concentration in alloys, for example in $Ti_{50-y}Ni_{50+y}$[16] [24], for small y.

Within the lower temperature region, there will be transitions from twinned to strain glass as the disorder concentration is varied. Again spin glasses can be used to guide expectations. Within the (soluble) SK model, this transition is at a constant $x = x_{\mathrm{c}1}$ for all T, but within the ferromagnetic region there are two sub-regions; for $x > x_{\mathrm{c}2}(T)$, with $x_{\mathrm{c}2}(T)$ increasing from $x_{\mathrm{c}1}$ as T is reduced, the ferromagnetic phase is ergodic, but for $x_{\mathrm{c}1} < x < x_{\mathrm{c}2}(T)$ the ferromagnetism is non-ergodic or "mixed" (ferromagnetic-spin glass).

---

[14] See also X Ren's chapter in this book [51].

[15] That is random-site Ising with the same $J(R)$.

[16] Note that y measures the density of defects compared with the pure case $Ti_{50}Ni_{50}$, whereas x earlier was the density of normal (host) atoms.

Fig. 10.4 Qualitative predicted phase diagram

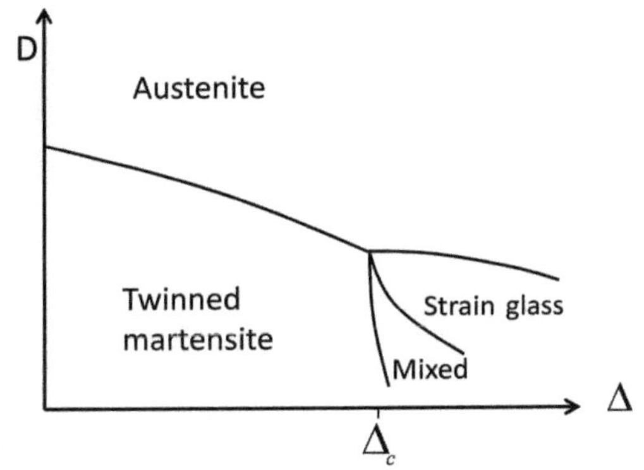

Turning to the martensitic systems with a quasi-continuous distribution of D, it can be noted that as D_0 is reduced more and more sites will pass the threshold of (10.8) and hence become magnetic sites in the effective Ising spin glass Hamiltonian. Correspondingly, the effective concentration of magnetic sites will increase as the temperature of the martensitic system decreases and the boundaries between ordered and strain glass regions will move further and further into the twinned region, yielding re-entrance, so that for Δ just greater than Δ_c, where Δ_c is the critical disorder at which the transition from austenite to lower symmetry occurs, one can anticipate a sequence of phases on lowering the temperature of austenite → strain glass → mixed twins/strain glass phase → ergodic twinned martensite.[17] Some features of an intermediate phase and re-entrance have been seen in experiments, although it is probable that the mixed phase is not truly equilibrium but rather only manifest on finite timescales. Figures 10.4 to 10.6 show the prediction and some experimental observations.[18]

For conceptual simplicity, the description above has been in terms of two-dimensional modelling. It can, however, be extended simply to three dimensions, for example by employing a lattice gas description $n_i = 0, 1$, to indicate whether a site is austenitic or martensitic, denoting the three orthogonal tetragonal variants by Potts "spins" $p_i = 1, 2, 3$, and writing the pseudo-spin Hamiltonian as

$$H = \sum_i D_i(1 - n_i) - \sum_{(ij)} n_i n_j J(\mathbf{R}_{ij})(\delta_{p_i p_j} - 1/3). \qquad (10.8)$$

Again $J(\mathbf{R})$ will have a short-range ferromagnetic part and a long-range part (but now, in three dimensions, going as R^{-3}), again with angular variation from negative

[17] Such "re-entrance" has been called "inverse freezing" and has been receiving much attention in other contexts; e.g. [55].

[18] Note: The transition curves shown in Fig 4.4 between pure twinned and mixed phase and between twinned and strain glass are qualitative, but both are swung towards the twinned state compared with their SK counterparts.

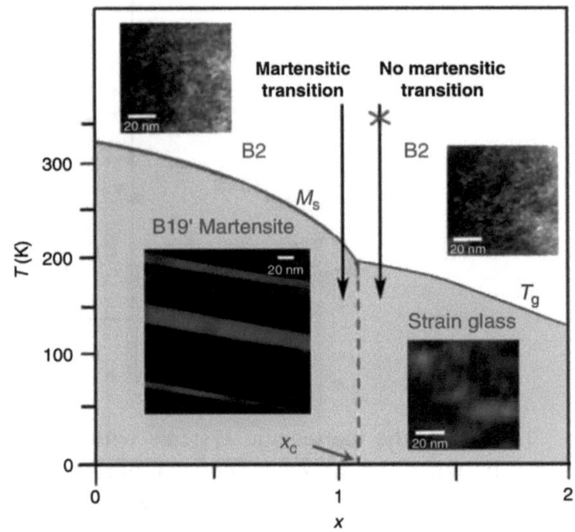

Fig. 10.5 Phase diagram of $Ti_{50-x}Ni_{50rx}$ reprinted with permission of MRS Bulletin from [25], Fig. 10.2

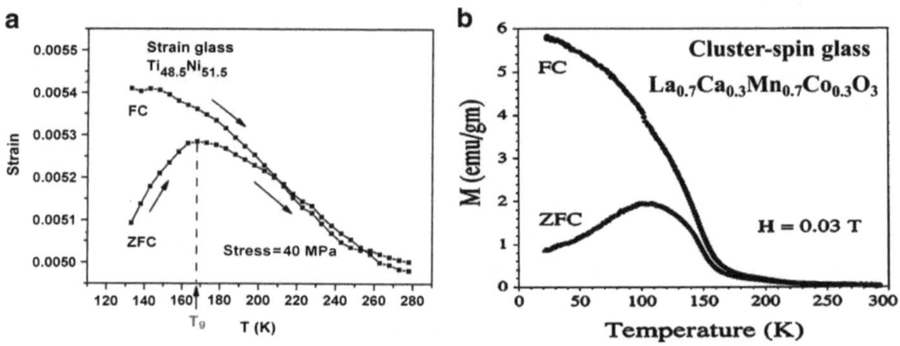

Fig. 10.6 (**a**) Compressibility of strain glass $Ti_{48.5}Ni_{51.5}$ (reprinted with permission from [24]; © (2007) American Physical Society; http://link.aps.org/abstract/PRB/v76/p132201), with, for comparison, (**b**) susceptibility of a cluster spin glass (reprinted with permission from [26]; © (2007) American Physical Society; http://link.aps.org/abstract/PRB/v56/p1345)

to positive and favouring the usual twin planes. The consequences will be similar to those discussed above.[19]

In fact, the original motivation for investigating a possible spin glass analogue in these systems [20] was to explain the possible origin of a different, apparently disordered state of martensitic materials that was observed as a precursor above the transition to the twin structure, exhibiting a mixture of regions of different

---

[19] Within mean field theory, Potts spin glasses of Potts dimension greater than 2 show additional re-entrance from spin glass to ferromagnet as the temperature is reduced [56].

tetragonal and austenite distortions and known as "tweed".[20] The authors recognized this behaviour as similar to the amorphous appearance of spin glasses and speculated that it might be an analogous glass, recognizing also that quenched randomness in the constitutive make-up was a necessary ingredient and assuming a model based on the SK spin glass with effective temperature-dependent bond randomness and an analogue of a spin glass phase between two ordered phases emulating austenite and twinned martensite.[21] In fact, it is now believed that the original tweed phase that is observed as precursor to the twinned martensitic phase is ergodic [27]. What was being anticipated theoretically was strain glass, as the above discussion demonstrates. It seems probable that tweed is actually a non-equilibrium precursor. Bounds for its existence follow qualitatively from the spinodals of the Landau–Ginzburg free energy obtained from the combination of the soft-spin local F_L and a soft-spin extension of H_I (with the discrete-valued S replaced by soft ϕ). It seems probable that a metastable tweed precursor will also occur at temperatures above the strain glass in its region of defect space, but it remains to consider it further theoretically.

One of the characteristic features of martensitic materials is one-way shape memory in which a shape that is imposed in the high-temperature austenitic phase, e.g. by plastic distortion or moulding at the time of preparation, is easily removed (or further distorted) by the application of only weak force in the twinned phase beneath the austenite–martensite transition, but reappears on heating back above that temperature. The usual pictorial description is in terms of (1) any high-temperature imposed macroscopic shape being maintained under cooling through the martensitic temperature while simultaneously the structure distorts microscopically into an equal mixture of twin types, with (2) further distortion in the twinned phase easily achievable by redistribution of weights among the twin types and (3) returning to austenite and the original shape on heating. One might wonder about the relationship of this effect with the memory effects observed in spin glasses. In fact, however, it is readily explainable without glassiness but does require going beyond hard pseudo-spins to deal with the plastic distortions in the martensite phase. The fact that martensitic twins are quite soft to distorting stresses that leave remanent strains demonstrates that the minima in the Landau–Ginzburg free energy are shallow, as also does the superelasticity exhibited above the martensitic phase transition temperature. Nevertheless, it would be interesting to look for analogues of spin glass non-ergodicity, such as the rejuvenation and memory effects shown in Fig. 10.3, which require a hierarchy of chaotically evolving metastable states. Two-way shape memory was demonstrated in a computer simulation [28] and argued to be such a spin-glass-like manifestation.

---

[20] This was also the initial motivation that led to [19].

[21] Reference [19] also assumed that the origin of tweed was quenched disorder but, as above, locally in a system with frustrated but not necessarily disordered exchange. In fact, the consequence is strain glass.

It should be emphasized that temperature has only been introduced implicitly through the variation of parameters in the Ginzburg–Landau free energy, particularly through the variation of the mean of the distribution of effective $\{D\}$. This would need to be noted if one wished to find the solutions of the model above by computer simulations; in particular, if one wished to investigate the ground state energy of $H_L + H_I$ by simulated annealing then one would need to employ another, artificial, annealing temperature T_A and reduce it to zero. A study of real thermal fluctuations would require modelling in terms of a real Hamiltonian, as opposed to this Ginzburg–Landau phenomenological emulation.

10.4 Relaxors

Another set of materials that undergo interesting structural deformation with apparently glass-like non-ergodicity are the so-called ferroelectric relaxors [29–31]. Here, we shall concentrate on systems epitomized by $PbMg_{1/3}Nb_{2/3}O_3$ (usually abbreviated as PMN)[22] and its alloys with $PbTiO_3$ (abbreviated as PT). They exhibit several features similar to spin glasses, including non-ergodicity as the temperature is lowered beneath a transition temperature [32]. Figure. 10.7 gives a comparison of the temperature and frequency dependence of the dielectric permittivity of relaxor PMN[23] and the magnetic susceptibility of spin glass $Pt_{0.975}Mn_{0.025}$, while Fig. 10.8 shows FC and ZFC measures for PMN to be compared with those of a simple spin glass shown in Fig. 10.2. Both of these comparisons suggest similarities and hence the existence of a multiplicity of metastable macrostates in the relaxors.[24]

PT is a member of a class of perovskite ionic crystals (see Fig. 10.9) characterized by the structure ABO_3 in which the A have charge +2, the B have charge +4 and the O have charge −2; typical examples of A are Sr, Ba and Pb and of B are Ti and Ta. Their more detailed structures are determined by the balance of their forces (short-ranged forces related to the sizes of the ions and longer-range Coulomb forces of both signs) and at lower temperatures they exhibit ferroelectric or anti-ferroelectric distortions to lower symmetry, depending on the specific system. In the case of PT, the low-temperature order is ferroelectric. PMN, however, is a substitutional alloy with the Ti ions (of charge +4) replaced by a mixture of 1/3 Mg (of charge +2) and 2/3 Nb (of charge +5), believed to be distributed quasi-randomly but maintaining

[22]Another example is $PbZn_{1/3}Nb_{2/3}O_3$ (abbreviated to PZN).

[23]Relaxors are so-called because of this significant frequency-dependent permittivity peak behaviour.

[24]The feature of frequency dependence of the peak in the real part of the dielectric permittivity or the magnetic susceptibility as a function of temperature decreasing with decreasing frequency is interpretable as reflecting the range of characteristic barrier penetration times, with higher barriers taking longer to surmount.

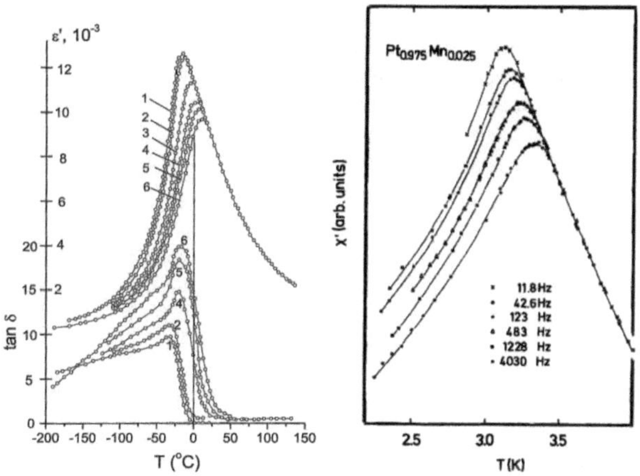

Fig. 10.7 (**a**) *Upper part*: Temperature dependence of real part of the dielectric permittivity of PMN for a range of frequencies, increasing through curves curves 1–6 (reprinted with permission from [33]; © (1961) American Institute of Physics); (**b**) frequency dependence of real part of magnetic susceptibility in $Pt_{0.975}Mn_{0.025}$ (reprinted with permission from [34])

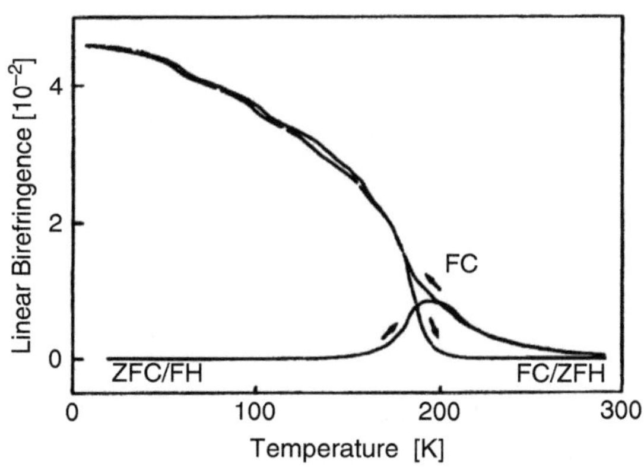

Fig. 10.8 Linear birefringence of PMN in the (001) plane induced by an electric field of 3 kV/cm along [110], under conditions of field cooling (FC) and zero-field cooling/field heating (ZFC/FC) (reprinted with permission from [32]; © (1992) American Physical Society; http://link.aps.org/abstract/PRL/v68/p847)

coarse charge neutrality. $PT_{1-x}PMN_x$ alloys[25] span the range from fully periodic to maximally random.

Let us now turn to modelling. Our philosophy will be to take a bare Hamiltonian H_0 to characterize minimally the displacement properties of pure PT, with a perturbation Hamiltonian H_1, again minimal, characterizing the perturbations caused by alloying with $Mg_{1/3}Nb_{2/3}$ in place of Ti.

In general ABO_3 systems, as the temperature is lowered, there can be displacement from the high-temperature pure perovskite structure of any of the constituent ions. There are competing forces at play with their relative strengths determining the actual low-temperature states, due to the mixture of signs of charges present as well as the normal short-range atomic forces, and this is evidenced by the fact that some ABO_3 are ferroelectric and some anti-ferroelectric. However, (1) our

[25]The usual convention in the field is to write PMN-PT, but we shall use PT-PMN here in keeping with our perspective of disordering a pure matrix.

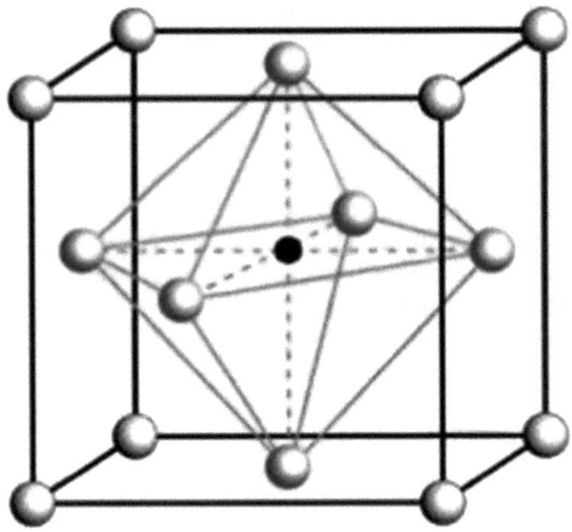

Fig. 10.9 Perovskite structure ABO_3. Here, A sites are at the corners of the cube, a B site is at the centre of the cube and the O sites are at the centre of the faces

principal interest is in the effects of substitutionally disordering the pure material by replacing B ions with a mixture of ions of different charges, (2) in PT and PMN the main displacement is observed to be of the Pb ions, and (3) it is known that PT is ferroelectric in a <111> direction. Hence, as a first approximation, we shall take the displacement structure of the pure low-temperature PT to be modelled simply in terms of Pb displacements via an idealized Hamiltonian

$$H_0 = \sum\nolimits_i (rS_i^2 + uS_i^4) - \sum\nolimits_{(ij)} J(\mathbf{R}_{ij}; \mathbf{S}_i, \mathbf{S}_j) \tag{10.9}$$

where the i label Pb ions, the $\{\mathbf{S}_i\}$ are their displacement vectors, allowed to vary continuously in length and direction, r is positive reflecting the elastic energy to move Pb atoms from their normal (high-temperature/high-symmetry) positions, u is positive[26], limiting displacement, and the $J(\mathbf{R}; \mathbf{S}, \mathbf{S})$ term is an anisotropic exchange term favouring ferroelectric ordering along the <111> directions and large enough to overcome the positive r and drive cooperative ferroelectric ordering when the temperature is reduced. In an obvious magnetic analogy, we shall refer to the $\{\mathbf{S}_i\}$ as (soft) spins.

Within this picture, the perturbation to PT caused by substitutionally alloying $Mg_{1/3}Nb_{2/3}$ in lieu of Ti can be considered by adding to H_0 an extra Hamiltonian contribution corresponding to extra charges -2 at locations occupied by Mg ions and charges $+1$ at locations occupied by Nb ions, with long-range consequences and together with short-range perturbations due to the different ionic radii of Ti, Mg and Nb. Let us concentrate on the effects of the charge perturbations. Since the Pb are charged, the perturbing extra charges on B sites will lead to additional

[26]Note that this is in contrast to the case of the martensitic materials discussed above, and implies a continuous transition, although one could easily modify to a negative u and include a positive sixth-order term to bound the Hamiltonian if one wished to allow the possibility of a first-order transition.

Coulomb forces trying to displace the Pb ions from their PT positions. Ignoring any displacements of the Mg or Nb themselves from the equilibrium Ti positions, these charges lead to a perturbing Hamiltonian $H_1 = \sum_{i\alpha} \mathbf{h}_{i\alpha}.\mathbf{S}_{i\alpha}$, where the $\mathbf{h}_{i\alpha}$ are effective "random fields" experienced at A-sites i due to the extra charges at B-sites α, taking the form $\mathbf{h}_{i\alpha} = -2\mathbf{g}_{i\alpha}$ if there is an Mg ion at α and $\mathbf{h}_{i\alpha} = +\mathbf{g}_{i\alpha}$ if there is a Nb ion at α, where $\mathbf{g}_{i\alpha}$ is a vector that points in the direction from i to α and whose magnitude scales with separation as $(R_{i\alpha})^{-2}$; there is no contribution ($\mathbf{h}_{i\alpha} = 0$) from α-sites where the Ti ions are not substituted.

There has been a lot of interest in the physics of random-field problems, paralleling the interest in random exchange problems typified by EA-like spin glasses [35].[27] For simple ferromagnets with continuous vector spins, it has been shown that uncorrelated random fields destroy the long-range order at dimensions less than four [36] due to the formation of domains. Much of the theoretical interest has, however, been in the random-field Ising model (RFIM), with uniaxially restricted spins, non-negative exchange interactions and uncorrelated random fields, for which the critical dimension for domain formation is 2. For this system, it was predicted [37] that in three dimensions and at intermediate temperatures there would be non-ergodic behaviour analogous to that of spin glasses, although a clear solution is still elusive and this non-ergodic state has been shown recently not to be the true equilibrium solution for a system with only ferromagnetic (or zero) interactions [38, 39].

But note that here (1) the fields are certainly not uncorrelated, both because of the need to maintain approximate local charge neutrality and because of the high correlation of the field directions experienced by pairs of Pb ions on either side of a Mg ion (both towards the Mg) or on either side of a Nb ion (both away from the Nb) and (2) while the "bare" system is ferromagnetic, the full interaction term in the effective spin Hamiltonian (10.9) also has anti-ferromagnetic elements as a function of $\{\mathbf{R}\}$ that could become relevant under inhomogeneous perturbation (as is the case for random-site spin glasses and martensitic systems). Furthermore, we know from conventional glasses that non-equilibrium glassy states can occur easily in practice even when the minimum energy state is crystalline.

Experimental implementation of uncorrelated random-signed fields in conventional magnetic systems has not proven possible, so experimental studies have studied instead random Ising anti-ferromagnets in the presence of uniform fields, utilizing a mapping to corresponding ferromagnets in random fields (directed oppositely on the sites of the two "hidden" anti-ferromagnetic sub-lattices). These have exhibited effects of non-ergodicity [40–42].

Relaxor ferroelectrics, such as PT–PMN, however, would seem to provide naturally effective random fields at random sites on the Ti lattice, subject to mesoscopic charge neutrality but not related to further implicit sub-lattices, on the top of a bare non-disordered effective Hamiltonian that leads to ferromagnetic ordering when all its sites are occupied, but with anti-ferromagnetic interactions too.

[27] Recall that most experimental spin glasses have site disorder but frustration in their exchange.

They therefore have the potential to be very interesting laboratories to study fundamentals of random-field problems. They do have vector 'spins' but also anisotropy that prefers the <111> orientations and can effectively change the spin character from continuous to quasi-discrete as the temperature is lowered. That quasi-discreteness is not, however, simple Ising (with two states) but rather has eight possible equivalent orientations, as also do the directions for the strongest (nearest-neighbour-effected) random fields.[28]

Experimentally, these relaxors have received extensive study with many interesting observations and deductions but without a consensus of understanding or detailed theory. Pure PT exhibits a ferroelectric phase transition at around 700 K, while PMN exhibits more than one "characteristic" temperature, a so-called Burns temperature at around 620 K marking an onset of deviations from a simple extrapolation of higher temperature properties, and two "transitions" [31], one around 420 K but maintaining ergodicity and another around 220 K [32] heralding the onset of non-ergodic behaviour. It has been suggested that these "transitions" indicate respectively, first, a random-field transition, but with the spins still having enough vector freedom to distort continuously transversely, and, second, behaviour analogous to that of RFIM as a consequence of freezing out of the transverse continuous freedom through the anisotropy favouring the <111> directions and hence the onset of spin angular discreteness [31]. Experiment indicates that (small) domains grow as the temperature is lowered in the higher of these regimes (and probably already starting from around 600 K), as might be anticipated from Imry–Ma theory for the regime where anisotropy is less effective in hindering angular deviation.

Some authors have suggested that these systems should be considered as random field. On the other hand, several authors have proposed that the non-ergodic phase of PMN and PMN–PT should be envisaged as a kind of spin glass freezing of the nano-domains through effective random domain exchange. In view of the large strength of the perturbing random fields (arising from their Coulomb origin), it is indeed likely that their effects will often be greater than that of some of the effective exchange terms and require account to be taken of longer range anti-ferromagnetic aspects of $\sum_{(ij)} J(\mathbf{R}_{ij}; \mathbf{S}_i, \mathbf{S}_j)$.[29] We have already noted in the examples of both spin glasses and martensitic alloys that the addition of local randomness to a pure but frustrated "bare" Hamiltonian can lead to behaviour analogous to a random-bond system and there are several other possible examples. Furthermore, there are often different ways to write Hamiltonians for different apparent emphasis and different ways to choose how to separate bare and perturbation parts of a full (even minimal) Hamiltonian. For example, the idealization of the PT–PMN system discussed above can be written to emphasize random spin correlation in the perturbation (restricted for illustration to nearest neighbour) as

---

[28] Effective random fields from further B atoms have intermediate orientations.

[29] Also, in this connection it is probably appropriate to draw attention to recent studies of dipolar glasses with local anisotropic preferred-axis disorder [57–59].

$$H_0 = \sum_i (rS_i^2 + uS_i^4) - \sum_{(ij)} J(\mathbf{R}_{ij}; \mathbf{S}_i, \mathbf{S}_j) - \sum_{(kl)} c_{kl}\, a_{kl}(\mathbf{S}_k - \mathbf{S}_l).(\mathbf{R}_k - \mathbf{R}_l) \tag{10.10}$$

where (kl) are pairs of near-neighbour Pb sites, $c_{kl} = 1$ if the Ti between k and l is replaced, otherwise $c_{kl} = 0$, and the a_{kl} have opposite signs (and different magnitudes) for replacement by Mg and Nb ions.

There are also several other examples of modified ABO_3 in which the randomizing or alloying is different and can be expected to lead to different pseudo-spin pictures; for example, a B can be replaced by a B′ with the same charge so there is no Coulomb contribution to the perturbation H_1, as in $PbZr_xTi_{1-x}O_3$, or even more perturbed by replacing some A by ions of higher charge together with free electrons, such as replacing some Pb by La. Thus, even the minimal model may need to be different in different cases. However, despite these different origins and details one is led to suspect that understanding of any non-ergodicity in these materials is to be found in terms of the multiple metastable state picture.

Finally, in this section, we might note that there are other types of relaxor ferroelectrics. For example, $Sr_{-0.61-x}Ce_xBa_{0.39}O_6$ has been proposed as a uniaxial relaxor realizing the RFIM in a materials analogue [43]. We shall not, however, pursue these further here.

10.5 Models, Simulations and Analysis

Minimalist models, clever computer simulation and often-subtle analytical studies have played important roles in helping understand spin glasses and generalize concepts. Real experimental systems have many controlling parameters and many variables, of different degrees of importance for determining behaviour, making complete modelling and simulation potentially confusing and difficult to perform on large enough systems, and hence the desire to simplify as much as possible while maintaining what are believed to be the most important ingredients, revising such beliefs in the light of comparison of predictions and observations. Simulations on such pared-down models enable larger systems to be studied under conditions that are known and also permit potentially instructive measurements for which no real experiments have yet been devised – an example of such a measurement that proved of great value in studies of spin glasses is of cross-correlations of two systems evolving simultaneously with the same control rules but different instances of stochastic noise.

Analytic solutions are essentially impossible for large disordered and frustrated systems, except for certain typical properties of systems with only infinite-range interactions drawn independently and identically from the same distribution, epitomized by the SK model [14] and its extensions, including local randomness also drawn independently from site-independent distributions. These latter have, however, been very instructive in forming conceptual pictures and devising new procedures.

The model of (10.1) is a simple emulation of a real experimental system, but in fact such site-disordered systems have received little simulation or analysis; these have instead concentrated on models with bond disorder, such as that of (10.2), because they are argued to have the same qualitative character within the spin glass phase. One could take the same perspective in discussing the martensites above if one is principally interested in the properties of the strain glass phase. Thus, one might employ the model Hamiltonian

$$H = \sum_i D_i S_i^2 + \sum_{(ij)} J_{ij} S_i S_j ; S = 0, \pm 1 \tag{10.11}$$

where the D and the J are drawn from simple distributions (such as Gaussian or top hat functions). This model has been studied analytically for the (soluble) infinite-ranged case [44–46], exhibiting the anticipated amorphousness of the strain glass state. A model combining this with a term of the form of (10.6) to emulate the transition from twinned martensite to strain glass has also been simulated [46]. These models, however, effectively put in ingredients guaranteed to yield strain glass. It would be interesting to minimize numerically $H = H_{\mathrm{L}} + H_{\mathrm{I}}$ of (10.4) and (10.5)[30] and show the emergence of the strain glass, but it is difficult to anticipate any result other than that phenomenologically deduced above. Indeed, simulations of the initial elasticity model have clearly demonstrated features analogous to tweed, as well as showing martensitic stripes [20, 23, 28, 47].

These considerations do suggest several other models and experimental systems as being of potential interest.[31] For example, there has been much interest in the cooperative formation of striped phases in many pure systems [48]. The above arguments suggest that dilution of such systems might often lead to transitions to "amorphous" glassy order. This could be the case with various "spin" types and one could consider examples with second-order or first-order transitions, driven from their high-symmetry phases by changes of anisotropy, thermal or quantum fluctuations.

The situation for the random-field model suggested for PT–PMN is more difficult to consider analytically. For example, replacement of the finite-range RFIM problem by one with uniform infinite-range exchange leads to triviality and thus mitigates against a useful simple mean-field solution. Nor has the RFIM received much simulation since this is made difficult by the discreteness of the spins. There is consequently much more uncertainty about glassiness in random-field problems. The observation of non-ergodicity in PMN and PT–PMN suggests, however, that it could be useful to consider minimal model simulation for these systems as random-field ones but appropriately correlated as discussed above. Details of the correct exchange and its anisotropy in (10.9) are not immediately obvious and their realistic inclusion would also complicate simulation, but it could be interesting to simulate

[30] For example, by simulated annealing or extremal optimization [60].

[31] Indeed some have been studied, but no attempt at a survey is made here.

a model with a simplified form, looking for an onset of non-ergodicity as the temperature is reduced and anisotropy frozen out. But this is beyond the scope of this article.

There has been an attempt to emulate the relaxor system with an infinite-range random-bond and random-field model [49], but again this effectively pre-empts the conclusion, while the even-qualitative validity remains undemonstrated.

10.6 Conclusion

In this chapter, it has been argued that experience in spin glasses and random magnets can provide useful perspectives for considering the origins of complex glassy behaviour in materials, although no claim is made of completeness of picture or of full originality. In keeping with the tradition of statistical physics, the approach has been to try to simplify the materials problems to provide minimal models for basic understanding, to consider the implications of such mapping through comparison with known spin glass systems, and to be followed eventually (not here) by extension to greater reality. The glassiness has consequently been identified in terms of the complex metastable state structure of spin glasses and high-disorder random-field systems.

Specifically, it has been argued that martensitic alloys with compositional defects behave like spin glass systems with first-order phase transitions as the temperature is lowered from the high-temperature high-symmetry austenite phase to a lower temperature periodic (twinned) phase for lower levels of defect concentration, or to a non-ergodic spin-glass-like phase for higher levels of defect concentration. Furthermore it has been suggested, from a mapping to a picture of bootstrapped effective Ising[32] spins, that if the quenched disorder has a quasi-continuous anisotropy strength distribution then the transition from twinned to strain glass phase as a function of temperature and defect concentration should be re-entrant. This feature implies that martensitic alloys could provide a useful "laboratory" for readily studying effective concentration variation across a periodic-spin glass transition as a function of temperature of the martensitic material rather than requiring the making of new alloys of different composition.

It has also been argued that PT–PMN relaxor ferroelectrics are most minimally modelled as random-field problems and might provide a useful experimental laboratory to study such model systems without needing to employ gauge mappings from random anti-ferromagnets in uniform fields, with their necessarily uniaxial fields of uniform strength and with the quasi-randomness tied to underlying sublattice structure. On the other hand, it has also been pointed out that these systems have important differences from the usual theoretical models with purely ferromagnetic exchange and uncorrelated random fields. Specifically, even a minimal

[32]Or Potts.

model has strong correlations in the random fields between neighbouring pseudo-spins (via intermediate random charges), while further correlations between those random charges imposed by mesoscopic charge neutrality and competitive effective exchange are also anticipated to be relevant. Another difference in the effective field character of PMN-like relaxors as compared with the effective random-field ferromagnets obtained by gauge transforming disordered anti-ferromagnets in applied fields is that the local random fields arising in relaxors through ionic replacement (such as Mg or Nb for Ti) are necessarily large and not tunable, whereas uniform applied fields can be readily modified in strength.

More generally, alloy-modified ABO_3 perovskite systems can involve also other random exchange effects due to ionic replacements. There is a need for further study to determine minimal features, but a starting point seems now reasonably clear.

In considering martensitic systems above the effect of disorder and defects in alloy constitution has been considered only via a phenomenological modification of the effective local temperature, or correspondingly the $\{a_i\}$ of (10.3) or the $\{D_i\}$ of (10.4). Defects can also lead to anisotropic strains yielding effective random-field terms $\sum_i h_i S_i$ or similar terms odd in $\{\phi_i\}$ [19]. Ren and co-workers have proposed that such random-field terms might be the origin of strain glass behaviour [50,51].[33]

One can certainly conclude that materials science can provide an extremely rich source for many-body systems exhibiting complex macroscopic behaviour in their local displacement correlation behaviour, due to quenched disorder and frustration. One might also note that the temperature range of behaviours of interest in this regard in these structural systems is much higher than those in conventional magnetic systems and the concentration of defects needed to induce strain glass behaviour is much smaller than those for either metallic or semiconducting spin glasses.

Finally, it is suggested that idealized models conceptually stimulated by these materials systems offer several interesting topics for further fundamental theoretical analysis and computer simulation [61].

Acknowledgments The author is grateful to Avadh Saxena and Turab Lookman for introducing him to martensitic shape-memory alloys and for useful discussions on the topic over many years of visits to Los Alamos National Laboratory, whose hospitality he also acknowledges. Also, in connection with martensitic alloys, he has appreciated correspondence with Jim Sethna and Xiaobing Ren. He thanks Roger Cowley, his colleague at Oxford, for introducing him to relaxor ferroelectrics, for informing him of several results and comparisons and for useful discussions on how to model and understand, and also Wolfgang Kleemann for very helpful comments on a draft of this paper and for drawing his attention to other relevant works on relaxors. Finally, he apologises again to the experts in martensites and relaxors whose work he has not acknowledged and indeed much of which he is insufficiently familiar with, but if he waited until he had had an opportunity to read, absorb and understand everything that has been done and written about, this article would not have been completed. Hopefully it will stimulate reactions, even if only of correction and objection.

[33] The present author is inclined to the belief that both effects are likely to be playing a role in martensitic systems, while random fields are dominant for relaxor ferroelectrics.

References

1. D. Sherrington, Physics and complexity. Philos.Trans. R. Soc. A **368**, 1175 (2010)
2. K. Binder, A.P. Young, Spin glasses: Experimental facts, theoretical concepts, and open questions. Rev. Mod. Phys. **58**, 801 (1986)
3. D. Sherrington, Spin glasses: a Perspective, ed by. E. Bolthausen, A. Bovier, *Spin Glasses* (Springer, Berlin, 2006)
4. M. Mézard, G. Parisi, M.A. Virasoro, *Spin Glass Theory and Beyond* (World Scientific, Singapore, 1987)
5. H. Nishimori, *Statistical Physics of Spin Glasses and Neural Networks* (Oxford University Press, Oxford, 2001)
6. M. Mézard, A. Montanari, *Information, Physics and Computation* (Oxford University Press, Oxford, 2009)
7. M. Talagrand, *Spin glasses: a Challenge for Mathematicians* (Springer, Berlin 2003)
8. A. Bovier, *Statistical Physics of Disordered Systems: a Mathematical Perspective* (Cambridge University Press, Cambridge, 2006)
9. J. Mydosh, *Spin Glasses: an Experimental Introduction* (Taylor and Francis, London, 1995)
10. H. Maletta, W. Zinn, Spin glasses, ed by K.A. Gschneider Jr., L. Eyring, *Handbook on the Physics and Chemistry of Rare Earths,* vol. **12**, (Elsevier, North-Holland, 1989), p. 213
11. M. Heimel, M. Pleiming, R. Sanctuary, (eds.), *Ageing and the Glass Transition,* (Springer, Berlin, 2007)
12. B.R. Coles, B. Sarkissian, R.H. Taylor, The role of finite magnetic clusters in Au-Fe alloys near the percolation concentration. Phil. Mag. B **37**, 489 (1978)
13. H. Maletta, P. Convert, Onset of ferromagnetism in EuxSr1-xS near x=0.5. Phys. Rev. Lett. **42**, 108 (1979)
14. D. Sherrington, S. Kirkpatrick, Solvable model of a spin glass. Phys. Rev.Lett. **35**, 1972 (1975)
15. V. Dupuis, E. Vincent, J.-P. Bouchaud, J. Hammann, A. Ito, H. Aruga Katori, Aging, rejuvenation, and memory effects in Ising and Heisenberg spin glasses. Phys. Rev. B **64**, 174204 (2001)
16. S.Nagata, P.H.Keesom, H.R.Harrison, Low-dc-field susceptibility of CuMn spin glass. Phys. Rev. B**19**, 1633 (1979)
17. S.F. Edwards, P.W. Anderson, Theory of spin glasses. J. Phys. F **5**, 965 (1975)
18. K. Bhattacharya, *Microstructure of Martensite* (Oxford University Press, 2003)
19. D. Sherrington: A simple spin glass perspective on martensitic shape-memory alloys, J. Phys. Cond. Mat. **20**, 304213 (2008)
20. S. Kartha, T. Castán, J.A. Krumhansl, J.P. Sethna, Spin-glass nature of tweed precursors in martensitic transformations. Phys. Rev. Lett. **67**, 3630 (1991)
21. S. Kartha, J.A. Krumhansl, J.P. Sethna, L.K. Wickham, Disorder-driven pretransitional tweed pattern in martensitic transformations. Phys. Rev. B**52**, 803 (1995)
22. S.R. Shenoy, T. Lookman, Strain pseudospins with power-law interactions: Glassy textures of a cooled coupled-map lattice. Phys. Rev. B **78**, 144103 (2008)
23. T. Lookman, S.R. Shenoy, K.Ø. Rasmussen, A. Saxena, A.R. Bishop, Ferroelastic dynamics and strain compatibility. Phys. Rev. B **67**, 024114 (2003)
24. S. Sarkar, X. Ren, K. Otsuka, Evidence for strain glass in the ferroelastic-martensitic system $Ti_{50-x}Ni_{50+x}$. Phys. Rev. Lett. **95**, 205702 (2005)
25. Y. Wang, X. Ren, K. Otsuka, A. Saxena, Evidence for broken ergodicity in strain glass. Phys. Rev. B **76**, 132201 (2007)
26. N. Gayathri, A.K. Raychaudhuri, S.K. Tiwary, R. Gundakaram, A. Arulraj, C.N.R. Rao, Electrical transport, magnetism, and magnetoresistance in ferromagnetic oxides with mixed exchange interactions: A study of the $La_{0.7}Ca_{0.3}Mn_{1-x}Co_xO_3$ system. Phys. Rev. B **56**, 1345 (1997)
27. X. Ren, Y. Wang, Y. Zhou, Z. Zhang, D. Wang, G. Fan, K. Otsuka, T. Suzuki, Y. Ji, J. Zhang, Y. Tian, S. Hoi, X. Ding, Strain glass in ferroelastic systems: Premartensitic tweed versus strain glass. Phil. Mag. **90**, 141 (2010)

28. J.P. Sethna, C.R. Myers, Martensitic tweed and the two-way shape-memory effect. arXiv:cond-mat/970203 (1997)
29. L.E. Cross, Relaxor ferroelectrics. Ferroelectrics **76**, 241 (1987)
30. W. Kleemann, Random fields in dipole glasses and relaxors. J. Non-Cryst. Solids **307–310**, 66 (2002); The relaxor enigma – charge disorder and random fields in ferroelectrics, J. Mater. Sci. **41**, 129 (2006)
31. R.A. Cowley, S.N. Gvasaliya, S.G. Lushnikov, B. Roessli, G.M. Rotaru, Relaxing with relaxors: a review of relaxor ferroelectrics, Advances in Physics **60**(2), 229 (2011)
32. V. Westphal, W. Kleemann, M.D. Glinchuk, Diffuse phase transitions and random-field-induced domain states of the "relaxor" ferroelectric $PbMg_{1/3}Nb_{2/3}O_3$. Phys. Rev. Lett. **68**, 847 (1992)
33. G.A. Smolenskii, V.A. Isupov, A.I. Agranoyskaya, S.N. Popov, Ferroelectrics with diffuse phase transitions. Sov. Phys. Solid State **2**, 2584 (1961)
34. G.V. Lecomte, H. von Löhneysen, E.F. Wassermann, Frequency dependent magnetic susceptibility and spin glass freezing in PtMn alloys. Z. Phys. B **50**, 239 (1983)
35. A.P. Young (ed.), *Spin Glasses and Random Fields* (World Scientific, Singapore, 1998)
36. Y. Imry, S-K. Ma, Random-field instability of the ordered state of continuous symmetry. Phys. Rev. Lett. **35**, 13909 (1975)
37. M. Mzard, R. Monasson, Glassy transition in the three-dimensional Ising model. Phys. Rev. B **50**, 7199 (1994)
38. F. Krzakala, F. Ricci-Tersenghi, L. Zdeborová, Elusive glassy phase in the random field Ising model. Phys. Rev. Lett. **104**, 207208 (2010)
39. F. Krzakala, F. Ricci-Tersenghi, D. Sherrington, L. Zdeborová, No spin glass phase in ferromagnetic random-field random-temperature scalar Ginzburg-Landau model. J. Phys. A: Math. Theor. **44**, 042003 (2011)
40. H. Yoshizawa, R. Cowley, G. Shirane, R.J. Birgenau, Neutron scattering study of the effect of a random field on the three-dimensional dilute Ising antiferromagnet $Fe_{0.6}Zn_{0.4}F_2$. Phys. Rev. B **31**, 4548 (1985)
41. P. Pollak, W. Kleemann, D.P. Belanger, Metastability of the uniform magnetization in three-dimensional random-field Ising model systems. II $Fe_{0.47}Zn_{0.53}F_2$. Phys. Rev. B **38**, 4773 (1988)
42. F.C. Montenegro, A.R. King, V. Jaccarino, S-J. Han, D.P. Belanger, Random-field-induced spin-glass behavior in the diluted Ising antiferromagnet $Fe_{0.31}Zn_{0.69}F_2$. Phys. Rev. B **44**, 2155 (1991)
43. W. Kleemann, J. Dec, P. Lehnen, R. Blinc, B. Zalar, P. Pankrath, Uniaxial relaxor ferroelectrics: The ferroic random-field Ising model materialized at last. Europhys. Lett. **57**, 14 (2002)
44. S.K. Ghatak, D. Sherrington, Crystal field effects in a general S Ising spin glass. J. Phys. C **10**, 3149 (1977)
45. A. Crisanti, L. Leuzzi, Thermodynamic properties of a full-replica-symmetry-breaking Ising spin glass on lattice gas: The random Blume-Emery-Griffiths-Capel model. Phys. Rev. B **70**, 014409 (2004)
46. R. Vasseur, T. Lookman, Effects of disorder in ferroelastics: A spin glass model for strain glass, Phys. Rev. B **81**, 094107 (2010)
47. S. Shenoy, T. Lookman, A. Saxena, A.R. Bishop, Martensitic textures: Multiscale consequences of elastic compatibility. Phys. Rev. B **60**, R12537 (1999)
48. See for example [17] and references therein; also M. Vojta, Lattice symmetry breaking in cuprate superconductors: stripes, nematics and superconductivity. Adv. Phys. **58**, 699 (2009)
49. R. Blinc, R. Pirc, Spherical random-bond–random-field model of relaxor ferroelectrics. Phys. Rev. Lett. **83**, 424 (1999)
50. D. Wang, Y. Wang, Z. Zhang, X. Ren, Modeling abnormal strain states in ferroelastic systems; the role of point defects. Phys. Rev. Lett. **105**, 205702 (2010)
51. X. Ren, *Strain Glass and Strain Glass* Transition, this book, chapter 11.
52. A. Guiliani, J.L. Lebowitz, E.H. Lieb, Ising models with long-range antiferromagnetic and short-range ferromagnetic interactions. Phys. Rev. B **74**, 064420 (2006)

53. M. Porta, T. Castán, P. Lloveras, T. Lookman, A. Saxena, S.R. Shenoy, Interfaces in ferroelastics: Fringing fields, microstructure, and size and shape effects. Phys. Rev. B**79**, 214117 (2009)
54. M.R. Garey, D.S. Johnson, *Computers and Intractability: a Guide to the Theory of NP-Completeness*, (W.H.Freeman, New York, 1979)
55. N. Schupper, N.M. Scherb, Inverse melting and inverse freezing: a spin model. Phys. Rev. E **72**, 046107 (2005)
56. D. Elderfield, D. Sherrington, The curious case of the Potts spin glass. J. Phys. C **16**, 4865 (1983)
57. S. Bedanta, W. Kleemann, Supermagnetism. J. Phys. D **42**, 013001 (2009)
58. J.F. Fernández, Equilibrium spin-glass transition of magnetic dipoles with random anisotropy axes. Phys. Rev. B **78**, 064404 (2008)
59. J.F. Fernandez, J.J. Alonso, Equilibrium spin-glass transition of magnetic dipoles with random anisotropy axes on a site diluted lattice. Phys. Rev. B **79**, 214424 (2009)
60. S. Boettcher, A.G. Percus, Optimization with extremal dynamics. Phys. Rev. Lett. **86**, 5211 (2001)
61. X. Ren, Y. Wang, K. Otsuka, P. Loveras, T. Castán, M. Porta, A. Planes, A. Saxena, Ferroelastic nanostructures and nanoscale transitions: ferroics with point defects. MRS Bull. **34**, 838 (2009)

Chapter 11
Strain Glass and Strain Glass Transition

Xiaobing Ren

Abstract Strain glass is a frozen disordered ferroelastic state with short-range strain order only. It is a conjugate state to the long-range ordered ferroelastic state or martensite. In this chapter, recent progress in strain glass and strain glass transition is reviewed. It is shown that a strain glass bears all the features of a glass, being parallel to other types of glasses such as relaxor ferroelectrics and cluster-spin glasses. Novel properties of strain glass are demonstrated. The origin of strain glass is discussed in terms of its relation to point defects. Finally, it is shown that the insight gained from strain glass may be able to solve a number of long-standing puzzles in ferroelastic community.

11.1 Disorder–Order and Disorder–Glass Transition in Nature: Anticipation of a Strain Glass Transition and Strain Glass

All kinds of matter tend to take a more ordered form at low temperature to reduce entropy, as required by the third law of thermodynamics [1]. This thermodynamic requirement is the origin of a great variety of disorder-to-order transitions observed in nature. The most familiar example is the liquid-to-crystal transition, which is an ordering of atomic configuration. Ordering of other physical quantities is also well known, such as the ordering of magnetic moment, electric dipole, or lattice strain. The corresponding disorder–order transitions are ferromagnetic transition, ferroelectric transition, and ferroelastic/martensitic transition, respectively (Fig. 11.1, the left arrow). These transitions play a central role in structural and functional materials and are also an important subject in materials science and in physics [3].

X. Ren (✉)
Ferroic Physics Group, National Institute for Materials Science, Tsukuba, Japan
e-mail: REN.Xiaobing@nims.go.jp

T. Kakeshita et al. (eds.), *Disorder and Strain-Induced Complexity in Functional Materials*, Springer Series in Materials Science 148,
DOI 10.1007/978-3-642-20943-7_11, © Springer-Verlag Berlin Heidelberg 2012

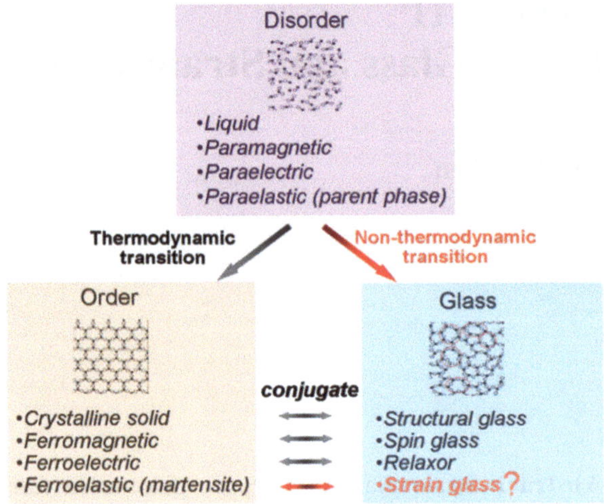

Fig. 11.1 Two classes of phase transitions in nature: disorder–order and disorder–glass (frozen disorder) [2]

Contrary to the above disorder–order transitions, which are driven by a thermodynamic requirement to reduce entropy, there exists another large class of transitions – the "glass transitions", where a disordered state is frozen into a statically disordered phase with local order only [4]. As the disordered glass state is not a favorable low-temperature state (i.e., low entropy state) from a thermodynamic consideration, disorder–glass transition is not a thermodynamic transition and it cannot be understood from thermodynamic principles.

Disorder–glass transitions are conjugate transitions of their corresponding disorder–order transitions, as shown in Fig. 11.1 (the right arrow). They are often formed by doping point defects into a system showing normal disorder–order transition. The most familiar disorder–glass transition is the structural glass transition, which is the conjugate transition of liquid to crystal transition. Structural glass transition can be formed by doping sufficient amount of point defects (or dopants) into a pure system that has a normal liquid-to-crystal transition. A daily life example is that dissolving gelatin into water can suppress the crystallization transition (ice formation) and instead the gelatin water transforms into a structural glass (the jelly!). Similarly, ferromagnetic transition has a conjugate glass transition – the cluster-spin glass transition; the latter can be formed by doping nonmagnetic defects (e.g., Zn) into a ferromagnetic system (e.g., $CoFe_2O_4$) [5], and the resultant cluster-spin glass is a frozen disordered arrangement of magnetic dipole clusters. Ferroelectric transition has a conjugate glass transition – the relaxor transition; the latter can be formed by doping point defects (e.g., La^{3+}) into a normal ferroelectric system (e.g., $BaTiO_3$), and the resultant relaxor (e.g., La–$BaTiO_3$) is a frozen disordered arrangement of electric dipoles [6].

By the same reasoning, it would be reasonable to expect a "strain glass transition", which is the conjugate glass transition of a ferroelastic/martensitic transition. Such a strain glass transition should be the transition from a dynamically disordered lattice strain state (the paraelastic or the parent phase) into a frozen disordered strain state (strain glass). However, such a possibility has remained obscure for a long time until very recently.

The first insightful suggestion of a possible strain glass was made in 1990s by two theoretical studies of Kartha et al. [7, 8] and Semenovskaya and Khachaturyan [9]. These studies suggested that the premartensitic tweed is a strain glass (they used the term "spin glass"). Premartensitic tweed, a cross-hatched microstructure, is known to appear prior to martensitic transformation in various systems, but its nature has remained unclear [10, 11]. A recent experiment [2] does not confirm the existence of a glass transition or frozen strain state in the tweed temperature regime.

The first direct evidence for strain glass and strain glass transition was found quite recently [12], in a well-studied system, the binary $Ti_{50-x}Ni_{50+x}$ system in the Ni-rich composition regime (Fig. 11.2). The near-stoichiometric $Ti_{50-x}Ni_{50+x}$ alloys have been known as the most important shape-memory alloys due to their martensitic transformation (long-range strain ordering) (left side of Fig. 11.2); however, with excess Ni doping above a critical concentration $x > x_c$, the system seems to exhibit no martensite but nanosized strain domains (right side of Fig. 11.2) down to 0 K. It is found that these "nanodomain" compositions exhibit clear features of a glass transition [13–17], which are detailed in the next section.

Now an increasing number of strain glass systems have been found and the number is increasing rapidly [18–25]. Strain glass seems to exist in most (if not all) ferroelastic systems with sufficient defect doping; recently, strain glass has been discovered even in a few ceramic ferroelastic systems [24, 25], suggesting that it is a very general phenomenon in ferroelastic systems.

In the following sections, we are going to show a generic phase diagram of strain glass and its relation with defect doping, key signatures of strain glass, some unique

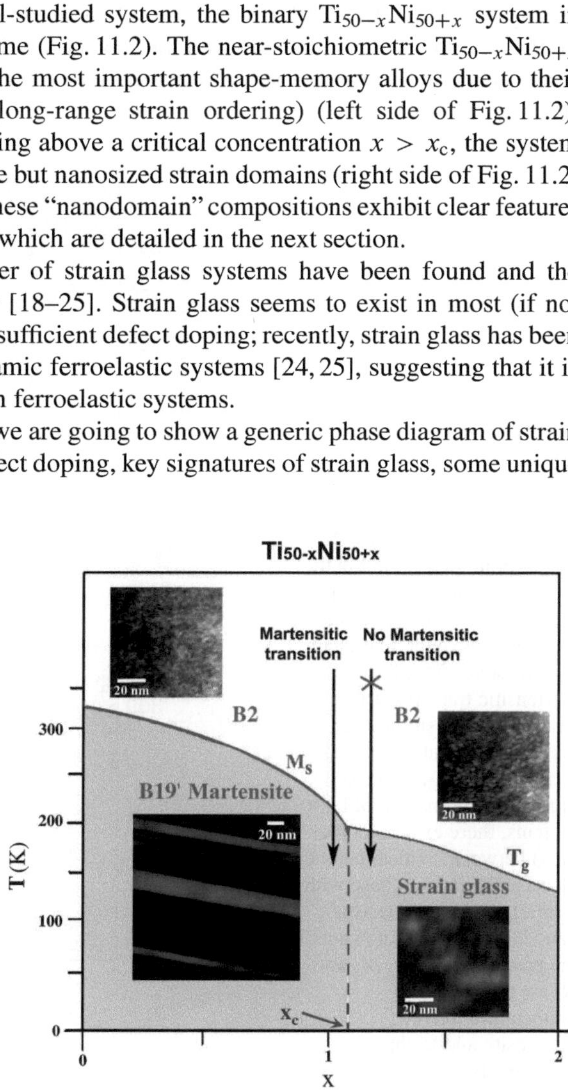

Fig. 11.2 Comparison between martensitic transformation ($x < x_c$) with strain glass transition ($x > x_c$) in $Ti_{50-x}Ni_{50+x}$ system [12]

properties of strain glass, and followed by a discussion about the origin of the strain glass and its models. Finally, we shall use the insight gained from strain glass to explain a few long-standing puzzles in ferroelastic/martensitic systems.

11.2 Phase Diagram of Strain Glass: Crossover from LRO to Glass Due to Point Defects

Figure 11.3a, b shows the new phase diagram of binary $Ti_{50-x}Ni_{50+x}$ [18] and ternary $Ti_{50}Ni_{50-x}Fe_x$ [19], respectively. Compared with the earlier phase diagrams of these two systems that showed only a strain-disordered phase (i.e., parent phase B2) and long-range strain-ordered phases (i.e., martensite phase B19′ and

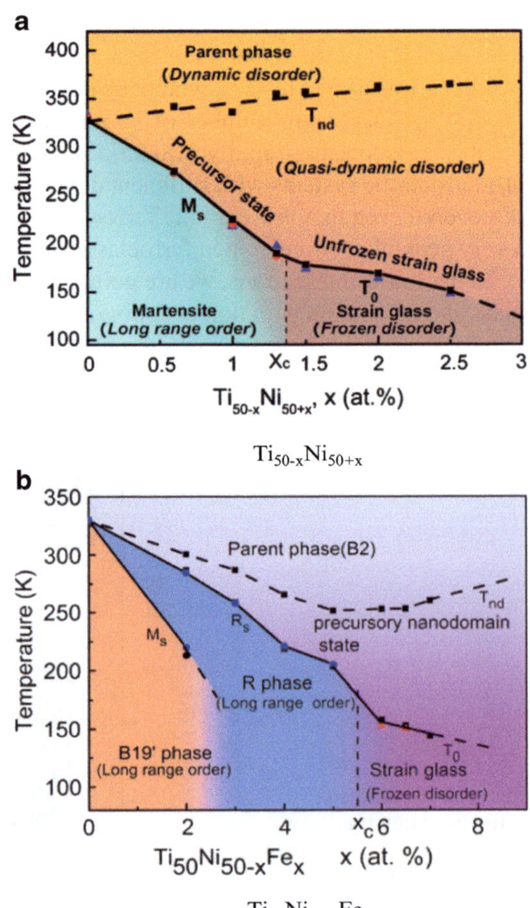

Fig. 11.3 Phase diagram of (**a**) binary $Ti_{50-x}Ni_{50+x}$ system [18] and (**b**) ternary $Ti_{50}Ni_{50-x}Fe_x$ system [19] In the binary system, excess Ni acts as point defect, and in the ternary system the dopant Fe acts as point defect. In both systems, it is found that there is a crossover from a martensitic transition to a strain glass transition when defect concentration exceeds a critical value x_c. On the high-temperature side of both systems, there exists a border (T_{nd}) between an ideal parent phase and a parent phase with a small portion of quasistatic nanodomains (also known as the precursor phase). M_s and R_s are the transition temperature to B19′ martensite and R phase, respectively

R), the new phase diagrams show two new phases characterized by short-range strain order. One is a so-called precursor phase, being essentially a dynamic strain-disordered B2 but embedded with some quasistatic strain nanodomains. Another is a strain glass phase, characterized by frozen disordered strains with nanosized strain domains. The precursor phase is a high-temperature phase above both a martensite phase (LRO) and a strain glass phase (frozen SRO).

Figure 11.3a, b shows an important common feature in the two different systems. There exists a critical point defect concentration x_c, at which a crossover from a long-range strain ordering (i.e., martensitic transformation) to a strain glass transition (i.e., a freezing or SRO) occurs. Here, point defect refers to excess Ni in $Ti_{50-x}Ni_{50+x}$ and dopant Fe in $Ti_{50}Ni_{50-x}Fe_x$. Figure 11.4 shows the distinction of various physical properties in $Ti_{50-x}Ni_{50+x}$ for $x < x_c \sim 1.3$ and $x > x_c$, respectively. Below the crossover limit ($x < 1.3$), the system undergoes a normal martensitic transformation, as characterized by a DSC peak, a hysteresis in resistivity curve, a frequency-independent elastic modulus dip, and mechanical loss (but the dip/peak height is frequency dependent due to a transient effect [26]). Above the crossover limit ($x > 1.3$), a very different transition behavior is found. It is characterized by a nearly vanishing DSC peak and the absence of transition hysteresis in the resistivity curve, as if there were no transition at all; however, a dip in the elastic modulus curve and a corresponding peak in the mechanical loss curve clearly suggest the existence of a transition. The dip/peak in the AC mechanical properties shows frequency dependence following Vogel–Fulcher relation, which is a key signature of a strain glass transition [12].

Similar to the two phase diagrams of $Ti_{50-x}Ni_{50+x}$ and $Ti_{50}Ni_{50-x}Fe_x$ shown in Fig. 11.3, other strain glass systems such as $Ti_{50}(Ni_{50-x}D_x)(D = Co, Cr, Mn)$ [20] and $Ti_{50}(Pd_{50-x}Cr_x)$ [21] also share a common feature: a crossover from a normal martensitic transformation to a strain glass transition at a critical defect doping level x_c. In these systems, the border separating an ideal parent phase and a precursor phase is yet to be determined.

In view of the commonality of defect-induced martensite (LRO) to strain glass (frozen SRO) crossover, it is possible to establish a generic phase diagram that captures the general features of all different systems. Such a phase diagram is shown in Fig. 11.5, which shows the relationship among a "strain liquid" (the ideal or normal parent phase), a "strain crystal" (martensite), a precursor or premartensitic phase (a sticky or less sticky strain liquid), and a strain glass (frozen strain liquid). The physical pictures of these states are as follows. At very high temperature ($T > T_{nd}$, refer to Fig. 11.3 for definition of T_{nd}), the system is in a strain liquid state or ideal parent phase state, where only dynamic distortion of the lattice is present. This is a completely disordered strain state. With the temperature decreasing to $T < T_{nd}$, the strain liquid gains some "stickiness" due to the appearance of some long life or quasistatic strain domains, as a result of defect doping (to be discussed later). This is the so-called precursor or premartensitic state. With the temperature decreasing further, this stickier strain liquid becomes more and more sticky. Depending on the concentration of defects, this liquid has two different destinies. For low-level doped strain liquid ($x < x_c$), the liquid cannot become

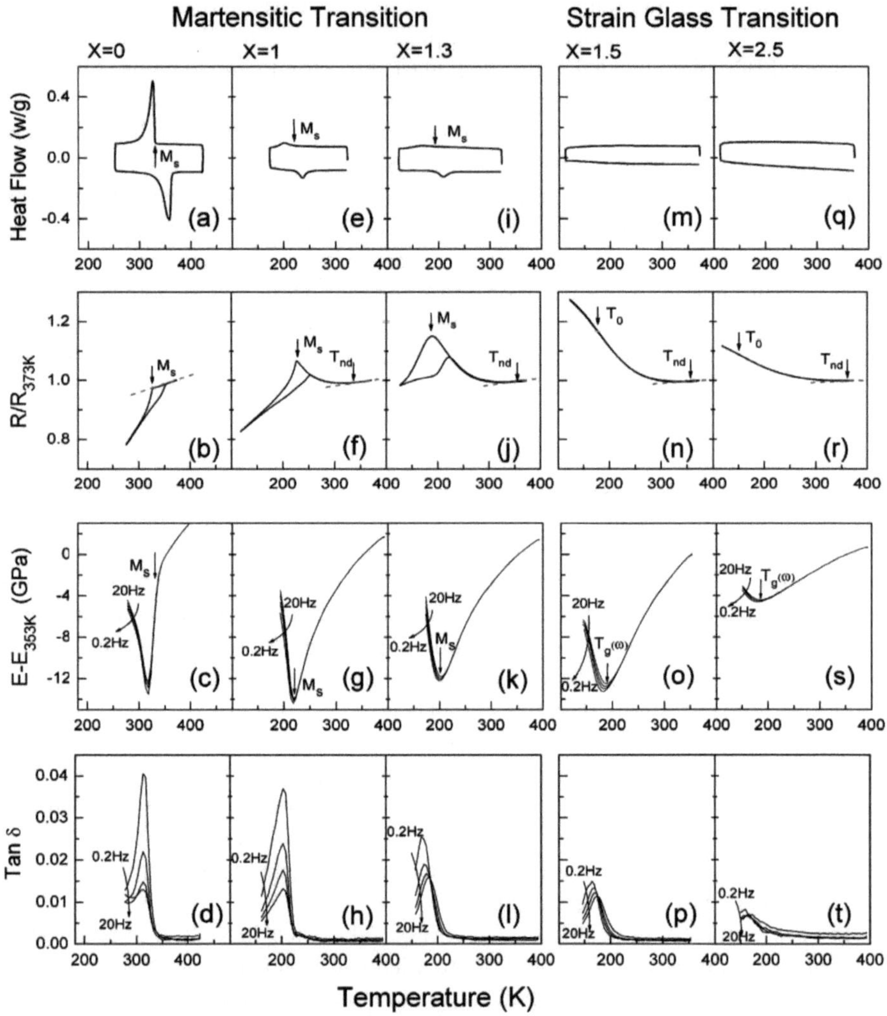

Fig. 11.4 Transition behavior of $Ti_{50-x}Ni_{50+x}$ as a function of point-defect concentration x [18]. There is a critical defect concentration $x_c \sim 1.3$, below which a martensitic transformation occurs and above which strain glass transition occurs

very sticky, and the less sticky liquid eventually transforms into a strain crystal – martensite. This is much the same as doping a small amount of gelatin into water will not stop the liquid from freezing into ice. At high-level doping ($x > x_c$), the liquid becomes very sticky so that the long-range ordering transition (martensitic transition) is suppressed and instead the liquid is frozen into a strain glass (or frozen strain liquid) at T_g. It is just like that doping sufficient amount of gelatin into water will make the solution very sticky and the solution eventually freezes into a jelly – a structural glass.

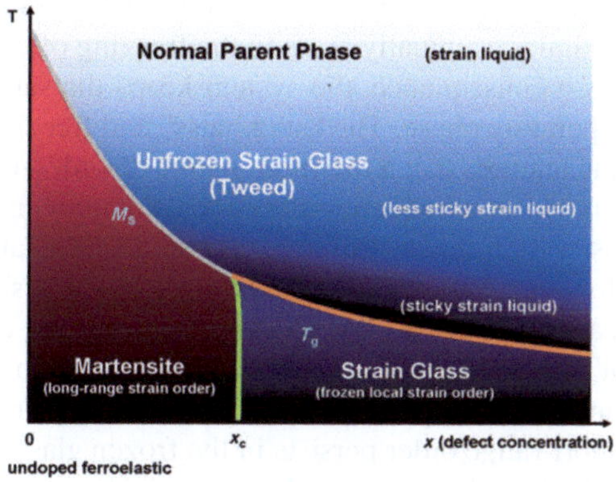

Fig. 11.5 A generic temperature vs. defect-concentration phase diagram for a defect-containing ferroelastic system. Relationship among four different strain states (normal parent phase martensite, tweed, and strain glass) is shown. A crossover from a martensitic transformation to a strain glass transition occurs when defect concentration x exceeds a critical value x_c

According to the above picture, an ideally pure system cannot develop into a sticky strain liquid state or precursor state, just like pure water cannot become sticky with cooling. Indirect evidence seems to support this scenario, as can be seen in Fig. 11.4b, where no anomaly in resistivity can be seen for a stoichiometric $Ti_{50-x}Ni_{50+x}(x = 0)$ (Fig. 11.4b), in contrast to the increasingly clear anomaly (an increase in resistivity prior to the martensitic or strain glass transition) with increasing x (excess Ni) (Fig. 11.4b, f, j, n, r). A more detailed discussion of the origin of precursor phase and the resistivity anomaly is given in Sects. 6.1 and 6.2.

Therefore, Fig. 11.5 provides a unified physical picture for all the four strain states observed in defect-containing ferroelastic systems and also for the crossover from martensitic transformation to strain glass transition.

11.3 Signatures of Strain Glass and Analogy with Other Glasses

A conventional disorder-to-order transition, such as liquid-to-crystal transition, ferromagnetic transition, or martensitic transformation, is a thermodynamic transition; it is characterized by the existence of a clear transition temperature, at which a sudden change in long-range order (atomic arrangement, spin, or lattice strain, etc.) occurs, and is accompanied by a large thermal effect. Therefore, such thermodynamic transitions can be easily identified by detecting the thermal effect, structural/symmetry change, and large domains of the low-temperature phase.

In contrast, a glass transition of any kind (either a structural glass transition or a spin glass transition) is characterized by a gradual slowing down of dynamics [4], or in simple words, the system becomes more and more "sticky" with cooling. This slowing down of dynamics eventually leads to the freezing of the disordered state – a glass transition. As a consequence, the system keeps the same symmetry as the disordered high-temperature phase. The "stickiness" and freezing of a glass system lead to several unique features that a thermodynamic disorder–order transition does not possess. (1) There exists a frequency-dependent anomaly in a relevant dynamic property (e.g., viscosity for liquid to structural glass transition) at the glass transition temperature T_g, following a Vogel–Fulcher relation. (2) The system shows a history dependence or so-called nonergodicity, as manifested by the deviation in a zero-field-cool/field-cool (ZFC/FC) curve. (3) There is no change in long-range order or equivalently no change in average symmetry or structure during a glass transition. (4) Local order or short-range order persists in the frozen glass state, manifested as nanosized domains. In the following, we shall present the recent evidence showing that the above essential features are satisfied by strain glass.

(1) A frequency-dependent AC modulus/loss anomaly at T_g following Vogel–Fulcher relation

Figure 11.6a shows a frequency-dependent anomaly in the dynamic mechanical properties of a strain glass $Ti_{48.5}Ni_{51.5}$ around the strain glass transition temperature T_g. There exist a dip in AC storage modulus curve and a corresponding peak in the mechanical loss curve, and the peak/dip temperature T_g is frequency-dependent, following the Vogel–Fulcher relation [12,16], $\omega = \omega_0 \exp[-E_a/k_B(T_g - T_0)]$, where T_g is the dip/peak temperature at frequency ω, T_0 is the "ideal" freezing temperature (T_g at 0 Hz), E_a is the activation energy, and k_B is the Boltzmann constant. It is interesting to note that very similar frequency-dependent anomaly at T_g has been observed in other two types of glasses: ferroelectric relaxor (freezing of local electric dipoles) (Fig. 11.6b) [27] and cluster-spin glass (freezing of magnetic moments) (Fig. 11.6c) [28].

(2) History dependence or nonergodicity of strain glass as manifested by the zero-field-cool/field-cool curve (ZFC/FC)

A critical proof for a glass transition is the existence of nonergodicity, or history dependence of the physical properties (like strain) in the glass state. This is usually done with a FC/ZFC experiment, which detects if there is a history dependence of the physical properties. Figure 11.7a shows the FC/ZFC curves of the strain glass $Ti_{48.5}Ni_{51.5}$ [14]. It is clear from the deviating FC and ZFC curves below T_g that the system is ergodic at $T > T_g$ and becomes nonergodic at $T < T_g$ (strictly speaking, small deviation actually begins above T_g, indicating a slight ergodicity-breaking). This proves that this system indeed undergoes a strain glass transition. It is noted that the FC/ZFC behavior of the strain glass shows a striking similarity with that of two other kinds of glass transitions, the relaxor ferroelectric transition (Fig. 11.7b) and the cluster-spin glass transition (Fig. 11.7c). As strain glass, relaxor, and cluster-spin glass are derived from three ferroic transitions (i.e., ferroelastic transition, ferroelectric transition, and ferromagnetic transition, respectively), these three glasses can be generalized into one bigger class of glasses – the ferroic

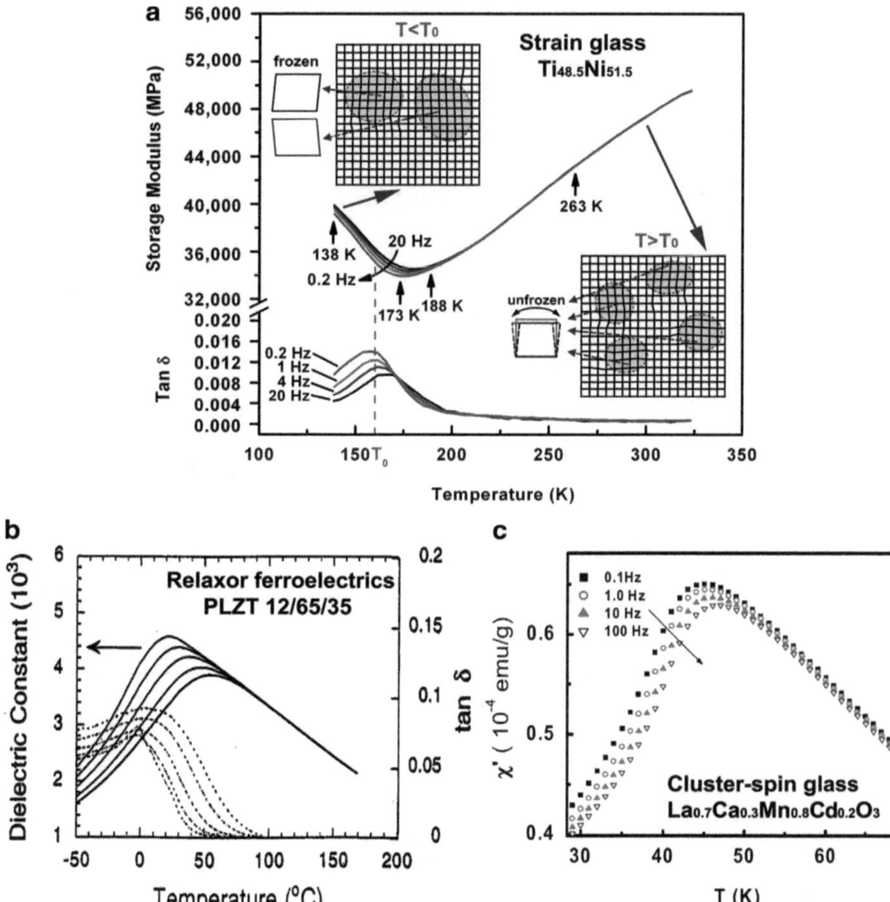

Fig. 11.6 (**a**) Strain glass transition is characterized by a frequency-dependent dip/peak in the AC elastic modulus/loss vs. temperature curve. The *inset* shows schematically the microscopic picture of the strain freezing process. This behavior bears a striking similarity with the frequency-dependent anomaly in (**b**) dielectric permittivity of a ferroelectric relaxor (PLZT 12/65/35 [27]) and (**c**) magnetic susceptibility of a cluster-spin glass ($La_{0.7}Ca_{0.3}Mn_{0.8}Cd_{0.2}O_3$ [28])

glasses [14]. They exhibit very similar behavior in their corresponding DC and AC properties.

(3) Invariance of average structure: no change in average symmetry or structure during a strain glass transition

As discussed above, a glass transition is a freezing of a certain disorder; thus, no change in long-range order or average structure occurs during the transition. Figure 11.8 shows that this is indeed the case. A strain glass alloy $Ti_{48.5}Ni_{51.5}$ exhibits no change in average structure during the whole temperature range encompassing the glass transition temperature $T_g \sim 168$ K.

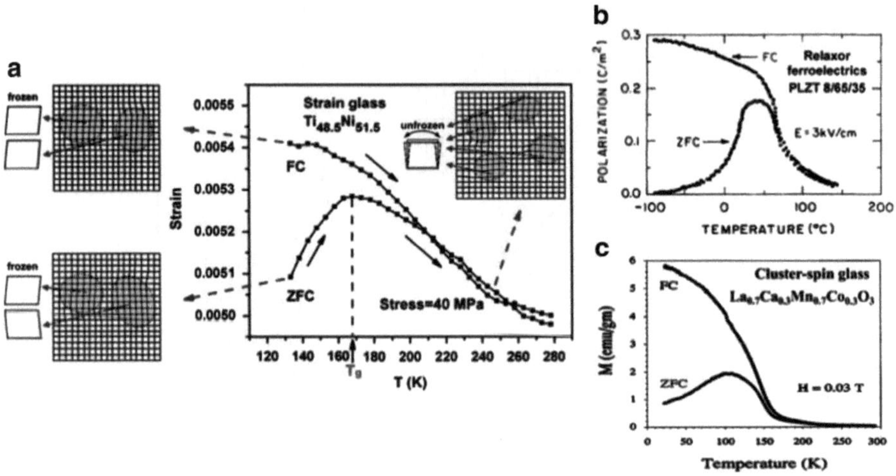

Fig. 11.7 (**a**) ZFC/FC curves of $Ti_{48.5}Ni_{51.5}$ strain glass show a large deviation below T_g (168 K) The schematics depict the unfrozen strain state above T_g and the two different frozen strain states below T_g under different thermal histories [14]. (**b**) The ZFC/FC curves of a ferroelectric relaxor (PLZT 8/65/35 [29]). (**c**) The ZFC/FC curves of a cluster-spin glass $La_{0.7}Ca_{0.3}Mn_{0.7}Co_{0.3}O_3$ [30])

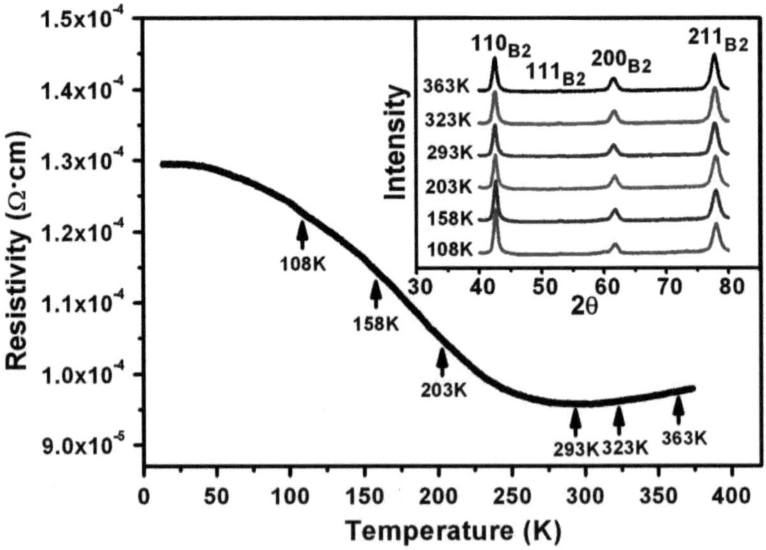

Fig. 11.8 Average structure remains invariant during a strain glass transition in at $T_g \sim 168$ K. The XRD pattern shows that the strain glass $Ti_{48.5}Ni_{51.5}$ alloy keeps an average "B2" structure and remains down to $T << T_g$. The freezing of nanodomains manifests as a rise in resistivity [16]

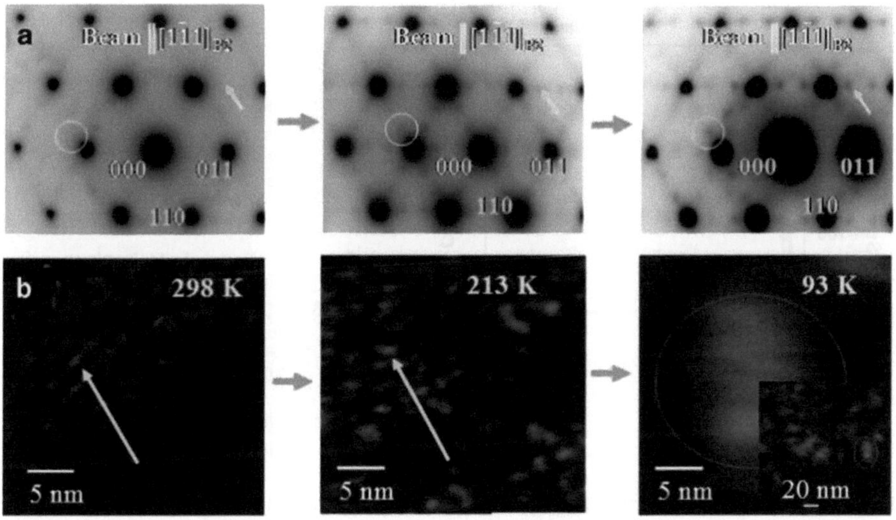

Fig. 11.9 Freezing process of the short-range ordered nanodomains in $Ti_{48}Ni_{52}$ strain glass alloy during cooling through the glass transition temperature $T_g \sim 158$ K. (**a**) The diffuse 1/3 reflections (marked by *yellow arrows*) correspond to the R-like local strain order, which strengthens during cooling. (**b**) The gradual growth and freezing of the nanodomains (SRO) with cooling; they never develop into a martensite (LRO) [12]

(4) Existence of local order or short-range order in the frozen glass state, as manifested by nanosized domains

Figure 11.9 shows the evolution of short-range strain order during cooling for a strain glass alloy $Ti_{48}Ni_{52}$. No long-range ordering (i.e., a martensitic transformation) is found over the whole temperature range encompassing its strain glass temperature $T_g \sim 158$ K, but the nanosized strain domains gradually grow and are finally frozen at a size of 20–25 nm. The diffuse 1/3 spots (Fig. 11.9a) suggest that the SRO of the nanodomains takes an R-like symmetry, being similar to one of the LRO structures of this system [12].

11.4 Novel Properties of Strain Glass

Strain glass exhibits a number of unexpected properties such as shape-memory effect and superelasticity [13, 15]. As a strain glass does not undergo a martensitic transition and looks like a nontransforming, "dead" material, there seems no reason to have a shape-memory effect, which characterizes only a martensitic system. However, as can be seen from Fig. 11.10a, a strain glass ($T_g \sim 168$ K, $T_0 = 160$ K) can also exhibit a shape-memory effect (when deformed at $T < T_g$ and heated up to $T > T_g$, the blue curve), very similar to a normal martensitic alloy. It is further found that at $T > T_g$, a strain glass can also exhibit superelasticity (the 173 K and

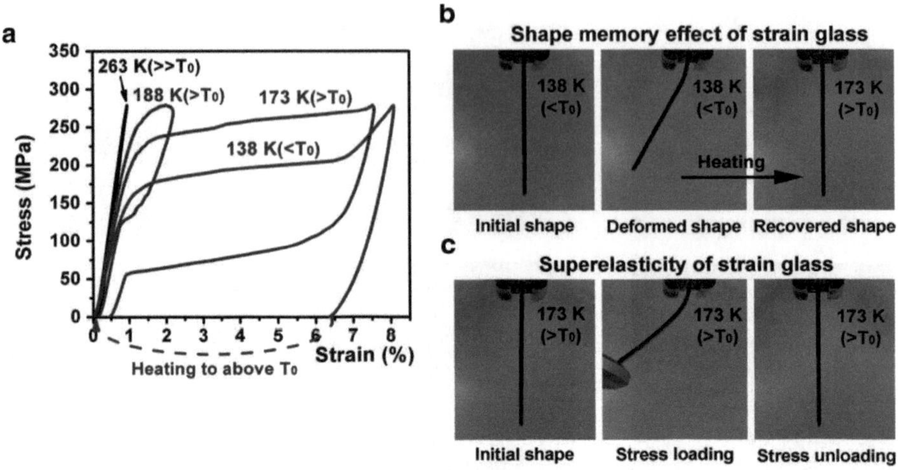

Fig. 11.10 The shape-memory effect and superelasticity of strain glass alloy $Ti_{48.5}Ni_{51.5}$. (**a**) The stress–strain curves over a wide temperature range spanning the freezing temperature ($T_g \sim$ 168 K, $T_0 = 160$ K) of strain glass transition. (**b**) and (**c**) show a visual evidence for shapememory effect and superelasticity, respectively [13]

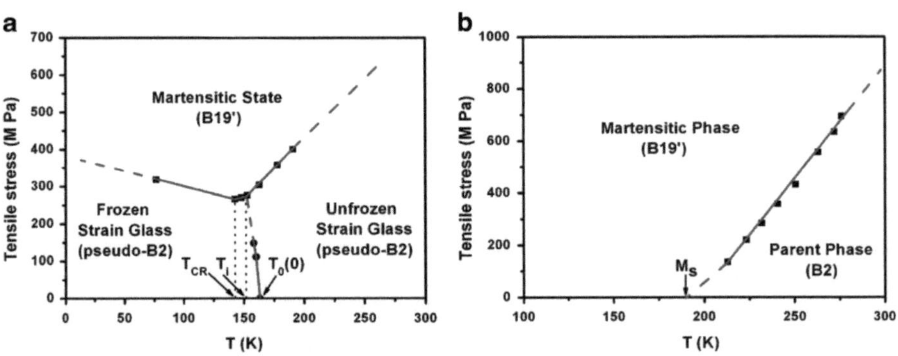

Fig. 11.11 Comparison between (**a**) the temperature–stress phase diagram of $Ti_{48.5}Ni_{51.5}$ strain glass alloy and (**b**) the temperature–stress phase diagram of $Ti_{48.4}Ni_{50.6}$ normal martensitic alloy [13]

188 K curves). The unexpected shape-memory effect of strain glass can be explained by the formation of a long-range strain order (martensite) by external stress and its recovery to the initial unfrozen strain glass state [13, 15]. Figure 11.5b, c shows the visual evidence for the shape-memory effect and superelasticity of the strain glass.

The shape-memory effect and superelasticity of strain glass can be explained by a stress–temperature phase diagram of a strain glass, as shown in Fig. 11.11a [15], and a comparison is made with a normal martensitic alloy $Ti_{48.4}Ni_{50.6}$ (Fig. 11.11b). In the strain glass alloy, application of sufficient stress at $T < T_g$ will force the frozen strain glass to transform into a long-range ordered form (i.e, martensite), and this results in a large plastic strain observed in Fig. 11.10a. Upon heating to

$T > T_g$, the martensite transforms back to the unfrozen strain glass and the shape is recovered. This is the shape-memory effect. At $T > T_g$, application of sufficient stress will drive the unfrozen (nearly ergodic) strain glass into a martensite and generate a large strain; upon unloading the martensite transforms back to the initial unfrozen strain glass state and this causes a spontaneous shape recovery. This is the superelasticity. The above mechanism has gained support from in situ XRD experiment [13]. On the other hand, a normal martensitic system does not exhibit a glass-to-martensite transition and the phase diagram is quite different at low-temperature side (Fig. 11.11b).

Another interesting property of strain glass is its broad damping peak [21], which may lead to some applications. This is because the peak is quite broad (spanning over 50 K) and sufficiently high compared with many other damping mechanisms. Compared with the unstable transient peak of a martensitic alloy [26], the damping peak of a strain glass is quite stable, being virtually independent of the cooling rate. Figure 11.12 shows that the damping peak of a $Ti_{50}Pd_{40}Cr_{10}$ strain glass has an

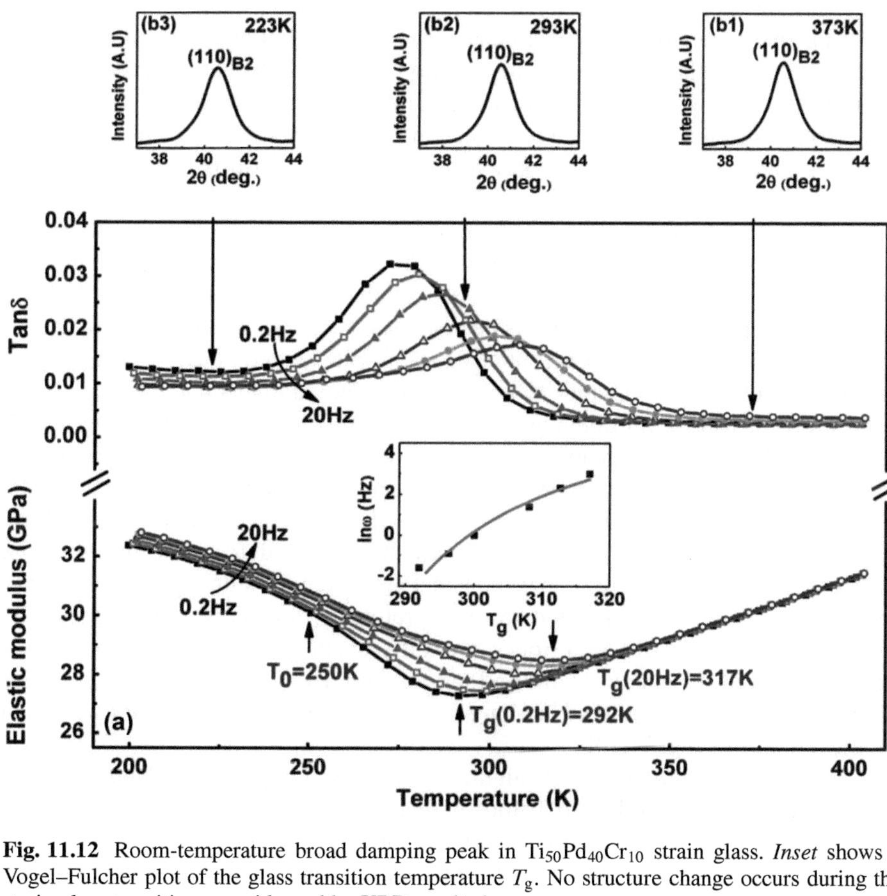

Fig. 11.12 Room-temperature broad damping peak in $Ti_{50}Pd_{40}Cr_{10}$ strain glass. *Inset* shows a Vogel–Fulcher plot of the glass transition temperature T_g. No structure change occurs during the strain glass transition, as evidenced by XRD result above

appreciable damping value of ~ 0.03 and the peak position is located around room temperature.

It is expected that strain glass may lead to many other interesting properties [22–25] and some may have significant technological implications.

11.5 Origin of Strain Glass and Theoretical Modeling/Simulations

Understanding the origin of strain glass starts from understanding the role of point defects or dopants, as strain glass appears only when sufficient amount of point defects is doped into a pure ferroelastic system, e.g., excess Ni into pure TiNi, Fe/Mn/Co/Cr into TiNi, and Cr into TiPd. There are two possible roles of point defects. One is an average effect, i.e., changing the global thermodynamic stability of martensite (reflected by a change in M_s temperature); another is a local effect, i.e., changing in the local stability around point defects.

So far there have been two different views on the local effect of point defects. One is to assume that the random distribution of point defects can cause a spatial distribution of M_s so that different locations have different M_s temperatures [7–9]. A suitable distribution of M_s can result in nanosized martensite domains that appear to be a strain glass [31, 32]. The key feature of this scenario is that point defects interact with strain order parameter only through composition effect (i.e., affecting the coefficient of harmonic terms in the Landau expansion), an isotropic effect. This type of model has shown to be able to reproduce some key features of strain glass [31, 32], but has yet to demonstrate a crossover phase diagram shown in Fig. 11.2.

Very recently a different view has been proposed [2, 33], which assumes that the local effect of point defects is essentially anisotropic; i.e., point defects create local anisotropic stresses that interact with the strain order parameter and thus dictate the local order of nanostrain domains. A cartoon of this idea is shown in Fig. 11.13. Figure 11.13a shows the analogy of a normal martensitic transformation in a pure system with the long-range propagation (ordering) of identical domino blocks. Clearly, the system can undergo a long-range ordering, because there are no defects to stop it. Figure 11.13b shows the situation with sufficient defect concentration. The "defects" in the domino array can be considered as irregular stones placed randomly in the array. Each stone has its own local preference for the way that the nearby blocks should fall, but different stones have different preference. As the result, when the dominos fall they cannot fall in a long-range ordered way as in the absence of defects; instead a local-ordered but long-range disordered pattern appears, as shown on the left side of Fig. 11.13b.

Strain glass formation stems from the same origin. Point defects (most probably defect pairs) create local lattice distortion that favors a particular strain domain for each defect pair, but as the defect pairs are randomly distributed, the system can order only into a locally ordered but long-range disordered state as shown on the

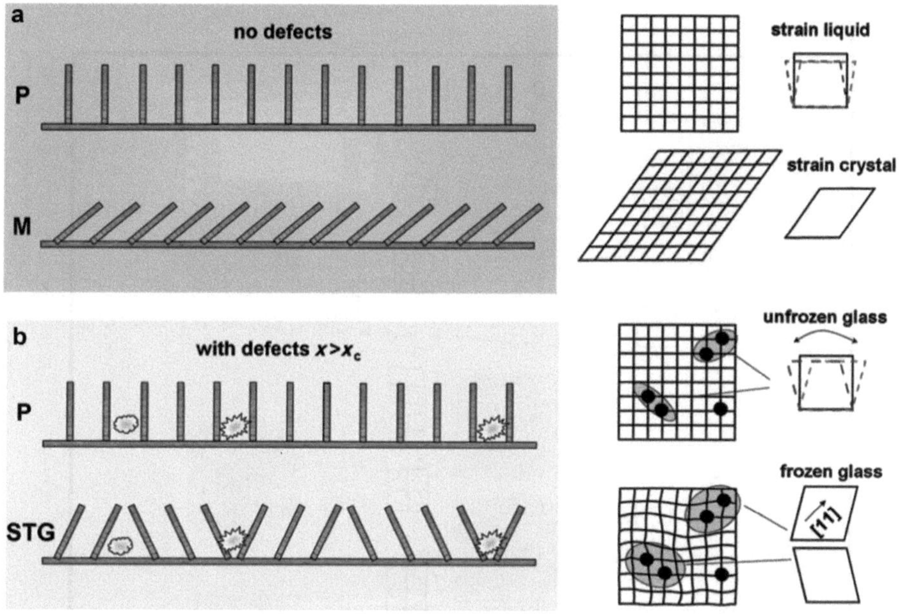

Fig. 11.13 A cartoon illustrating the microscopic origin of strain glass: destruction of long-range strain ordering by point defects. (**a**) Long-range strain ordering in the absence of defects. (**b**) Formation of short-range ordering due to point defects [2]

right side of Fig. 11.13b. This is analogous to the "domino array with stones" case. We believe that Fig. 11.13 reveals the essential physics of a strain glass; it may be a starting point for a quantitative theory of strain glass.

A qualitative explanation of the martensite to strain glass crossover is given in Fig. 11.14 [18]. It shows that with increasing defect concentration, the local anisotropic stresses of the defect pairs become dominant and the long-range strain ordering (i.e., formation of martensite) becomes more and more difficult; this is reflected by a decreasing martensite domain size. When defect concentration is above a critical value x_c, long-range ordering is fully suppressed; instead the system can be frozen into a strain glass state where local strain order or nanosized martensite domains still persist (Fig. 11.15).

The formation of strain glass can be phenomenologically described by the existence of many local minima in the phase-space [15]. These local minima are considered to be formed by the point defects. Then strain glass can be described by a competition between the local barriers and thermal activation. When thermal activation is insufficient to carry the system over the local barriers, the system becomes frozen and a strain glass is formed.

Based on the local anisotropic stress effect of point defects (also called "local field effect"), a phase field model has been proposed [33]. Figure 11.16 shows a comparison of the experimental phase diagram (Fig. 11.16a) and a simulated phase

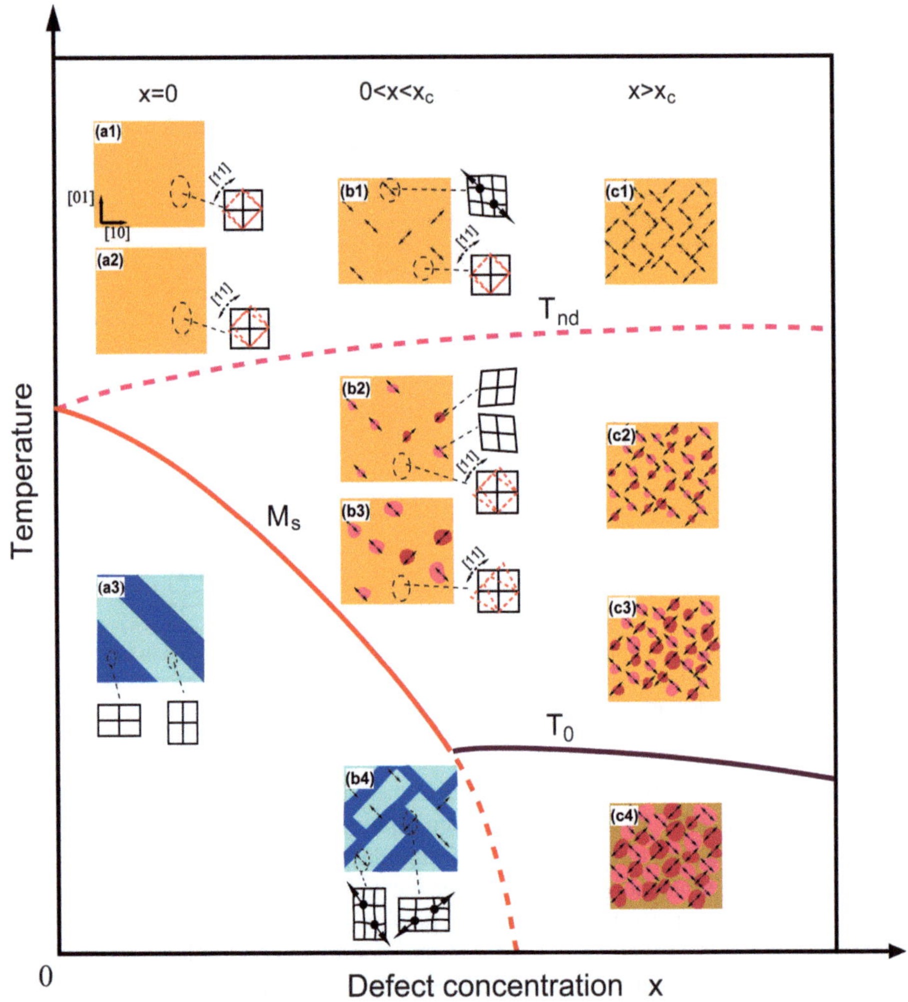

Fig. 11.14 Microscopic picture about the crossover behavior from normal martensitic transition to strain glass transition as a function of defect concentration. The point-defect-induced random stress fields are expressed by *arrows* [18]

diagram (Fig. 11.16b) based on this model. The simulated phase diagram well reproduces the most important feature of a strain glass system: a crossover from a normal martensitic transformation to a strain glass transition at a critical defect concentration. It also reproduces the existence of a precursor state or tweed state below a critical temperature T_{nd}.

Figure 11.17a shows the simulated microstructure evolution as a function of temperature at various defect concentrations. It is found that above a critical doping level ($c \sim 0.125$), no martensitic transformation can be identified and instead the

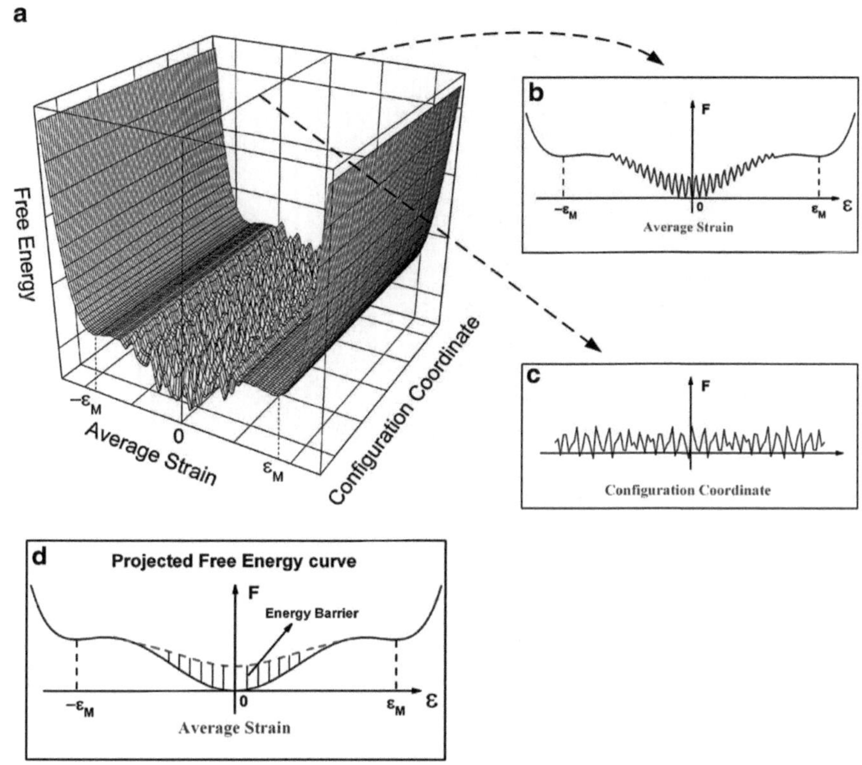

Fig. 11.15 The free energy landscape of a strain glass in a phase space, which is characterized by numerous local minima. (**a**) The 3D free energy landscape of a strain glass in the microscopic configuration coordinateaverage strain space. (**b**) and (**c**) are two sectional views of the 3D free energy landscape. (**d**) Projected free energy curve in free energy vs. average strain plane [15]

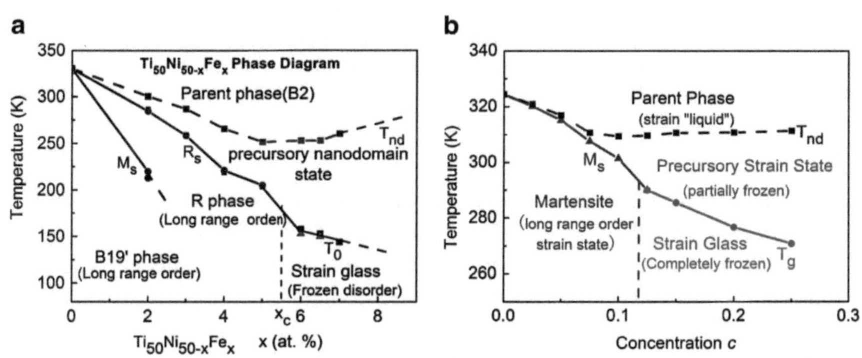

Fig. 11.16 (**a**) Experimental phase diagram of $Ti_{50}Ni_{50-x}Fe_x$. (**b**) The calculated phase diagram [33]

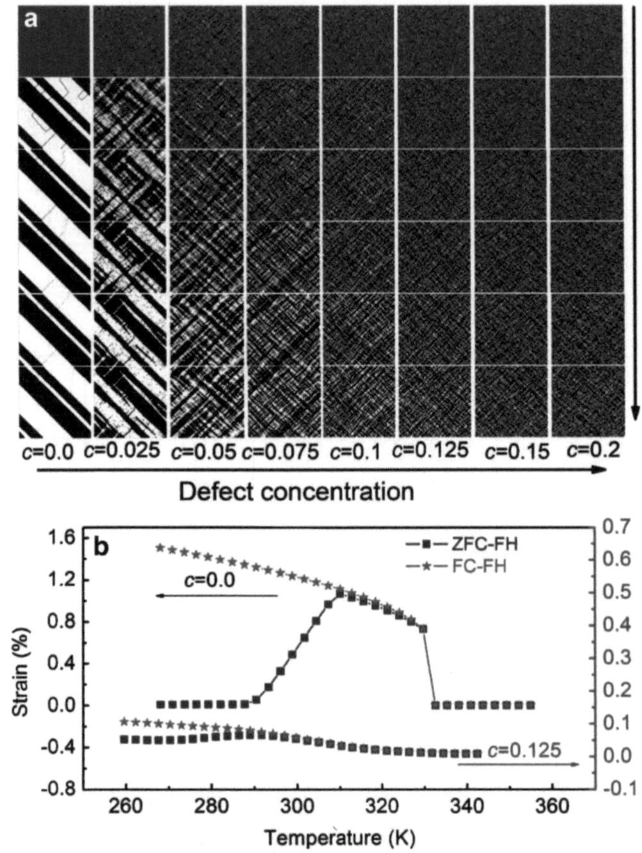

Fig. 11.17 (**a**) Simulated martensitic microstructure at different defect concentrations. The defect concentration c increases from left to right and the temperature decreases from top to bottom. *Gray color* describes the parent phase; *white* and *black color* describes the two martensitic variants. (**b**) Zero-field cooling or field cooling of normal martensitic transformation ($c = 0.0$) and strain glass transition ($c = 0.125$). The jump in strain for $c = 0.0$ corresponds to the martensitic transformation

system is frozen into a nanodomain microstructure, the strain glass. Figure 11.17b shows different ZFC/FC curves of a martensite ($c = 0.0$) and a strain glass ($c = 0.125$). For a normal martensitic system, the system remains ergodic during the martensitic transition, and significant nonergodicity appears in fully martensitic state. For a strain glass, detectable nonergodicity starts well above the freezing temperature, as the system is easy to become "sticky" for a strain glass system; significant ergodicity-breaking occurs at T_g.

Very recently, ideas borrowed from spin glass theories [34, 35] have been applied to explore the origin of strain glass [36, 37] and to try to unify different types of glasses [Sherrington, private discussion, 2010]. A detailed discussion about such models can refer to another chapter of this book written by Prof. Sherrinton [private discussion, 2010].

11.6 Strain Glass as a Solution to Several Long-Standing Puzzles About Martensite

Several puzzles in our martensite community have remained unsolved for many decades. Now it seems that they have a common solution, if viewed from the angle of strain glass.

(1) Origin of precursor or premartensitic tweed

A long-standing puzzle in martensite community is why quasistatic ferroelastic domains can exist well above the onset of martensite transformation (M_s) (or glass transition temperature T_g for strain glass compositions), where ferroelastic state is apparently unstable thermodynamically. These quasistatic strain domains are usually known as the "precursor tweed" or "precursor nanodomains", as they appear as cross-hatched or dotted nanodomains under TEM. The appearance of the quasistatic nanodomains above M_s or T_g can be well explained by the role of point defects shown in Fig. 11.14. We first consider the tweed state in a strain glass composition $x > x_c$ (see Fig. 11.14). When $T > T_{nd}$, the system is perfectly dynamic or ergodic, at which strong thermal vibration destroys any static strain domains. At $T_g < T < T_{nd}$, some static nanodomains can exist and be stabilized owing to the anisotropic stresses of the point defects; this is the "precursor tweed", but these quasistatic domains have only a small volume fraction so that as a whole the system still appears quasidynamic or "sticky strain liquid". With the temperature decreasing further to $T < T_g$ (or T_0, the ideal glass temperature), the volume fraction of the quasistatic nanodomains increases and the whole system appears frozen.

Recent experimental measurement of the relaxation spectrum of a strain glass [17] seems to support this view. Figure 11.18a shows the relaxation spectrum of a strain glass over a wide temperature range spanning its ideal freezing temperature T_0. An important feature is that a quasistatic (or slowly varying) component can exist well above the glass transition temperature T_0, but its volume fraction is small (Fig. 11.18b, $T > T_0$). This means the existence of quasistatic nanodomains in a "strain liquid"; this corresponds to a "sticky strain liquid" state. This is the nature of the precursor tweed. With the temperature decreasing to $T < T_0$, the volume fraction of the quasistatic component increases dramatically and this causes the whole system to freeze.

For a martensitic composition, the picture is essentially the same, but with a difference that the volume fraction of the quasistatic nanodomains is smaller, due to its lower point defect concentration, as shown in Fig. 11.14 (for $x < x_c$). As a result, the strain liquid (the parent phase) is not sticky enough to undergo a freezing

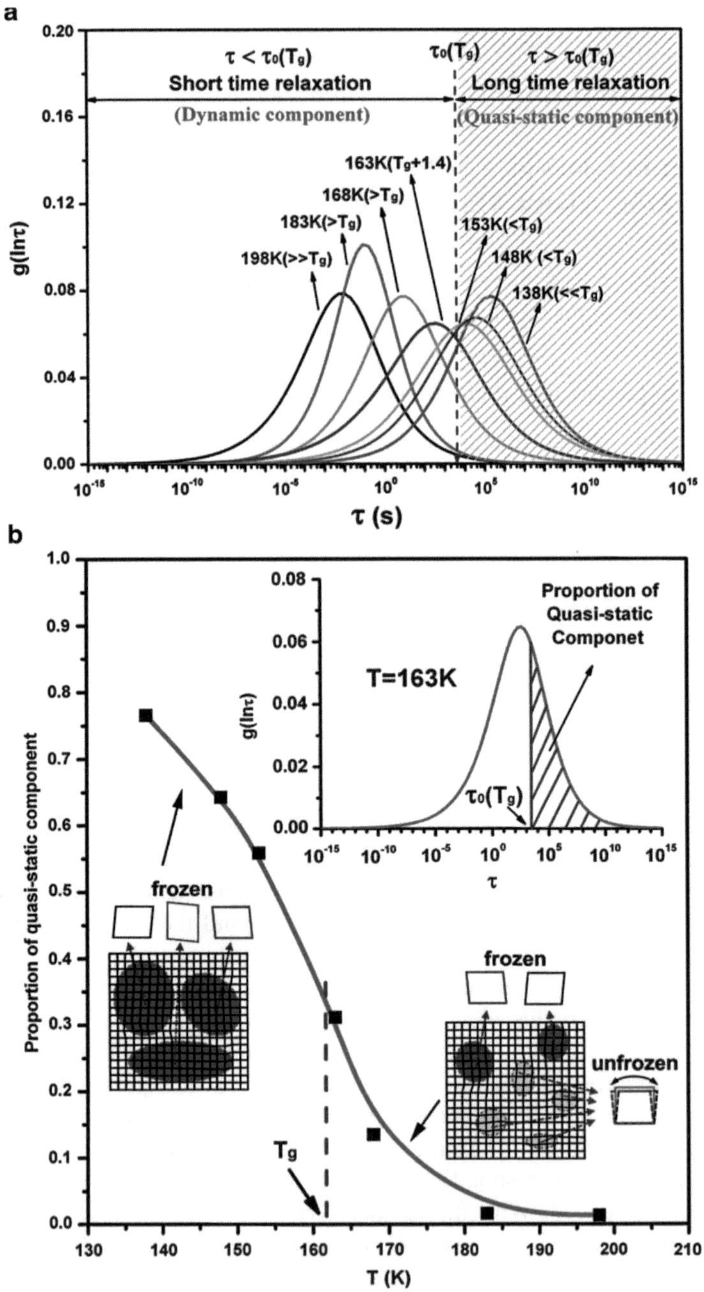

Fig. 11.18 (**a**) Evolution of the relaxation spectrum of $Ti_{48.5}Ni_{51.5}$ strain glass with temperature. (**b**) The fraction of quasistatic nanodomains increases continuously and shows a rapid increase around the freezing temperature T_g on cooling. The schematic *insets* illustrate that a small fraction of static nanodomains can exist even above T_g and almost all the nanodomains are frozen well below T_g [17]

transition; instead it undergoes a long-range ordering of the strains (i.e., martensitic transformation).

(2) Negative temperature dependence of electrical resistivity in TiNi-based shape-memory alloys

A negative temperature dependence (NTD) of electrical resistivity is characteristic of insulators rather than metals. However, Ti–Ni–Fe and Ni-rich Ti–Ni systems exhibit NTD in the martensitic alloys prior to the martensitic transformation and also appear in their strain glass alloys [38]. Now with the microscopic mechanism of the tweed and strain glass understood, it is possible to provide a simple answer to this long-standing puzzle.

The origin of NTD in TiNi-based strain glass can be derived from the following three important facts: (1) ferroelastic nanodomains start to appear at T_{nd} and gradually increase in size and volume fraction with decreasing temperature, as shown in Fig. 11.19d; (2) the nanodomains have an R-phase structure, as suggested by the 1/3 superlattice diffraction spots (Fig. 11.9a); (3) the R structure has a higher specific electrical resistivity than that of the B2 parent phase, as having been well known in the literature. At $T > T_{nd}$, the system exhibits a normal metallic behavior, i.e., having a positive temperature dependence of resistivity. Below T_{nd}, nanosized R-domains with higher specific resistivity start to form and increase gradually in volume fraction. This leads to an opposite effect to counteract the normal phonon scattering effect in metals, as shown in Fig. 11.19b. With further decrease in temperature, the volume fraction of the R-like nanodomains further increases and their positive contribution to the resistivity eventually exceeds the phonon scattering effect. As a result, the resistivity increases with decreasing temperature when the temperature is well below T_{nd}, as shown in Fig. 11.8b.

Similar NTD phenomenon has been reported for Ni-rich Ti–Ni and Ti–Ni–Fe martensitic systems (i.e., $x < x_c$) prior to their martensitic transformation temperatures [38]. The physical origin is the same as the above for strain glasses, because these compositions also produce nanosized R-domains well above the martensitic transformation temperature [i.e., $T_{nd} >> M_s$ (or R_s)] and the NTD is caused by gradual increase in volume fraction of the R-like nanodomains prior to the martensitic transformation.

(3) Decrease in transformation latent heat with increasing defect concentration.

Another long-standing puzzle about the martensitic transition of $Ti_{50-x}Ni_{50+x}$ system (or TiNi–Fe/Cr/Co/Mn as a whole) is the drastic decrease of the transition latent heat or entropy as defect content x increases, and the latent heat eventually vanishes at $x \geq 1.5$, as shown in Fig. 11.4i. As transition entropy reflects the jump in order parameter (lattice strain) at M_s, there seems no reason why such a jump must decrease with increasing defect concentration and must vanish at high defect concentration. In the following, we show that this puzzle can be well explained by the microscopic picture in Fig. 11.14.

The transition entropy or latent heat reflects the difference in strain order between the parent phase and martensite phase at transition temperature. For a

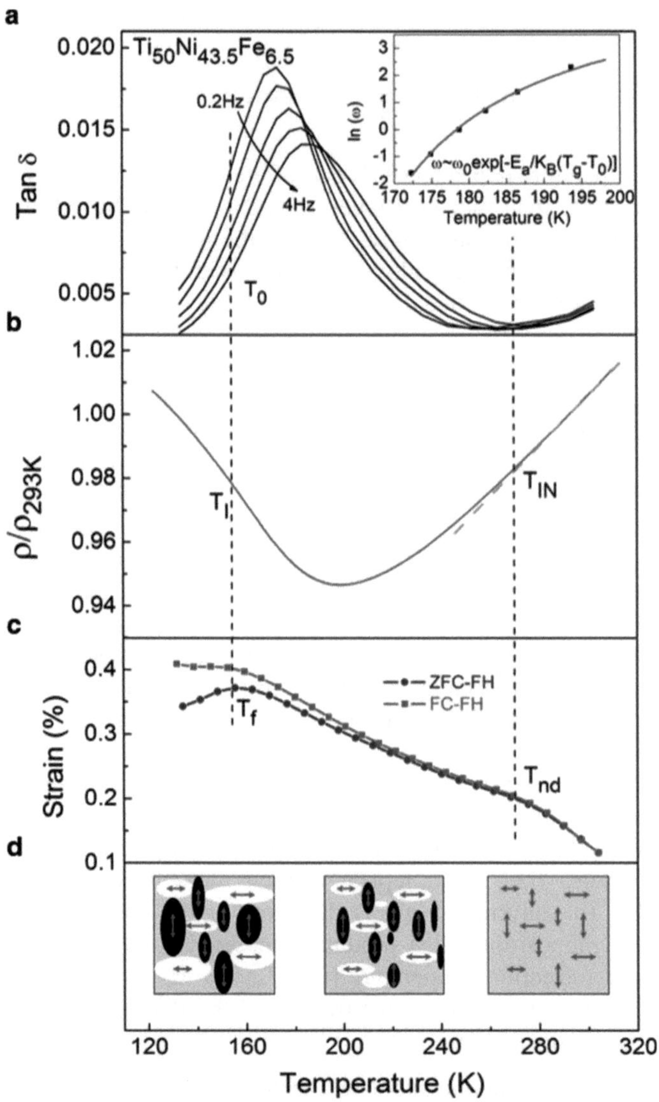

Fig. 11.19 Correlation among different characteristics of a strain glass transition. (**a**) Freezing behavior seen from internal friction; (**b**) negative temperature dependence of electrical resistivity; (**c**) ZFC/FC curves (**d**) Schematic microscopic illustration of strain glass freezing process. The *arrows* represent the local strain states caused by point defect. The *white* and *black colors* represent different martensitic variants [19]

pure $Ti_{50-x}Ni_{50+x}(x = 0)$ martensitic system, the transition is between a fully strain-disordered parent phase [Fig. 11.14(a2)] and a fully long-range strain-ordered martensite [Fig. 11.14(a3)]; thus, the transition entropy $\Delta S_{x=0}$ is the largest. For slightly doped alloys ($x < x_c$), the transition is between a partially ordered (locally ordered) parent phase [Fig. 11.14(b3)] and an imperfectly long-range strain-ordered martensite [Fig. 11.14(b4)]; thus, the difference becomes less. This results in a decrease in transition entropy or latent heat with increasing doping level.

At high defect content ($x > x_c$), the unfrozen strain glass state with quasidynamically disordered strain [Fig. 11.14(c3)] cannot form long-range strain ordering but instead gradually freezes into a frozen strain glass state with static disordered strain [Fig. 11.14(c4)]. Since there is no obvious change in the degree of strain order during a strain glass transition, the transition entropy or latent heat becomes negligible in the strain glass composition regime ($x > x_c$).

11.7 Summary

Recent progress (mostly experimental) about an emerging field – strain glass – is reviewed in this chapter, aiming at delineating a consistent picture of strain glass. Strain glass is a glass form of a ferroelastic system due to point defect doping. Strain glass not only provides interesting properties, some not seen in martensite, but also provides important clues to understanding the role of point defects in ferroelastic systems and consequently resolves a number of long-standing puzzles. Comparison with ferroelectric relaxor and cluster-spin glass leads to the concept of "ferroic glass" [39], which may lead to a unified understanding of the glass phenomena in ferroic materials. Owing to the unique way that nanosized microstructure of strain glass responds to external stimuli, there is a good reason to expect that strain glass will provide new opportunities for novel technological applications.

Acknowledgments The author thanks T. Lookman, A. Saxena, D. Sherrington, Y. Wang, D. Wang, Z. Zhang, Y.M. Zhou, J. Zhang, Y.Z. Wang, T. Suzuki, and K. Otsuka for discussions. He also acknowledges the financial support from Kakenhi of JSPS.

References

1. G. Careri, *Order and Disorder in Matter* (AddisonWesley, Massachusetts 1984)
2. X Ren, Y. Wang et al., Strain glass in ferroelastic systems: Premartensitic tweed versus strain glass Philos. Mag **90**, 141 (2010)
3. V.K. Wadhawan, *Introduction to Ferroic Materials* (Gordon and Breach, Amsterdam, 2000)
4. K. Binder, W. Kob, *Glassy Materials and Disordered Solids* (World Scientific, London, 2005)
5. R.N. Bhowmik, R. Ranganathan, Anomaly in cluster glass behaviour of $Co_{0.2}Zn_{0.8}Fe_2O_4$ spinel oxide. J. Magn. Magn. Mater. **248**, 101 (2002)
6. R.T. Zhang, J.F. Li, D. Vieland, Effect of aliovalent substituents on the ferroelectric properties of modified barium titanate ceramics – Relaxor Ferroelectric Behavior. J. Am. Ceram. Soc. **87**, 864 (2004)
7. S. Kartha, T. Castan, J.A. Krumhansl, J.P. Sethna, Spin-glass nature of tweed precursors in martensitic transformations. Phys. Rev. Lett. **67**, 3630 (1991)

8. S. Kartha, J.A. Krumhansl, J.P. Sethna, L.K. Wickham, Disorder-driven pretransitional tweed pattern in martensitic transformations. Phys. Rev. B **52**, 803 (1995)
9. S. Semenovskaya, A.G. Khachaturyan, Coherent structural transformations in random crystalline systems. Acta Mater. **45**, 4367 (1997)
10. A. Planes, L. Manosa, Vibrational properties of shape-memory alloys. Solid State Phys. **55**, 159 (2001)
11. K. Otsuka, X. Ren, Physical metallurgy of Ti-Ni-based shape memory alloys. Prog. Mater. Sci. **50**, 511 (2005)
12. S. Sarkar, X. Ren, K. Otsuka, Evidence for strain glass in the ferroelastic-martensitic system $Ti_{50-x}Ni_{50+x}$. Phys. Rev. Lett. **95**, 205702 (2005)
13. Y. Wang, X. Ren, K. Otsuka, Shape memory effect and superelasticity in a strain glass alloy. Phys. Rev. Lett. **97**, 225703 (2006)
14. Y. Wang, X. Ren, K. Otsuka, A. Saxena, Evidence for broken ergodicity in strain glass. Phys. Rev. B. **76**, 132201 (2007)
15. Y. Wang, X. Ren, K. Otsuka, A. Saxena Temperature-stress phase diagram of strain glass $Ti_{48.5}Ni_{51.5}$ Acta Mater. **56**, 2885 (2008)
16. Y. Wang, X. Ren, K. Otsuka Strain glass: glassy Martensite, Mater. Sci. Forum. **583**, 67 (2008)
17. Y. Wang et al., Evolution of the relaxation spectrum during the strain glass transition of $Ti_{48.5}Ni_{51.5}$ alloy. Acta Mater. **58**, 4723 (2010)
18. Z. Zhang et al., Phase diagram of $Ti_{50-x}Ni_{50+x}$: Crossover from martensite to strain glass. Phys. Rev. B **81**, 22402 (2010)
19. D. Wang et al., Strain glass in Fe-doped Ti-Ni Acta Mater **58** 6206 (2010)
20. Y.M. Zhou et al., Strain glass in doped $Ti_{50}(Ni_{50-x}D_x)$ (D = Co, Cr, Mn) alloys – Implication for the generality of strain glass in defect-containing ferroelastic systems, Acta Mater. **58**, 5433 (2010)
21. Y.M. Zhou et al. High temperature strain glass in $Ti_{50}(Pd_{50-x}Cr_x)$ alloy and the associated shape memory effect and superelasticity Appl. Phys. Lett. **95**, 151906 (2009)
22. Y.C. Ji et al., to be published (2010)
23. S. Ren, BS thesis, Xi'an Jiaotong University, 2007
24. P. Zhang, BS thesis Xi'an Jiaotong University, 2009
25. Y. Ni, BS thesis, Xi'an Jiaotong University, 2009
26. J. Van Humbeeck The Martensitic Transformation, Mechanical Spectroscopy Q^{-1} **382** (TransTech, Zurich 2001)
27. Q. Tan, J.F. Li, D. Viehland, Role of potassium comodification on domain evolution and electrically induced strains in La modified lead zirconate titanate ferroelectric ceramics. J. Appl. Phys. **88**, 3433 (2000)
28. S. Karmakar, S. Taran, B.K. Chaudhuri, H. Sakata, C.P. Sun, C.L. Huang, H.D. Yang, Disorder-induced short-range ferromagnetism and cluster spin-glass state in sol-gel derived $La_{0.7}Ca_{0.3}Mn_{1-x}Cd_xO_3(0 = x = 0.2)$. Phys. Rev. B **74**, 104407 (2006)
29. D. Viehland, J.F. Li, S.J. Jang, L.E. Cross, M. Wuttig, Glassy polarization behavior of relaxor ferroelectrics. Phy. Rev. B **46**, 8013 (1992)
30. N. Gayathri, A.K. Raychaudhuri, S.K. Tiwary, R. Gundakaram, A. Arulraj, C.N.R. Rao, Electrical transport, magnetism, and magnetoresistance in ferromagnetic oxides with mixed exchange interactions: A study of the $La_{0.7}Ca_{0.3}Mn_{1-x}Co_xO_3$ system. Phys. Rev. B **56**, 1345 (1997)
31. P. Lloveras, T. Castan et al., Influence of elastic anisotropy on structural nanoscale textures. Phys. Rev. Lett. **100**, 165707 (2008)
32. P. Lloveras, T. Castan et al., Glassy behavior in martensites: Interplay between elastic anisotropy and disorder in zero-field-cooling/field-cooling simulation experiments. Phys. Rev. B **80**, 054107 (2009)
33. D. Wang, Y. Wang, Z. Zhang X. Ren Modeling abnormal strain states in ferroelastic systems: the role of point defects Phys. Rev. Lett. **105**, 20570 (2010)
34. D. Sherrington, S. Kirkpatrick Solvable model of a spin-glass Phys. Rev. Lett. **35**, 1792 (1975)

35. S. Kirkpatrick, D. Sherrington Infinite-ranged models of spin-glasses Phys. Rev. B **17**, 4384 (1978)
36. D. Sherrington A simple spin glass perspective on martensitic shape-memory alloys J. Phys.: Condens. Matter. **20**, 304213 (2008)
37. R. Vasseur T. Lookman Effects of disorder in ferroelastics: A spin model for strain glass. Phys. Rev. B **81**, 094107 (2010)
38. T. Kakeshita, T. Fukuda, H. Tetsukawa et al., Negative temperature coefficient of electrical resistivity in B2-Type Ti–Ni Alloys. Jpn. J. Appl. Phys. **37**, 2535 (1998)
39. X. Ren, et al., Ferroelastic nanostructures and nanoscale transitions: ferroics with point defects. MRS Bull. **34**, 838 (2009)

Chapter 12
Precursor Nanoscale Textures in Ferroelastic Martensites

Pol Lloveras, Teresa Castán, Antoni Planes, and Avadh Saxena

Abstract This chapter deals with nanoscale spatially inhomogeneous states that occur as precursors of martensitic/ferroelastic transitions in many off-stoichiometric shape-memory alloys. We show that these states are a result of the competition between elastic anisotropy and disorder arising, for instance, from compositional fluctuations. In the limit of high disorder and/or low elastic anisotropy, we show that nanoscale inhomogeneities give rise to glassy behaviour, while the structural transition is inhibited.

12.1 Introduction

Systems exhibiting spatially inhomogeneous states are a subject of increasing interest because of their potential importance in engineering functional materials [1]. A special situation is that of nanoscale textures originating as precursors to phase transitions in ferroic and multiferroic materials. Such states consist of self-organized multiphase structures of coexisting regions with properties varying over nanometer distances [2]. The understanding of these complex structures is a challenging nonlinear problem usually involving the interplay of disorder and (multiple) long-range interactions. Regarding engineering applications, the control

P. Lloveras · T. Castán (✉) · A. Planes
Facultat de Física, Departament d'Estructura i Constituents de la Matèria, Universitat de Barcelona, Diagonal 647, 08028 Barcelona Catalonia, Spain

Institut de Nanociència i Nanotecnologia, Universitat de Barcelona, Catalonia, Spain
e-mail: pol@ecm.ub.es; teresa@ecm.ub.es; toni@ecm.ub.es; antoniplanes@ub.edu

A. Saxena
Theoretical Division, Los Alamos National Laboratory, Los Alamos, NM 87545, USA

Institut de Nanociència i Nanotecnologia, Universitat de Barcelona, Barcelona, Spain
e-mail: avadh@lanl.gov

T. Kakeshita et al. (eds.), *Disorder and Strain-Induced Complexity in Functional Materials*, Springer Series in Materials Science 148,
DOI 10.1007/978-3-642-20943-7_12,

of such nanostructures opens a new route to fabricate self-organized functional structures within a homogeneous medium [3].

Here, we focus on precursor nanostructures revealed by high-resolution imaging techniques already well above a ferroelastic phase transition temperature. This type of precursor, historically termed tweed [4], was first observed in cubic materials [5–10] with underlying modulations in strain. The corresponding pattern exhibits anisotropic cross-hatched correlations that in real-space strain contrast TEM images resemble tweed textiles. More recently, precursor structures with modulations in the magnetization giving rise to stripe-like patterns have been experimentally observed [11] and theoretically described [12]. Also, precursor structures have been observed in ferroelectrics [13–15], with modulations in polarization and in magnetoelastic materials with modulations in magnetization and/or the strain [16–18]. We now have enough experimental evidence to substantiate the earlier suggestion [19, 20] that the concept of tweed-like precursors is rather universal and applicable to a broad class of ferroic materials. In this sense, it was anticipated that the occurrence of this type of precursors requires very general conditions: (1) High sensitivity (in the sense of response functions) to local symmetry breaking perturbations arising from disorder in the form of inhomogeneities, resulting in long-lived fluctuations of the low-temperature phase, (2) long-range interactions in order to induce a global response of the system and (3) anisotropy that determines the morphology and symmetry of the resulting pattern. The conjunction of these three requirements gives rise to the formation of self-organized precursor microstructures as the natural cooperative response of anisotropic long-range dipolar interactions (elastic, magnetic and electric) to local inhomogeneities that couple to the order parameter of the transition (strain, magnetization and polarization). It is worth mentioning that the actual observation of real-space precursor nanostructures requires that the timescale associated with the disorder be large compared to that of the order parameter relaxation.

In numerous systems showing nanoscale textures, even not being a primary order parameter, strain plays a crucial role since it couples to the relevant degrees of freedom of the system. This, for instance, seems to be the case of manganites [21], where magnetic and electronic properties interact with crystal structure giving rise to textures with coexisting metallic and insulating phases. Also, in the case of magnetoelectrics [1], which are simultaneously ferromagnetic and ferroelectric, the magnetoelectric coupling in some cases is mediated by the strain. As mentioned above, strain, and thus elasticity, plays a crucial role in determining the actual symmetry properties of ferroic and multiferroic nanostructures. From this point of view, ferroelastic materials offer a unique scenario where purely structural (strain) textures can be studied [22]. Among ferroelastics, materials termed as martensites [23] include a large number of metals and alloys with the common feature of undergoing a diffusionless first-order structural phase transition from an open bcc-β high-temperature phase to a more close-packed structure upon cooling. In the absence of an external mechanical stress, the product phase consists of a coherent mixture of twin-related orientation variants. Associated with this ferroelastic character, some materials exhibit shape-memory effect and superelasticity [24, 25].

Indeed, the most relevant aspect related to ferroelasticity [25] is the ability to recover its original macroscopic shape upon heating after being significantly deformed. This is the shape-memory effect that for decades has attracted a great deal of interest due to a broad variety of technological applications. The main cause of such behaviour is the underlying solid-to-solid ferroelastic phase transition that can be induced by tuning an external control parameter such as temperature or stress. Upon cooling, the parent phase becomes unstable giving rise to a spontaneous ordering deformation that involves a loss of symmetry. As a consequence, the product phase may take multiple equivalent (degenerate) states, or variants, that while having the same crystal structure, differ in their mutual orientation. In order to ensure coherence at the interfaces and to minimize the total energy, the different equivalent ordered states accommodate by forming bands consisting of alternating twin-related variants. In response to the induced long-range elastic interactions, such long-range modulations in the strain result in a multidomain anisotropic pattern that preserves the original macroscopic shape. It is worth mentioning that the anisotropy is related to the existence of well-defined soft crystallographic directions of the parent phase, which as a matter of fact operate as easy channels for energy relaxation.

Actually, the microstructure of a given ferroelastic material depends on several variables such as the specimen size, thermal/stress history, specific alloy composition and transformation path. In particular, it is well established that the presence of disorder may significantly affect the final microstructure and, more interestingly, the technological operative regimes [26, 27]. Indeed, alloys are one of the most prominent examples containing inherent point defects, which come from sample preparation, in the form of statistical compositional fluctuations. Furthermore, disorder may be easily modified by means of changing the percentage of each element either through an off-stoichiometry composition or by doping with an extra element. These procedures of engineering disorder are currently used to design materials with specific functional properties [28, 29].

As indicated previously, precursor textures in ferroelastic martensitic materials are intimately related to the presence of disorder. Indeed, point defects break the crystal symmetry of the parent phase and create an inhomogeneous strain field. As a consequence, the material proceeds by relaxing the excess energy along the easy directions dictated by the properties of its corresponding elastic response, in particular, by the values of the elastic moduli. This results in an anisotropic textured pattern with tendency to correlate along the soft direction(s). Interestingly, this tendency is more enhanced with lower values of the corresponding elastic modulus. In addition, the correlation length might be shortened by the disorder. The interplay between anisotropy and the disorder determines the morphology and final microstructure of the precursor pattern. Special attention is required in the two limiting cases of very low anisotropy and very high disorder. In the case of very low anisotropy, the morphology of the precursor pattern is mottled-like with domains of almost spherical shape and size of a few nanometers [30]. In the other limiting case of high disorder concentration, the martensitic transition is arrested [31, 32] and instead a glassy state is found at very low temperatures [31, 33]. This is experimentally supported by the history dependence observed in the zero-field-cooling/field-cooling (ZFC/FC) strain curves [33], with the applied external field being the stress. Such a non-transforming frozen state has the same morphology as the precursor state indicating a close relationship between the two states. Although we shall discuss this behaviour in detail in this chapter, we wish to indicate that numerical simulations of an appropriate model agree satisfactorily with the general scenario suggested above [34–36].

This chapter is intended to provide a unified framework for the different morphologies observed in nanostructural precursors in cubic shape-memory alloys. We start by providing a brief overview of the actual experimental situation. Next we introduce a model and present simulations appropriate to describe and reproduce the different observed patterns and finish by outlining our main conclusions.

12.2 Structural Precursor Textures in Cubic Ferroelastics

Nanoscale textures appearing prior to the martensitic transformation have been observed as anisotropic tweed [5–10] or isotropic mottled [30] morphology depending on the elastic anisotropy of the material [30]. Indeed, highly anisotropic systems such as Ni–Al give rise to cross-hatched tweed, whereas more isotropic systems such as Ti–Ni-based alloys result in a mottled microstructure of almost spherical nanodomains. We suggest [37] that both are caused by a similar mechanism that is some incipient instability of the parent phase and the presence of some kind of disorder, while the different morphology is simply due to a definitively different value in the elastic anisotropy. This section is devoted to providing experimental evidence to support the previous statement.

12.2.1 Tweed Textures

The high-temperature phase of most ferroelastic materials undergoing a martensitic transition exhibits a cubic bcc-based open structure with an unusual low value of the shear elastic constant $C' = (C_{11} - C_{12})/2$ and positive variation with temperature. This, as pointed out by Zener long time ago [38], is indicative of an increasing instability with decreasing temperature against distortions of the $\{110\}$ planes along the <1–10> directions. Later on Friedel [39] suggested that the, somehow unexpected, stability of the bcc phase at high temperatures is due to a large vibrational entropy arising from the low energy of the whole TA_2-phonon branch, which, moreover, softens with temperature. In particular, this is the case of Cu-based shape-memory alloys, usually known as entropy-driven alloys [40].

This intrinsic instability associated with the cubic character of the high-temperature parent phase is of crucial importance in the observation of precursor microstructures, eventually already well above the phase transition. Indeed, the signature of the Zener instability in diffraction patterns is seen in the form of anisotropic thermal scattering consisting of strong diffuse streaks running along the <1–10> directions [41]. Sometimes, the corresponding real-space image exhibits tweed. Tweed, a name already coined in the 1960s [4], refers to the fine-scale, diffuse, striated microstructure observed in TEM of the parent cubic phase. The linear striations are of typically ~3–6 nm periodicity and lie, on average, parallel to the traces of the {110} planes of the parent phase. They have been identified [42,43] to arise from static (or quasi-static) <110><1–10> displacement waves with wavelengths up to ~100 nm. Such a periodic (or quasi-periodic) strain modulation is the natural lattice accommodation in response to any perturbing field sensitive to couple locally with the strain. The upper panel of Fig. 12.1 displays the tweed striations in Ni–Al as experimentally observed in TEM. In the left side, the cross-hatched correlations along both diagonals are visible, whereas in the right side the superposition disappears after tilting the sample slightly. The lower panel

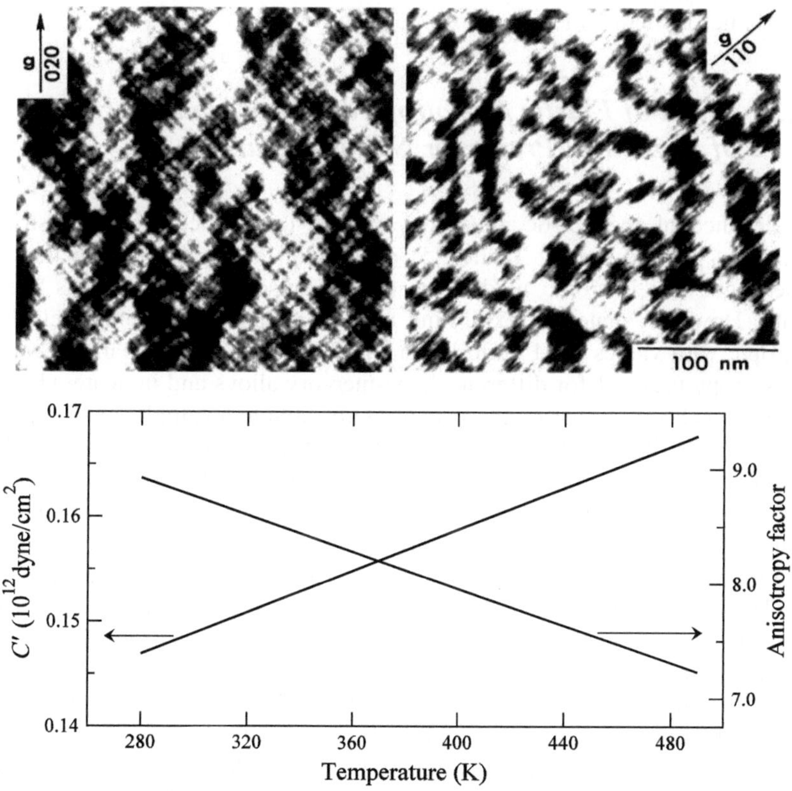

Fig. 12.1 Pretransitional behaviour of Ni–Al. The *upper panel* shows the tweed precursor textures as seen in TEM (adapted from [44]). The *lower panel* exhibits the temperature behaviour of the elastic constant C' and the elastic anisotropy factor A for the $Ni_{63,2}$ $Al_{36.8}$ alloy with $M_s \sim 280\,K$ (adapted from [6, 7])

exhibits the softening behaviour in Ni–Al for the relevant elastic constant C' and the corresponding increase in the elastic anisotropy factor A. For a cubic crystal, such a factor is defined as the ratio $A = C_{44}/C'$. Figure 12.1 has been adapted from References [44] (top) and [6, 7] (bottom), respectively.

Tweed is neither restricted nor common to all martensites. On the one hand, tweed textures have been observed in cubic materials that do not transform martensitically [5] and, on the other, there exist martensitic materials where tweed textures have never been observed. We wish to stress that tweed is a generic behaviour likely to occur in any highly anisotropic cubic alloy [45, 46]. Whether it can be observed in TEM images or not, it may depend on different conditions such as the resulting timescale. Moreover, it seems now clear that the softening of C' is not enough to observe tweed textures, but additionally one requires a high value of the elastic anisotropy factor A. In the context of martensites, given that tweed is triggered by the incipient instability of the parent phase, it can be considered to be a genuine precursor of the incoming phase. Furthermore, once these structures are created in the parent phase, they definitively affect the transformation path and the obtained microstructure of the martensitic phase.

12.2.2 *Effect of Elastic Anisotropy on the Morphology of Structural Precursor Nanostructures*

The relevance of the elastic anisotropy on the occurrence of tweed striations was already pointed out by Tanner [5] and Enami et al. [6, 7] more than three decades ago. For isotropic media, the elastic anisotropy factor A is identical to unity. However, materials exhibiting tweed precursors are definitively highly anisotropic with values of A quite larger. In Table 12.1, we compare the value of the anisotropy factor A for different shape-memory alloys and indicate whether or not tweed has been observed. Of special interest among examples in Table 12.1 is the case of TiNi-based alloys, where values of A are rather small compared with the other materials. In spite of the significant softening of C,′ no tweed precursors have been observed in the alloy family. Certainly, this is due to the simultaneous softening of C_{44}, which in turn prevents the elastic anisotropy to increase. In fact, the simultaneous softening of both elastic shears C' and C_{44} is a general feature of all Ti–Ni-based alloys [24]. Such behaviour is illustrated in Fig. 12.2 for the case of $Ti_{49}Ni_{51}$ alloy. Notice the variation of the elastic anisotropy factor A with temperature (lower panel). It never rises significantly and reaches the minimum value at the transition temperature M_s. This behaviour is clearly in contrast to the case of the highly anisotropic Ni–Al alloy discussed before.

Associated with the particular softening behaviour shown in the upper panel of Fig. 12.2, TiNi-based alloys also exhibit structural precursors, but in this case under a completely different morphology that was discussed in the previous section for Ni–Al. In Fig. 12.3, we show energy-filtering dark-field images of $Ti_{50}Ni_{48}Fe_2$ in the parent phase. One observes that the microstructure is mottled-like (each bright

Table 12.1 Comparative pretransitional behaviour in several shape-memory alloys. The value of the elastic anisotropy factor A and the eventual observation of tweed are included

Alloy	Softening of C'	A	Tweed	References
$Fe_{70}Pd_{30}$	Yes	~ 15	Yes	[47]
$Ni_{62.5}Al_{37.5}$	Yes	~ 9	Yes	[48]
Cu–Zn	Yes	~ 11	Yes	[8–10]
$Cu_{68}Zn_{16}Al_{16}$	Yes	~ 14	Yes	[49, 50]
Cu–Al–Ni	Yes	~ 13	–	[51]
Au–Zn–Cu	Yes	$\sim 12-20$	–	[51]
Au–Cd	Yes	$\sim 11-14$	–	[51]
Ni_2MnGa	Yes	~ 23	Yes	[52, 53]
$Ti_{49.7}Pd_{43.8}Cr_{6.5}$	–	~ 3.6	Weak	[54–56]
$Ti_{50}Ni_{50}$	Yes	~ 2	No	[51]
$Ti_{50}Ni_{48}Fe_2$	Yes	~ 2	No	[30, 32]
$Ti_{50}Ni_{40}Cu_{10}$	Yes	~ 2.4	–	[51]
$Ti_{50}Ni_{30}Cu_{20}$	Yes	~ 2.8	–	[51]

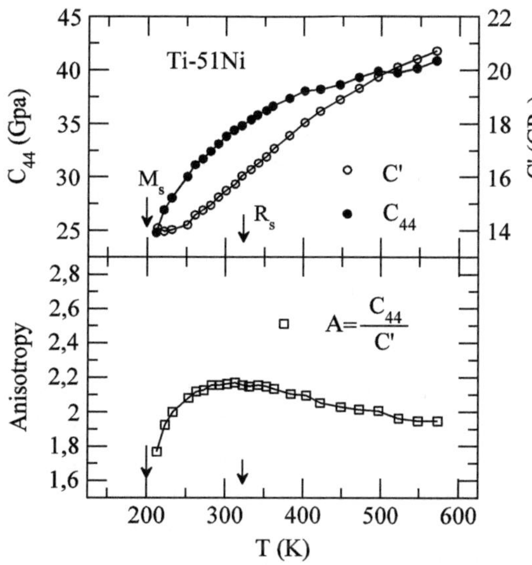

Fig. 12.2 Temperature behaviour of elastic constants in a Ti_{49}–Ni_{51} shape-memory alloy. The *upper panel* shows the simultaneous softening of both shear moduli C' and C_{44}, while the *lower panel* depicts the corresponding temperature evolution of the elastic anisotropy factor. The figure has been adapted from reference [51]

dot corresponds to a distorted domain) and comprises domains of almost spherical shape and size of 5 nm or smaller.

Murakami and Shindo [30, 57] were the first to describe such precursors in $Ti_{50}Ni_{50-x}Fe_x (x = 2)$ and to relate the isotropic morphology to the low value of A. Interestingly, similar morphology has been also observed at higher content of Fe. In fact, Choi et al. [32] have performed recently an extensive study of the precursor phenomena and transformation behaviour for increasing content of iron with $x = 2, 4, 6$ and 8 at. % by means of optical and TEM observations and X-ray diffraction. Dark-field images of the parent phase revealed the same nanoscale domain-like structure as in Fig. 12.3. Moreover, for $x > 4$, both cubic-to-R and R-to-B19′ structural transitions were found to be suppressed.

Fig. 12.3 Dark-field image of the parent phase in $Ti_{50}Ni_{48}Fe_2$. Courtesy of Prof. Y. Murakami

Indeed, similar behaviour is observed in the self-doping $Ti_{50-x}Ni_{50+x}$ alloy [58]. As the concentration of point defects (excess of Ni) increases from the stoichiometric value, one observes a precursor mottled-like nanostructure already in the parent phase. If one increases the doping of excess Ni above a critical value ($x >\sim 1.3$), the martensitic transition disappears. Interestingly, it has been shown that this non-transforming state is a frozen disordered state of local lattice strains with properties characteristic of a glassy state that survives down to very low temperatures [31, 33, 34].

A standard experiment for detecting signatures of glassy behaviour is the so-called Zero-Field-Cooling (ZFC)/Field-Cooling (FC) measurements. Similarly to the magnetic case, the method consists of comparing the two strain curves obtained by applying a small stress field upon heating, following a cooling under zero stress (ZFC) in one case and under nonzero (but low) stress (FC) in the other. Deviations between both curves are indicative of history dependence, commonly associated with glassy behaviour, as typically observed in spin glasses [59], relaxor ferroelectrics [60] and recently in $Ti_{50-x}Ni_{50+x}$ [61]. This behaviour is illustrated in the upper panel of Fig. 12.4 for $x = 1.5$. It is worth mentioning that additionally, the material exhibits frequency-dependence anomaly in AC mechanical susceptibility experiments obeying Vogel–Fulcher relation [34, 58]. The lower panel of Fig. 12.4 shows the temperature-composition phase diagram of $Ti_{50-x}Ni_{50+x}$. In the left side, it illustrates schematically the morphology of the two low-temperature phases that can be obtained by self-doping with excess of Ni. At the right side, the phase diagram includes not only the parent and the martensitic (B19′) phases, but also the regions of occurrence of mottled-like precursors and the frozen glassy states. The upper curve T_{nd} is the onset temperature of precursor nanodomain state, whereas T_0 is the freezing temperature and both have been determined from the anomalies that

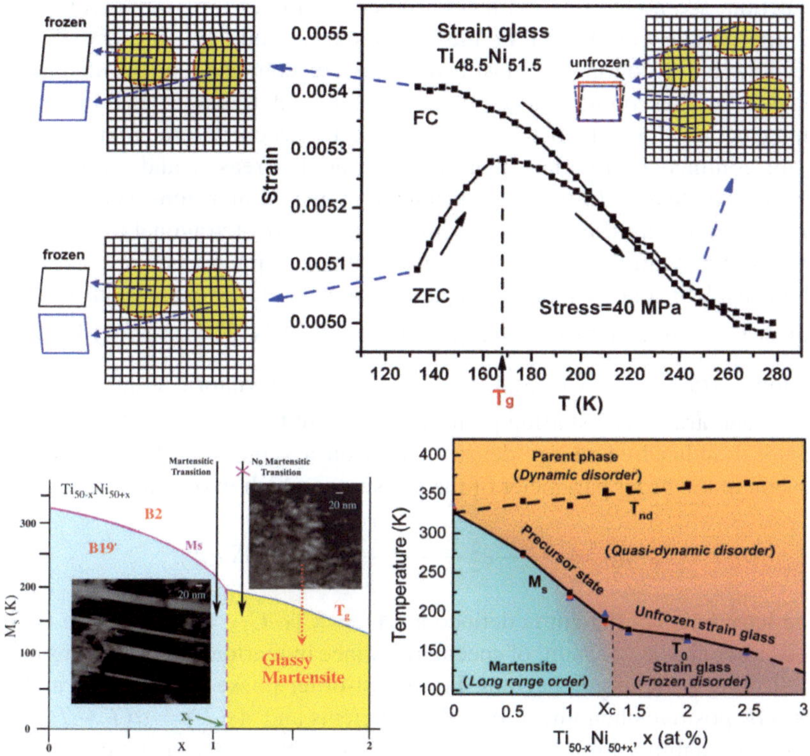

Fig. 12.4 (*Upper panel*) History dependence of the field(stress)-cooling/zero-field(stress)cooling (FC/ZFC) strain curves for the frozen glassy system $Ti_{48.5}Ni_{51.5}$. (*Lower panel*) Temperature-composition phase diagram for $Ti_{50-x}Ni_{50+x}$ as a function of the excess Ni content. For details, see the text. Courtesy of Prof. X. Ren

show up in electrical resistivity measurements [58]. It is very important to notice that the microstructure of the glassy state is essentially that of the precursor state and therefore consists of almost spherical nanodomains. In other words, the frozen glassy state at low temperatures inherits the same morphology of the precursor state at high temperature. This reinforces the idea that precursor textures, once formed, definitively affect the final microstructure of the low-temperature state.

12.3 Phenomenological Modeling

To understand the pretransitional and transitional behaviours described in the previous sections, we need a model suitable to reproduce the thermodynamics of the ferroelastic phase transition. The model used here is a strain-based formalism within a coarse-grained Ginzburg–Landau framework. It allows analysing separately the effect of the main ingredients necessary for tweed to occur and listed in the introduction. We notice that the natural order parameter for a structural transition is an appropriate component of the strain tensor, although the actual degrees of freedom for a continuum elastic medium are contained in the displacement variables. Another point that should be mentioned refers to the fact that the inhomogeneous strain fluctuations giving rise to structural precursors of interest here are confined into certain planes [25]. For this reason and without loss of generality, we shall illustrate the problem in two dimensions and focus on a square-to-rectangular transition (that mimics a cubic-to-tetragonal transition). The corresponding symmetry adapted tensor has three components that correspond to the three available elastic modes in square symmetry. Their expressions in terms of the linear Lagrangian tensor ε_{ij} are given by $e_1 \equiv \left(1/\sqrt{2}\right)(\varepsilon_{xx} + \varepsilon_{yy})$ for the bulk dilatational strain, $e_2 \equiv \left(1/\sqrt{2}\right)(\varepsilon_{xx} - \varepsilon_{yy})$ for the deviatoric strain and $e_3 = \varepsilon_{xy}$ for the shear strain. The starting point in the assembly of the model is a triple well Ginzburg–Landau free-energy density based on a symmetry-allowed polynomial expansion in the order parameter of the transition, which in our case is e_2:

$$f_{\mathrm{GL}} = \frac{A_{2(T)}}{2} e_2^2 - \frac{\beta}{4} e_2^4 + \frac{\gamma}{6} e_2^6 + \frac{\kappa}{2} |\vec{\nabla} e_2|^2. \tag{12.1}$$

The harmonic coefficient is defined as $A_2 = C_{11} – C_{12} = 2C'$, where C' is the second-order elastic constant, of special relevance in martensitic transitions because of its striking softening behaviour. For a first-order phase transition, β and γ are taken to be positive phenomenological coefficients and $A_2(T) = \alpha(T - T_c)$, where T_c is the low stability limit of the high-temperature phase; κ is a positive coefficient that accounts for the interfacial energy.

As we already pointed out, disorder is an essential ingredient to induce heterogeneous textures. In the present model, it is introduced taking into consideration the following two facts: (1) in any alloy, the nominal composition is a statistical variable and therefore fluctuates from one point to another of the material and (2) in martensitic transitions the transition temperature is very sensitive to alloy composition. Consequently, it is reasonable to assume that statistical compositional fluctuations give rise to a certain distribution of local transition temperatures. This may be accomplished via a local coupling to disorder in the harmonic coefficient $A_2(T)$ that now depends linearly on both temperature and disorder; that is $A_2(T, \eta) = \alpha[(T - T_c) + \eta(\mathbf{r})]$. The variable $\eta(\mathbf{r})$ is a random variable; Gaussian distributed around zero (standard deviation ζ) and spatially correlated using an exponential pair correlation function (correlation length ξ). Notice that the distribution of local transition temperatures $T_o(\mathbf{r})$ will cause that, in some regions, the transformation into the product phase be well above (or below) the nominal transition temperature T_0. As a consequence, and given that the transition is of first order, there will be a spatial distribution of local states with different degree of metastability, separated among them by energy barriers. This turns out to be very important to understand the formation of frozen disordered states.

The free-energy density in (12.1) is enough to obtain the phase diagram but not for textures. The reason is that in expression (12.1) it is implicitly assumed that the non-order-parameter strain components, e_1 and e_3, are identically zero. Nevertheless, the coherent matching of nearby transformed cells may well lead to effective contributions of non-order-parameter strain components, which are allowed by symmetry. To be sure, this is the origin of the long-range strain interaction. To satisfy this requirement, we append the simplest expansion in e_1 and e_3 to expression (12.1). That is:

$$f = f_{\mathrm{GL}} + f_{\mathrm{NOP}} = f_{\mathrm{GL}} + A_1 e_1^2 + A_3 e_3^2. \tag{12.2}$$

Once more the harmonic coefficients are merely second-order elastic constants given by $A_1 = C_{11} + C_{12}$ and $A_3 = 4C_{44}$. Notice that the augmented free-energy density $f(\mathbf{r})$ is a functional of the three strain fields $e_1(\mathbf{r})$, $e_2(\mathbf{r})$ and $e_3(\mathbf{r})$, which is not very convenient. In order to get rid of the non-order-parameters dependence, we proceed by minimizing $f(\mathbf{r})$ with respect to e_1 and e_3, but taking into account that all these strains e_1, e_2 and e_3 are obtained as derivatives of the same underlying displacement field and are therefore not independent. The linking is provided by the Saint-Vénant compatibility condition, which ensures the integrity of the lattice. Its expression in 2D is given by [44]:

$$\nabla^2 e_1 - \sqrt{8}\partial_{xy} e_3 - (\partial_{xx} - \partial_{yy}) e_2 = 0. \tag{12.3}$$

Performing standard Euler–Lagrange minimization, one obtains that f_{NOP}, when expressed in terms of the order parameter, is nonlocal and can be written as follows:

$$f(\boldsymbol{r}) = f_{\mathrm{GL}}(\boldsymbol{r}) + \int e_2(\boldsymbol{r}) U(\boldsymbol{r}, \boldsymbol{r}') e_2(\boldsymbol{r}') d\boldsymbol{r}'. \tag{12.4}$$

The potential $U(\mathbf{r}, \mathbf{r}')$ is anisotropic and in real space decays as $\sim 1/r^2$ [62], implying that the response to local perturbations has large-scale effects. Especially appealing is the expression for the potential in k-space [44, 62],

$$V(k_x, k_y) = A_3 \frac{(k_x - k_y)^2}{(A_{3/} A_1) k^4 + 8(k_x k_y)^2}. \tag{12.5}$$

This expression makes clear why cross-hatched correlations along $k_x = \pm k_y$ are favoured. Figure 12.4 shows a plot of $V(k_x, k_y)$ revealing the valleys along the diagonals (Fig. 12.5).

12.4 Numerical Simulation Results

This section is devoted to presenting and discussing the results obtained by solving numerically the model described above and contained in (12.1)–(12.5). The model parameters have been taken from [44] and have been converted into reduced units

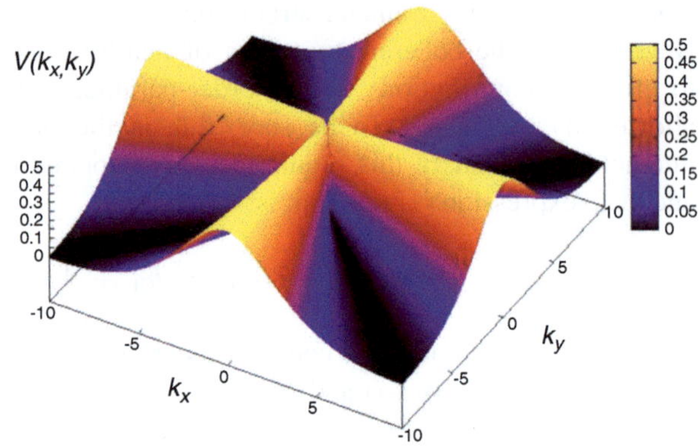

Fig. 12.5 3D plot of the Fourier kernel $V(k_x, k_y)$ of the repulsive anisotropic potential

by setting α, κ and T_c equal to unity so that the unit length is $l_0 = 0.24$ nm. The remaining model parameters take the following values: $\beta = 275\kappa/l_0^2$, $\gamma = 4.86 \times 10^5\kappa/l_0^2$, $A_3 = 4.54\kappa/l_0^2$ and the ratio A_3/A_1 is set to $A_3/A_1 = 2$. The correlation length and the standard deviation of the disorder are chosen to be $\xi = 20l_0$ and $\zeta = 0.32T_c$, respectively, in order to reproduce the typical length scale for tweed textures and its onset temperature.

It can be shown [63] that the Landau free energy (12.1) can be rescaled in such a way that no free parameters remain in the model besides the control parameters, external stress field σ and reduced temperature T. Moreover, it has been verified that changes in the ratio A_3/A_1 and/or in the correlation length ξ do not produce any additional physical insight [36]. Thus, the model parameters of physical relevance for this study can be approximately reduced to A_3, which is the weight of the long-range anisotropic potential (12.5) (A_3/A_1 is kept constant), and to the standard deviation of the disorder ζ. We notice that the elastic anisotropy factor $A = C_{44}/C'$ can also be expressed in terms of the model parameters, namely $A = A_3/2A_2$. Consequently, at constant T, the anisotropy factor will behave as $A \sim A_3$ and variations in the anisotropy A can be accomplished by simply modifying the value of A_3. Concerning variations in the amount of disorder they are performed through changes in the value of ζ. Numerical results presented in this section have been obtained on a simulation cell of size $(10^3 \times 10^3)l_0^2$, discretized onto a 512×512 mesh and subjected to standard periodic boundary conditions. The final (stabilized) configurations are obtained by using a purely relaxational dynamics.

12.4.1 Effect of the Elastic Anisotropy on Structural Precursors: From Cross-Hatched to Mottled Morphology

In Fig. 12.6, we show selected strain-field configurations for different values of the temperature T and the anisotropy $A(\sim A_3)$. For the highest value of A_3 (top row), strongly anisotropic cross-hatched structures can be observed (well) above the

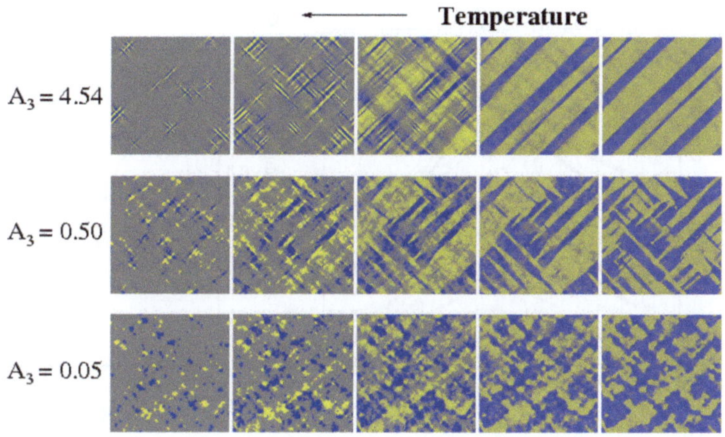

Fig. 12.6 Snapshots of stabilized strain-field configurations for three different values of the elastic anisotropy ($\sim A_3$) as a function of temperature. The two equivalent low-temperature variants are denoted in *blue* and *yellow*, whereas *grey* represents the parent phase. *Arrow* denotes the increasing direction of the temperature. For details, see the text

transition temperature T_0, (located in the middle of the row), which have features similar to tweed precursors in Fig. 12.1. When the temperature is decreased below T_0, ferroelastic twins that correlate along the entire system with well-defined long-range diagonal interfaces appear.

Regarding variations of A_3 at a given temperature, the configurations clearly show that when A_3 is decreased the structures gradually lose directionality. This behaviour is consistent with decreasing anisotropy. In particular, for the lowest value of $A_3 = 0.05$, at high temperature (precursor regime), instead of tweed, a droplet-like structure is obtained, reminding the mottled-like structures observed in systems with low values of the anisotropy A such as Ti–Ni-based alloys mentioned previously (see Fig. 12.3). Interestingly, for this low value of A_3 twins are suppressed and the surviving low-temperature structures are strongly correlated with the mottled precursors, with very similar morphologies and characteristic lengths. This behaviour is in agreement with experimental observations in isotropic Ti–Ni [58] and Ti–Ni(Fe) [32] shape-memory alloys.

To further characterize the textures in Fig. 12.6, it is interesting to look for anomalies in the thermodynamic properties. The behaviour of the heat capacity C (which can be computed from the thermodynamic relation $C = -Td^2F/dT^2$) is shown in Fig. 12.7a for different values of A_3. For the highest values, a sharp peak is obtained around $T_0 \sim 1$, indicating the ferroelastic transition. Moreover, a bump can be observed slightly above T_0 due to the effect of heterogeneities, which in this case correspond to tweed precursors. As the value of A_3 is gradually decreased, the peak (indicated by arrows) softens and shifts to lower temperatures. For values of $A_3 < 1$, the peak first splits apart from the bump and next disappears leaving the bump as the only surviving anomaly, which has been identified to be

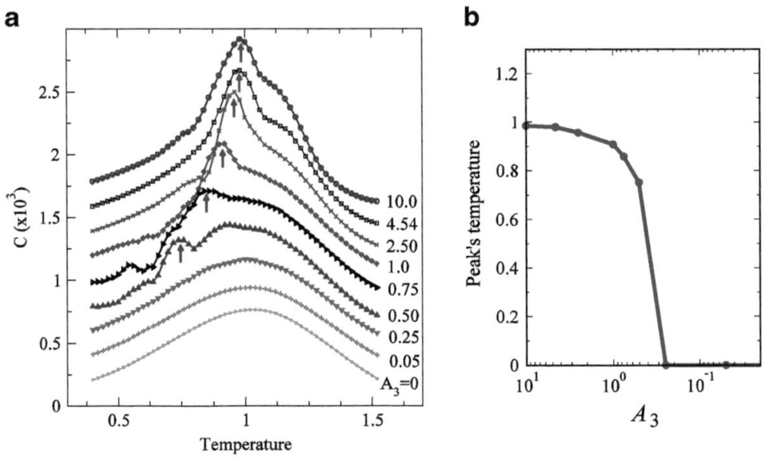

Fig. 12.7 (**a**) Heat capacity for different values of A_3. The position of the peak for each curve is indicated by an *arrow*. For the sake of clarity, the curves have been shifted conveniently. (**b**) Dependence of the temperature of the peak in C on the value of A_3

associated with the heterogeneities. Precisely, the suppression of the peak coincides with the suppression of twins (see Fig. 12.6) and therefore with the inhibition of the ferroelastic transition, whereas the remaining bump is associated with mottled-like textures. In Fig. 12.7b, we have plotted the temperature of the peak as a function of A_3. The emerging curve separates the transforming (inside the curve) and non-transforming regions in this space of parameters. We would like to mention that this behaviour is consistent with that shown in the phase diagram of Fig. 12.4 [29].

From the results above, it becomes clear that for low values of the parameter A_3, long-range anisotropic (elastic) interactions are weak and may actually be screened by the free-energy barriers erected by local disorder. Consequently, the long-range correlations shrink and local regions behave independently from each other.

Indeed, experiments are not performed as a function of the elastic anisotropy, which additionally does not change significantly inside a given alloy family. Usually, experimental studies are performed for a particular alloy family as a function of the doping or self-doping content. The increase in doping content implies an increase in the degree of compositional fluctuations (i.e. disorder) present in the system. Thus, it seems appropriate to analyse the effect of disorder on the behaviour of the system. This will be done in the next subsection.

12.4.2 *Effect of the Disorder: Frozen Glass State*

Figure 12.8 shows low-temperature strain-field configurations obtained for different values of both anisotropy ($\sim A_3$) and disorder ($\sim\zeta$). It can be observed that twins are obtained only in the range of low disorder. Moreover, the higher the anisotropy,

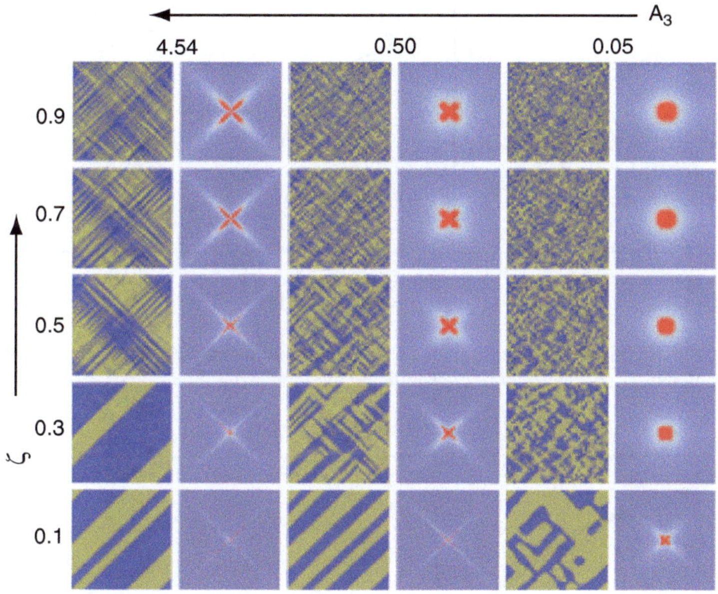

Fig. 12.8 Low-temperature structures obtained for different values of A_3 and ζ. We have also included the Fourier transform $(|F(e_2)|^2$

the larger the disorder required to inhibit the formation of twins. For illustration purposes, at the right side of each real-space configuration, we have included the corresponding intensity of the Fourier transform $(|\mathrm{F}(e_2)|^2)$, which is related to the diffraction pattern. They highlight the fact that directionality increases with increasing anisotropy and/or decreasing disorder. Moreover, the longitudinal length of the reciprocal pattern goes as the inverse of the characteristic length of the real-space structures, which confirms that an increase of the disorder entails a decrease in the characteristic length scale of the microstructures.

We notice that the suppression of the ferroelastic transition by means of increasing the amount of disorder has been experimentally observed in many different materials such as Ti–Ni [31], Ti–Ni(Fe) [32, 64], Ti–Pd(Cr) [54, 65], YBaCuO [66, 67] and Fe–Pd [68].

Motivated by recent zero-field-cooling/field-cooling (ZFC/FC) experiments in self-doping $Ti_{50-x}Ni_{50+x}$ shape-memory alloy [33], we have performed analogous simulation experiments, but with the applied external field being the stress. The corresponding contribution to the free energy is taken into account by appending a term $(-\sigma e_2)$ to the expression (12.1). The results are shown in Fig. 12.9 as a function of disorder and for three different values of anisotropy [35]. Indeed, deviation between ZFC and FC strain curves is obtained for the three values of anisotropy. Moreover, the deviation depends on the disorder in such a way that it occurs at higher temperatures with increasing disorder.

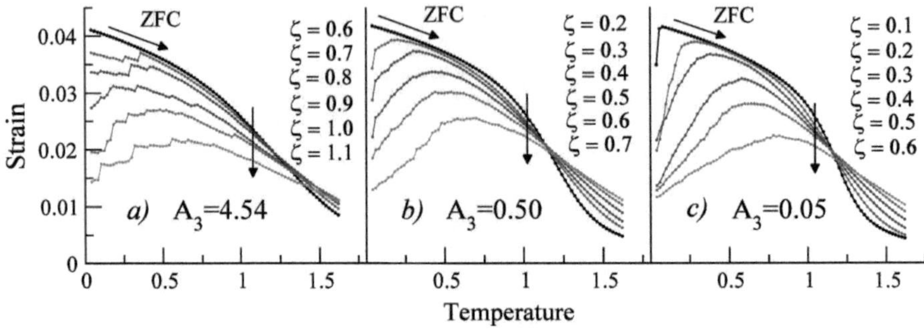

Fig. 12.9 ZFC/FC simulation experiments as a function of disorder ζ and for three different values of the parameter A_3. The *vertical arrows* indicate the increasing direction of disorder

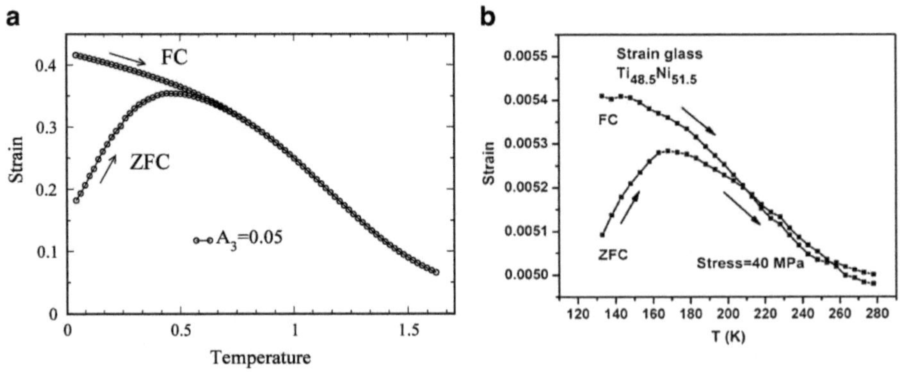

Fig. 12.10 ZFC/FC strain curves in the isotropic case. (**a**) from numerical simulations and (**b**) experiments

In summary, results in Fig. 12.9 corresponding to $A_3 = 0.05$ are in agreement with the experiments performed on the isotropic Ti–Ni shape-memory alloy [34]. Furthermore, they strongly suggest that the glassy frozen states are also likely to exist for highly anisotropic materials, provided the disorder present in the system is high enough.

Next, in Fig. 12.10 we compare [34] ZFC/FC strain curves obtained from numerical simulations (a) and from experiments in $Ti_{48.5}Ni_{51.5}$ strain glass (b). The agreement is indeed remarkable.

12.4.3 Thermomechanical Behaviour

It is well known that stoichiometric Ti–Ni, commonly known as Nitinol among engineers, is one of the most used shape-memory alloys for technological applications. Recently [69], it has been reported that the non-transforming, strain glass Ti–Ni also shows superelasticity and the shape-memory effect (see Fig. 12.11a). This certainly widens the horizons for new non-transforming shape-memory alloys.

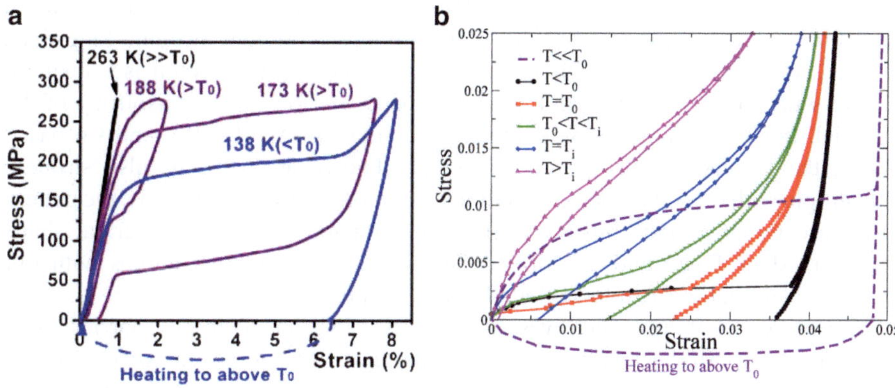

Fig. 12.11 (**a**) Stress–strain curves in glassy $Ti_{48.5}Ni_{51.5}$ ($T_0 \sim 160$ K) showing superelasticity and shape-memory effect (courtesy of Prof. X. Ren). (**b**) Stress–strain simulation results for the non-transforming glassy system, which also displays both superelasticity and shape-memory effect

We have used the model presented in Sect. 12.3 to study the effect of applying an external stress field in systems displaying transforming and non-transforming behaviour. We have obtained that, regardless of amount of disorder and value of the elastic anisotropy factor, the application of an external stress enables to induce a single martensitic variant, in agreement with experiments [69]. Depending on temperature either superelastic or pseudoplastic behaviour is found. In the latter case, upon unloading, the strain is not recovered, but instead a single variant state remains stable. However, upon heating the reverse transition occurs and the initial undeformed shape is recovered. Therefore, even if the martensitic transition is inhibited, shape-memory behaviour is operative. The results are shown in Fig. 12.11b. The dashed line corresponds to the shape-memory effect. Moreover, we also obtain (partial) superelastic effect at higher temperatures.

Although it is not shown here, it is worth mentioning that variations in the specific amount of disorder (and/or the anisotropy) lead to changes in the transition stress, hysteresis area, percentage of recovered strain and operative ranges. This is in agreement with experimental observations in a wide range of alloys [26,27,70,71]. In particular, results from the present model indicate that in the glassy frozen state these systems show large thermomechanical response in the sense that, for a given amount of disorder, at all temperatures, the stress necessary to reach a given strain decreases with decreasing anisotropy. This behaviour is certainly interesting from a practical point of view since it opens up new possibilities for designing more efficient shape-memory actuators.

12.5 Conclusions

Precursor nanoscale textures in ferroelastic martensites originate from the incipient instability intrinsic to the cubic character of the high-temperature parent phase. Such an instability (Zener instability) reflects in a low value of the elastic constant

C', which provides large <110><1–10> strain fluctuations. Eventually, when such fluctuations couple to local disorder (point defects), they become long-lived or even completely pinned. The cooperative response of the crystal is a textured pattern with a particular morphology that depends on the elastic anisotropy of the material. For highly anisotropic materials, such as Ni–Al-, Fe–Pd- and Cu-based alloys, the arising microstructure, commonly known as tweed, is also strongly anisotropic with cross-hatched correlations. For more isotropic materials, the pattern loses directionality and consists of tiny nanodomains of almost spherical shape as observed in Ti–Ni and Ti–Ni(Fe). This scenario completely agrees with our numerical simulations results included in this chapter.

The presence of disorder gives rise to a distribution of energy barriers that, above a certain critical amount (that depends on the elastic anisotropy), succeeds in screening the long-range anisotropic potential (12.5), breaking correlations and thus suppressing the transition to the martensitic twinned structure. In this case, the system shows glassy behaviour originating from kinetic freezing associated with long-time relaxation effects and characterized by the splitting of zero-field-cooling and field-cooling strain vs. temperature curves. Simulations suggest that frozen glassy states may exist independently of the elastic anisotropy, provided that the amount of disorder is appropriate. Actually, for low values of anisotropy, the required critical amount of disorder is also low and the system becomes short-ranged and purely disorder driven. However, for high values of anisotropy, the critical amount of disorder is high. Although in this case the twins are broken as well, glassy tweed textures are predicted to persist at low temperature, as a successful compromise between disorder and anisotropy. In this situation, the system is expected to behave both anisotropy driven and disorder driven (whereas twinning is only anisotropy driven). The general trends deduced from the simulations are also in very good qualitative agreement with recent experiments in Ti–Ni-based alloys where it has been shown that increasing point defect concentration prevents the martensitic transition to occur and instead the system shows glassy features. However, more experiments are needed in systems with selected values of the elastic anisotropy and disorder in order to confirm the specific features of the competition between these two parameters in determining the characteristics of martensitic transitions in ferroelastic materials. In any case, the concepts and results presented here are quite general and may apply to a wide variety of ferroic as well as multiferroic materials.

Acknowledgments The authors are grateful for fruitful, insightful and stimulating discussions with a number of researchers including Professors T. Kakeshita, K. Otsuka, X. Ren and T. Fukuda. We are also indebted to Prof. Y. Murakami and Prof. X. Ren for providing some of the pictures shown in this chapter. This work was supported by CICyT (Spain) Project No. MAT2007–61200 and the U.S. Department of Energy.

References

1. E. Dagotto, Complexity in strongly correlated electronic systems. Science **309**, 257 (2005)
2. T. Lookman, P. Littlewood, Nanoscale heterogeneity in functional materials. MRS Bull. **34**, 822 (2010)

3. N. Mathur, P. Littlewood, The third way. Nat. Mater. **3**, 207 (2004)
4. The name *tweed* was first coined by W.K. Armitage in his PhD. Thesis, University of Leeds, UK (1963)
5. L.E. Tanner, Diffraction contrast from elastic shear strains due to coherent phases. Philos. Mag. **14**, 111 (1966)
6. K. Enami, J. Hasunuma, A. Nagasawa, S. Nenno, Elastic softening and electron-diffraction anomalies prior to the martensitic transformation in a Ni-Al β_1 alloy. Scr. Metall. **10**, 879 (1976)
7. K. Enami, A. Nagasawa, S. Nenno, On the premartensitic transformation in the Ni-Al β_1 alloy. Scr. Metall. **12**, 223 (1978)
8. I.M. Robertson, C.M. Wayman, Tweed microstructures I. Characterization in ß-NiAl. Philos. Mag. A **48**, 421 (1983)
9. I.M. Robertson, C.M. Wayman, Tweed microstructures II. In several phases of the Ni-Al system. Philos. Mag. A **48**, 443 (1983)
10. I.M. Robertson, C.M. Wayman, Tweed microstructures III. Origin of tweed contrast in β and γ Ni-Al Alloys. Philos. Mag. A **48**, 629 (1983)
11. Y. Murakami, D. Shindo, K. Oikawa, R. Kainuma, K. Ishida, Magnetic domain structures in Co–Ni–Al shape memory alloys studied by Lorentz microscopy and electron holography. Acta Metall. **50**, 2173 (2002)
12. M. Porta, T. Castan, A. Planes, A. Saxena, Precursor nanoscale modulations in ferromagnets: Modelling and thermodynamic characterization. Phys. Rev. B **72**, 054111 (2005)
13. X. Meng, K.Z. Baba-Kishi, G.K.H. Pang, H.L. Chan, C.L. Choy, Tweed domains in ferroelectrics with compositions at the morphotropic phase boundary. Philos. Mag. Lett. **84**, 191 (2004)
14. O. Tikhomirov, H. Jiang, J. Levy, Local ferroelectricity in $SrTiO_3$ thin films. Phys. Rev. Lett. **89**, 147601 (2002)
15. Z. Xu, Myung-Chul Kim, Jie-Fang Li, Dwight Viehland, Observation of a sequence of domain-like states with increasing disorder in ferroelectrics. Philos. Mag. A **74**, 395 (1996)
16. Y. Kishi, M. De Graef, C. Craciunescu, T.A. Lograsso, D.A. Neumann, M. Wuttig, Microstructures and transformation behavior of CoNiGa ferromagnetic shape memory alloys. J. Phys. IV (France) **112**, 1021 (2003)
17. M. De Graef, Y. Kishi, Y. Zhu, M. Wuttig, Lorentz study of magnetic domains in Heusler-type ferromagnetic shape memory alloys. J. Phys. IV (France) **112**, 993 (2003)
18. H.S. Park, Y. Murakami, D. Shindo, V.A. Chernenko, T. Kanomata, Behavior of magnetic domains during structural transformations in Ni_2MnGa ferromagnetic shape memory alloy. Appl. Phys. Lett. **83**, 3752 (2003)
19. A. Saxena, T. Castán, A. Planes, M. Porta, Y. Kishi, T.A. Lograsso, D. Viehland, M. Wuttig, M. De Graef, Origin of magnetic and magnetoelastic tweed-like precursor modulations in ferroic materials. Phys. Rev. Lett. **92**, 197203 (2004)
20. T. Castán, E. Vives, L. Mañosa, A. Planes, A. Saxena, Disorder in Magnetic and Structural Transitions: Pretransitional Phenomena and Kinetics, in ed by A. Planes, Ll. Mañosa, A. Saxena. *Magnetism and Structure in Functional Materials,* Springer Series in Materials Science, vol 79. (Springer, Berlin 2005)
21. N. Mathur, P. Littlewood, Mesoscopic texture in manganites. Phys. Today **56**, 25 (2003)
22. K. Bhattacharya, *Microstructure of Martensite* (Oxford University Press., New York, 2003)
23. A.G. Khachaturyan, *The Theory of Structural Transformations in Solids* (Wiley, New York, 1983)
24. K. Otsuka, X. Ren, Physical metallurgy on Ti-Ni-based shape memory alloys. Prog. Mat. Sci. **50**, 511 (2005)
25. E.K.H. Salje, *Phase Transitions in Ferroelastic and Co-elastic Crystals* (Cambridge University Press., Cambridge, 1990)
26. R. Kainuma, Y. Imano, W. Ito, Y. Sutou, H. Morito, S. Okamoto, O. Kitakami, A.F.K. Oikawa, T. Kanomata, K. Ishida, Magnetic-field-induced shape recovery by reverse phase transformation. Nature Lett. **439**, 957 (2006)

27. Y. Wang, X. Ren, K. Otsuka, A. Saxena, Temperature-stress phase diagram of strain glass $Ti_{48.5}Ni_{51.5}$. Acta Mater. **56**, 2885 (2008)
28. J. Cui, Y. Chu, O. Famodu, Y. Furuya, J. Hattrick-Simpers, R. James, A. Ludwig, S. Thienhaus, M. Wuttig, Z. Zhang, I. Takeuchi, Combinatorial search of thermoelastic shape-memory alloys with extremely small hysteresis width. Nature Mat. **5**, 286 (2006)
29. R. Delville, S. Kasinathan, Z. Zhang, J.Van Humbeeck, R.D. James, D. Schryvers, Transmission electron microscopy study of phase compatibility in low hysteresis shape memory alloys. Philos. Mag. **90**, 177 (2010)
30. Y. Murakami, H. Shibuya, D. Shindo, Precursor effects of martensitic transformations in Ti-based alloys studied by electron microscopy with energy filtering. J. Microscopy **203**, 22 (2001)
31. S. Sarkar, X. Ren, K. Otsuka, Evidence for strain glass in the ferroelastic martensitic system $Ti_{50-x}Ni_{50+x}$. Phys. Rev. Lett. **95**, 205702 (2005)
32. M.-S. Choi, T. Fukuda, T. Kakeshita, H. Mori, Incommensurate- commensurate transition and nanoscale domain-like structure in iron doped Ti-Ni shape memory alloys. Philos. Mag. **86**, 67 (2006)
33. X. Ren, Y. Wang, Y. Zhou, Z. Zhang, D. Wang, G. Fan, K. Otsuka, T. Suzuki, Y. Ji, J. Zhang, Y. Tian, S. Hou, X. Ding, Strain Glass in ferroelastic systems: Premartensitic tweed versus strain glass. Philos. Mag. **90**, 141 (2010)
34. X. Ren, Y. Wang, K. Otsuka, P. Lloveras, T. Castan, M. Porta, A. Planes, A. Saxena, Ferroelastic nanostructures and nanoscale transitions: Ferroics with point defects. MRS Bull. **34**, 838 (2009)
35. P. Lloveras, T. Castan, M. Porta, A. Planes, A. Saxena, Glassy behavior in martensites: Interplay between elastic anisotropy and disorder in ZFC/FC simulation experiments. Phys. Rev. B **80**, 054107 (2009)
36. P. Lloveras, T. Castan, M. Porta, A. Planes, A. Saxena, Thermodynamics of stress-induced ferroelastic transitions: Influence of anisotropy and disorder. Phys. Rev. B **81**, 214105 (2010)
37. P. Lloveras, T. Castan, M. Porta, A. Planes, A. Saxena, Influence of elastic anisotropy on nanoscale textures. Phys. Rev. Lett. **100**, 165707 (2008)
38. C. Zener, Theory of strain interaction of solute atoms. Phys. Rev. **74**, 639 (1948)
39. J. Friedel, On the stability of the body centred cubic phase in metals at high temperatures. J. Physique Lett. **35**, L59 (1974)
40. A. Planes, Ll. Mañosa, Vibrational properties of shape-memory alloys. Solid State Phys. **55**, 159 (2001)
41. B.E. Warren, *X-Ray Diffraction* (Dover Publications, NewYork, 1990) p. 174
42. L.E. Tanner, A.R. Pelton, R. Gronsky, The characterization of pretransformation morphologies: periodic strain modulations. J. Physique *Colloque C4* **43**, 169 (1982)
43. D. Schryvers, L.E. Tanner, G. Van Tendeloo, Premartensitic Microstructures as seen in the High Resolution Electron Microscope: A Study of a Ni-Al Alloy, NATO ASSI on *Phase Stability,* Crete (1987)
44. S. Kartha, J.A. Krumhansl, J.P. Sethna, L.K. Wickham, Disorder-driven pretransitional tweed pattern in martensitic transformations. Phys. Rev. B **52**, 803 (1995)
45. S.M. Shapiro, J.Z. Larese, Y. Noda, S.C. Moss, L.E. Tanner, Neutron-scattering study of premartensitic behaviour in Ni-Al alloys. Phys. Rev. Lett. **57**, 3199 (1986)
46. S. Muto, S. Takeda, R. Oshima, Analysis of lattice modulations in the tweed structures of an Fe-Pd alloy by image processing of a high-resolution electron micrograph. Jap. J. Appl. Phys. **29**, 2066 (1990)
47. S. Muto, R. Oshima, F.E. Fujita, Elastic softening and elastic strain energy consideration in the FCC-FCT transformation of Fe-Pd alloys. Acta Metal. Mater. **38**, 685 (1990)
48. D. Schryvers, D.E. Lahjouji, B. Slootmarkers, P.L. Potapov, HREM investigations of martensite precursor effects and stacking sequences in Ni-Mn-Ti alloys. Scr. Mater. **35**, 1235 (1996)
49. G. Van Tendeloo, M. Chandrasekaran, F.C. Lovey, Modulated microstructures in β Cu-Zn-Al. Metall. Trans. A **17**, 2153 (1986)
50. A. Planes, Ll. Mañosa, E. Vives, Vibrational behavior of bcc Cu-based shape-memory alloys close to the martensitic transition. Phys. Rev. B **53**, 3039 (1996)

51. X. Ren, N. Miura, J. Zhang, K. Otsuka, K. Tanaka, M. Koiwa, T. Suzuki, Y. Chumlyakoz, M. Asai, A comparative study of elastic constants of Ti-Ni-based alloys prior to martensitic transformation. Mat. Sci. Eng. A **312**, 196 (2001).
52. J. Worgull, E. Petit, J. Trevisono, Behavior of the elastic properties near an intermediate phase transition in Ni_2MnGa. Phys. Rev. B **54**, 15695 (1996)
53. E. Cesari, V.A. Chernenko, V.V. Kokorin, J. Pons, C. Seguí, Internal friction associated with the structural phase transformations in Ni-Mn-Ga alloys. Acta Mater. **45**, 999 (1997)
54. S.M. Shapiro, G. Xu, B.L. Winn, D.L. Schlagel, T. Lograsso, R. Erwin, Anomalous phonon behaviour in the high-temperature shape-memory alloy $Ti_{50}Pd_{50-x}Cr_x$. Phys. Rev. B **76**, 054305 (2007)
55. S.M. Shapiro, G. Xu, B.L. Winn, D.L. Schlagel, T. Lograsso, R. Erwin, Phonon precursors to the high temperature martensitic transformation in $Ti_{50}Pd_{48}Cr_2$. J. Phys. IV France **112**, 1047 (2003)
56. A.J. Schwartz, L.E. Tanner, Phase transformations and phase relations in the TiPd-Cr pseudobinary system I. Experimental observations. Scr. Metall. Mater. **32**, 675 (1995)
57. D. Shindo, Y. Murakami, Advanced transmission electron microscopy study on premartensitic state of $Ti_{50}Ni_{48}Fe_2$. Sci. Technol. Adv. Mater. **1**, 117 (2000)
58. Z. Zhang, Y. Wang, D. Wang, Y. Zhou, K. Otsuka, X. Ren, Phase diagram of $Ti_{50-x}Ni_{50+x}$: Crossover from martensite to strain glass. Phys. Rev. B **81**, 224102 (2010)
59. S. Nagata, P. Keesom, H.R. Harrison, Low-dc-field susceptibility of CuMn spin glass. Phys. Rev. B **19**, 1633 (1979)
60. D. Viehland, J.F. Li, S.J. Jang, L.E. Cross, M. Wuttig, Glassy polarization behaviour of relaxor ferroelectrics. Phys. Rev. B **46**, 8013 (1992)
61. Y. Wang, X. Ren, K. Otsuka, A. Saxena, Evidence for broken ergodicity in strain glass. Phys. Rev. B **76**, 132201 (2007)
62. T. Lookman, S.R. Shenoy, K.Ø. Rasmussen, A. Saxena, A.R. Bishop, Ferroelastic dynamics and strain compatibility. Phys. Rev. B **67**, 02411 (2003)
63. See for instance, F. Falk, Model free energy, mechanics, and thermodynamics of shape memory alloys. Acta Metall. **28**, 1773 (1980)
64. M.-S. Choi, T. Fukuda, T. Kakeshita, Anomalies in resistivity, magnetic susceptibility and specific heat in iron-doped Ti-Ni shape memory alloys, Scr. Mater. **53**, 869 (2005)
65. Y. Zhou, D. Xue, X. Ding, K. Otsuka, J. Sun, X. Ren, High temperature strain glass in $Ti_{50}(Pd_{50-x}Cr_x)$ alloy and the associated shape memory effect and superelasticity. Appl. Phys. Lett. **95**, 151906 (2009)
66. Y. Xu, M. Suenaga, J. Tafto, R.L. Saba, A.R. Moodenbaugh, P. Zolliker, Microstructure, lattice parameters and superconductivity of $YBa_2(Cu_{1-x}Fe_x)O_{7-\delta}$ for $0 < x < 0.33$. Phys. Rev. B **39**, 6667 (1989)
67. W.W. Schmahl, A. Putnis, E.K.H. Salje, P. Freeman, A. Graeme-Barber, R. Jones, K.K. Singh, J. Blunt, P.P. Edwards, J. Loran, K. Mirza, Twin formation and structural modulations in orthorhombic and tetragonal $YBa_2(Cu_{1-x}Co_x)_3O_{7-\delta}$. Philos. Mag. Lett. **60**, 214 (1989)
68. R. Oshima, M. Sugiyama, F. Fujita, Tweed structures associated with Fcc-Fct transformations in Fe-Pd alloys. Metall. Trans. A **19**, 803 (1988)
69. Y. Wang, X. Ren, K. Otsuka, Shape memory effect and superelasticity in a strain glass alloy. Phys. Rev. Lett. **97**, 225703 (2006)
70. N. Nakanishi, T. Mori, S. Miura, Y. Murakami, S. Kachi, Pseudoelasticity in Au-Cd thermoelastic martensite. Philos. Mag. **28**, 277 (1973)
71. V.A. Chernenko, V. L'vov, J. Pons, E. Cesari, Superelasticity in high temperature Ni-Mn-Ga alloys. J. Appl. Phys. **93**, 2394 (2003)

Chapter 13
Metastability, Hysteresis, Avalanches, and Acoustic Emission: Martensitic Transitions in Functional Materials

Martin-Luc Rosinberg and Eduard Vives

Abstract We review several aspects of the dynamics of first-order phase transitions in functional materials. In particular, we focus on recent models of the athermal evolution in driven ferromagnets that provide a global picture of metastability and hysteresis, and show that first-order phase transitions in these systems proceed by avalanches. Within this theoretical framework, we discuss recent experiments on acoustic emission avalanches in structural phase transitions.

13.1 Introduction

Functional materials are based on the interplay of different ferroic properties such as ferroelasticity, ferromagnetism, and ferroelectricity. In order to obtain a sufficiently large response to the external excitation, these materials are typically tuned so as to cross a first-order phase transition (FOPT), in which one or several order parameters (strain, magnetization, polarization, etc.) exhibit a macroscopic discontinuity. It is, thus, important to understand the dynamics of FOPTs in such materials for applications.

M.-L. Rosinberg (✉)
Laboratoire de Physique Théorique de la Matière Condensée, Université Pierre et Marie Curie, 4 Place Jussieu, 75252 Paris, France
e-mail: mlr@lptmc.jussieu.fr

E. Vives
Facultat de Física, Departament d'Estructura i Constituents de la Materia, Universitat de Barcelona, Martí i Franquès 1, 08028 Barcelona, Catalonia, Spain

Institut de Nanociència i Nanotecnologia (IN2UB), Universitat de Barcelona, Barcelona, Catalonia, Spain

Department of Physics, University of Warwick, Coventry CV4 7AL, UK
e-mail: eduard@ecm.ub.es

T. Kakeshita et al. (eds.), *Disorder and Strain-Induced Complexity in Functional Materials*, Springer Series in Materials Science 148,
DOI 10.1007/978-3-642-20943-7_13,

FOPT in solids hardly occurs at thermal equilibrium. Typically, the energy barriers involved in the transition are very large compared to thermal fluctuations so that the order parameters evolve following metastable trajectories. The transitions are then called athermal, and instead of being sharp like in ideal FOPTs, they extend over a broad range of the driving parameters and show hysteresis. In many cases, the hysteresis (or at least a part of it) cannot be suppressed by driving the system more slowly because it is not related to the fact that the system cannot respond instantaneously due to relaxational delay. This kind of hysteresis is usually called rate-independent hysteresis.

The high energy barriers have two origins: on the one hand, real materials always exhibit some amount of quenched disorder that determines the nucleation sites and can thus strongly affect the metastable path. On the other hand, when one of the order parameters involved in the transition is strain (like in martensitic transformations), a complex microstructure naturally arises at the FOPT due to the symmetry differences between the parent and product phases. This also strongly affects the metastable trajectory and the hysteresis.

In Sect. 13.2, we shall introduce very simple models of athermal evolution in driven ferromagnets. They give us a global picture of the relationship between metastability and hysteresis and show that athermal FOPT in the presence of disorder proceeds via avalanches. This means that the response of the system to a smooth driving consists in a sequence of discontinuous jumps of the order parameter separated by periods of inactivity. Microscopically, avalanches are associated with the motion of an interface and/or with the nucleation of a domain of the new phase. These models also describe how the statistical distribution of the avalanche sizes changes with the amount of disorder and how avalanches and hysteresis depend on the driving mechanism, temperature, driving rate, number of cycles through the transition, etc. In magnetic materials, the signature of the avalanche dynamics is the so-called Barkhausen noise, which can be monitored by using a pick-up coil. Similar phenomenology is observed in other disordered systems, for instance, in ferroelectrics or superconductors.

In Sect. 13.3, we focus on the ferroelastic case of structural phase transitions where avalanches can be recorded as acoustic emission (AE) events. These are produced when an interface separating two different crystallographic structures advances producing an elastic wave (typically with frequencies in the ultrasonic range) which propagates through the material and can be recorded at the surface by an appropriate transducer. Although a satisfactory theoretical framework to interpret the results of AE experiments is still lacking, the interaction between experiments and theory over the past 15 years has been intense and fruitful. We hope that this short review will help to clarify the status of some recent advances and contribute to further progress.

13.2 What Can We Learn from Simple Models?

On general grounds, one can relate the intermittent and hysteretic response of a disordered system to slowly changing external conditions to the existence of a corrugated (free) energy landscape. At low enough temperature this landscape is indeed characterized by an enormous number of local minima (or metastable states) and the energy barriers are so large that thermally activated processes play a negligible role. As stressed above, true equilibrium is then never reached on experimental timescales and the system can only move from one metastable state to another as the external control parameter (e.g., strain, magnetic field, pressure, or temperature) is changed and the initial state in which the system was trapped becomes unstable. This collective event (avalanche) is usually very fast, at least compared to the rate of variation of the external parameter, and this results in a jump discontinuity in the nonequilibrium response. One then often considers the so-called adiabatic limit in which the rate is merely taken to zero.

Can we go beyond such general statements? For instance, what can be said about the number of metastable states, their energy, or their magnetization (in the remainder of this section, we shall most often refer to magnetic systems as illustration)? What is the relationship between the organization of the states and the shape of the saturation hysteresis loop obtained by cycling the field between large negative and positive values? What is the influence of the driving mechanism on the dynamical response (although one often controls an intensive external force or field, in other situations, e.g., in experiments with ferroelastic materials, one usually controls the strain, which is an extensive quantity, instead of the stress). What is the statistical distribution of the size and duration of the avalanches? Why is a power-law (scale-free) behavior extending over several decades so often observed? Is this associated with a nonequilibrium critical point and then what is the range of the critical regime? In which cases do microscopic details affect large-scale events and in which cases are they irrelevant?

Such questions clearly touch fundamental issues in the theory of disordered systems and to answer them (or at least some of them), it has proven useful to consider models that are simple enough to allow for a partial analytical description and for extensive numerical studies. Perhaps the simplest (and yet not fully understood) prototype is the zero-temperature nonequilibrium random-field Ising model (RFIM) proposed in 1993 by J.P. Sethna, J.A. Krumshansl, and their collaborators as a model for hysteresis and crackling noise in disorder-driven first-order phase transformations [1]. This model, which contains the most important physical ingredients (quenched-in disorder, interactions, and external control parameter), has been intensively studied over the past 15 years and has been applied to many different physical situations, from fluid invasion inside porous solids to group decision making (we refer the reader to [2] for a comprehensive review). In particular, the RFIM appears to be the convenient theoretical framework to understand the hysteresis behavior associated with the capillary condensation of gases in amorphous porous solids. In this case, the driving force is the gas pressure in the external reservoir, the order parameter is the amount of adsorbed gas inside the solid, and temperature is just a parameter that changes the topology of the free-energy landscape (but is still too low for making activated processes efficient and inducing rate-dependent hysteresis effects) [3]. A nice example coming from low-temperature physics is the condensation of helium in silica aerogels where a description in terms of random (but correlated) fields gives a rationale to the changes in the shape of the hysteresis loop with porosity and temperature [4, 5].

The RFIM is defined by the Hamiltonian

$$\mathcal{H} = -J \sum_{<i,j>} s_i s_j - \sum_i (H + h_i) s_i, \tag{13.1}$$

where $\{s_i\}$ are N Ising spins placed on the sites of a lattice (e.g., a cubic lattice), $J > 0$ is a ferromagnetic coupling between nearest-neighbor spins, H is the external field, and $\{h_i\}$ is a set of uncorrelated random fields usually drawn from a Gaussian distribution probability with zero mean and standard deviation Δ.

The zero-temperature metastable evolution induced by the external field consists in a single-spin-flip dynamics: metastable states are thus defined by the condition

$$s_i = \mathrm{sign}(f_i), \tag{13.2}$$

where $f_i = J \sum_{j/i} s_j + H + h_i$ is the effective local field, and a spin flips when its local field changes sign. To stay in the adiabatic limit, the external field H is kept constant during the propagation of an avalanche. (Note that a two-spin-flip dynamics has also been considered recently to test the robustness of the model behavior with respect to an additional relaxation process [6].) The most salient feature of the model is the existence of two regimes of avalanches depending on the amount of disorder (i.e., the value of Δ). In strong disorder, spins mostly flip individually so that avalanches are of microscopic size and the magnetization curve is smooth macroscopically. On the other hand, in the low disorder regime, spins tend to flip collectively, which results in a system spanning avalanche seen as a macroscopic jump in the magnetization curve. In between, there is a critical disorder Δ_c and a critical field H_c at which avalanches of all sizes are observed. The avalanche size distribution then follows on long length scales a power-law behavior as $p(S) \propto S^{-(\tau+\sigma\beta\delta)} \mathrm{d}S$, where τ, σ, β, and δ are universal critical exponents [2].

We now focus on two issues that were not discussed in [2].

13.2.1 Relationship Between Hysteresis and the Distribution of Metastable States

Let us first discuss the issue of the number and distribution of metastable states in the field-magnetization plane. Since we are interested in the relationship to the hysteresis loop induced by a variation of the magnetic field H, the relevant quantity is not the total number of metastable states, but the number of states with a given magnetization at a given external field $N(m, H)$ (determining the *full* topology of the energy landscape is also a very interesting and challenging issue that has received some attention in recent years [7]). On general grounds, one expects $N(m, H)$ to scale exponentially with the system size, say the number N of elementary domains, spins, etc. It is, thus, the logarithm of $N(m, H)$, which is an extensive quantity like the free energy. Since this quantity is sample-dependent, the physically relevant quantity is $\overline{\log N(m, H)}$, where the average, denoted here by the overbar, is taken over a representative set of disorder realizations. This leads to define the magnetization-dependent *quenched complexity* as

$$\Sigma_Q(m, H) = \lim_{N\to\infty} \frac{1}{N}\overline{\log N(m, H)}. \tag{13.3}$$

Note that we consider the average of the logarithm and not the logarithm of the average (the so-called *annealed* average) because the hysteresis loop is a self-averaging quantity (sample-to-sample fluctuations vanish in the thermodynamic limit) and we thus need to describe the behavior of a *typical* sample. Considering the annealed average is misleading since there exist a certain number of atypical samples that give a finite contribution to $\log \overline{N(m, H)}$ [8, 9] (on the other hand, computing the quenched complexity is much more difficult task).

The crucial point is that the hysteresis loop in the strong disorder regime (i.e., when the loop is smooth) is just the convex envelope of the set of metastable states in the field-magnetization plane and identifies with the contour $\Sigma_Q(m, H) = 0$. This is still a kind of conjectural statement, but it is strongly supported by (a) an exact theorem and (b) analytical and numerical calculations. The exact theorem is the so-called *no-passing* rule [1] which applies to systems with ferromagnetic interactions only (or to elastic media with a convex elastic potential). The no-passing rule states that the $T = 0$ metastable dynamics conserves the partial ordering of the states: in other words, a spin never flips back when the field is varied monotonically. This is sufficient to prove the remarkable property of return point memory [1], which is observed in many experimental systems, and this also implies that there are no metastable states outside the hysteresis loop in a given disorder sample (and therefore, on average, the density of metastable states scales to zero exponentially outside the loop [8]). On the other hand, the no-passing rule does not imply that the number of metastable states is exponentially large (and therefore comparable to the total number of states) everywhere *inside* the loop. Of course, it is known experimentally that there are many metastable states inside the loop, as illustrated by the so-called "scanning" curves obtained by reversing the field (or the stress) before reaching saturation (or complete phase transformation). However, these states that are reachable by a field history starting from one of the two saturation states only represent a negligible subset of the whole set of metastable states, albeit probably the most interesting one. In fact, very little is known about their actual number [10]. Therefore, in principle, a region could exist in the vicinity of the hysteresis loop where the number of metastable states is only subexponential (and thus $\Sigma_Q(m, H) = 0$). Such a region, however, does not exist in the 1D RFIM for which the quenched complexity has been computed analytically [8]. There is also good numerical evidence that the same is true in three dimensions [9] and we believe that this is a general feature.

The situation in the low disorder regime is different as one does not expect the envelope of the metastable states to be convex anymore. This conjecture is again based on numerical and analytical calculations of the complexity [9, 11], but it also quite naturally explains the presence of a finite jump in the magnetization curve because m must always be a monotonic function of H (on general grounds, the "susceptibility" $\mathrm{d}m/\mathrm{d}H$ must be a positive quantity (see also [12])).

These predictions are nicely illustrated by the soft-spin version of the RFIM in the infinite range limit where each spin is now a continuous variable taking values between $-\infty$ and $+\infty$ and is coupled to all other spins with coupling J/N, as described by the Hamiltonian

$$\mathcal{H} = -\frac{J}{2N}\sum_{i\neq j} s_i s_j - \sum_i (H + h_i)s_i + \sum_i V(s_i) \tag{13.4}$$

where $V(s) = (k/2)[s - \mathrm{sign}(s)]^2$ is a double-well potential that mimics the two states of the hard-spin model. The metastable states are now solutions of the equation

$$s_i - \mathrm{sign}(s_i) = \frac{Jm + H + h_i}{k}. \tag{13.5}$$

The mean-field character of the model allows one to compute the hysteresis loop and the complexity $\Sigma(m, H)$ analytically [13]. Some typical results are shown in Figs. 13.1 and 13.2 (note that the magnetization does not saturate when $H \to \pm\infty$ because the spins are unbounded variables).

In the small disorder regime, the shape of the curve Σ vs.m changes drastically with H. For $H = 0$, the complexity varies continuously with m, reaches a maximum at $m = 0$ (which is thus the most probable magnetization of the metastable states), and vanishes at $m \approx \pm 1.5$, which are exactly the values of the magnetization along the two branches of the hysteresis loop. On the other hand, for a larger field (e.g., $H = 2$), the accessible magnetization domain breaks into two disjoint intervals with no metastable states in between (in the interval $1 \leq m \leq 2.35$ for $H = 2$): this is at the origin of the finite jump in the ascending branch of the hysteresis loop, as can be seen in Fig. 13.1. One may notice some resemblance of this scenario (a phase transition induced by a disconnected order parameter space) with the ergodicity-breaking scenario observed in systems with long-range interactions (see e.g., [14]). However, in the present case, we believe that this is not a consequence of the mean-field character of the model and that this scenario is very general.

These results have some interesting consequences. First, on the theoretical side, because they bring up the possibility of studying the hysteresis loop without following the dynamical evolution, which may prove useful to resolve the pending

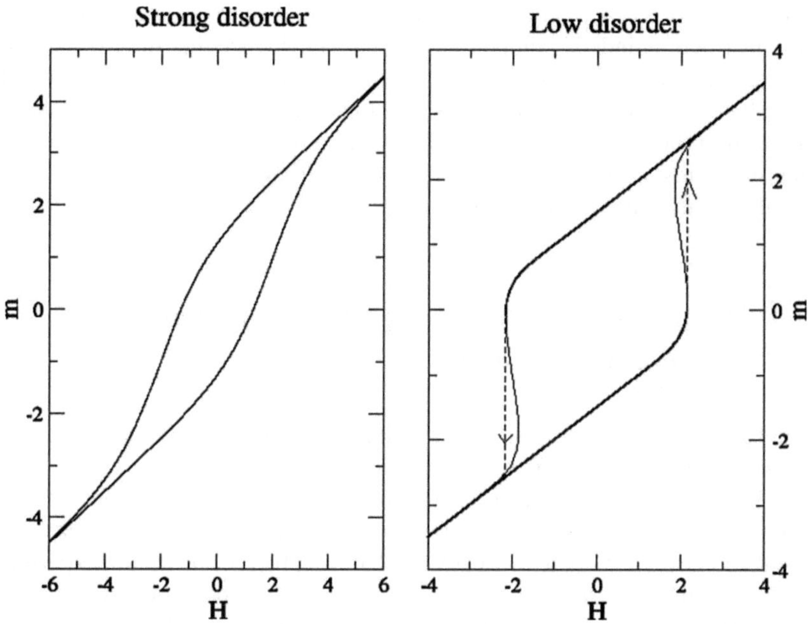

Fig. 13.1 The soft-spin version of the mean-field RFIM: hysteresis loop (*dashed line*) and contour $\Sigma_Q(m, H) = 0$ (*solid line*) for $\Delta = 0.8$. For strong disorder, the two curves coincide. For small disorder, the contour $\Sigma_Q(m, H) = 0$ is reentrant and the magnetization curve has a finite jump

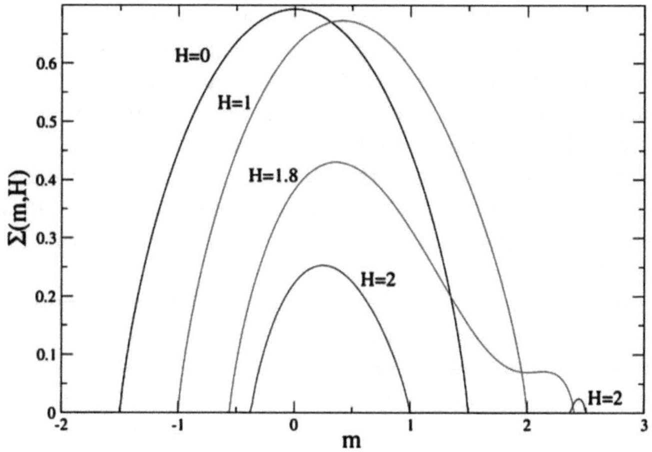

Fig. 13.2 Complexity vs. magnetization for different values of the magnetic field in the low disorder regime

issue of the universality of equilibrium and nonequilibrium disorder-induced phase transitions (there is indeed compelling numerical evidence that the two transitions belong to the same universality class [15], but there is no convincing theoretical proof so far). Secondly, because the distribution of the metastable states in the field-magnetization (or strain–stress) plane also gives a rationale for the influence of the driving mechanism on the nonequilibrium hysteretic response. This is an issue of practical relevance, as we shall now discuss.

13.2.2 Influence of the Driving Mechanism and the Effect of Long-Range Forces

A solid bar can be put into tension by specifying either the load that is placed upon it (or hangs from it) – a "soft" loading device, or its elongation – a "hard" loading device. In a ferromagnetic material, one usually measures the magnetic flux as a function of the applied magnetic field, but one can also make the field slave of the magnetization by using some feedback mechanism [16]. In adsorption experiments, depending on the size of the gas reservoir connected to the experimental cell, the isotherms could in principle evolve from a "grand-canonical" to a "canonical" type [17]. More generally, depending on the system under consideration, one may control either the externally applied field (stress, magnetic field, gas pressure, etc.) or the thermodynamically conjugated variable (strain, magnetization, mass of the adsorbed gas, etc.). At equilibrium the response does not depend on which is the control variable, but what happens far from equilibrium when the response to a smooth external driving is a sequence of avalanches that reflect irreversible transitions between metastable states? What are the differences between the two situations? Since there are very few examples in which the two experimental setups have been used with the same disordered sample, the hysteresis loops shown in Fig. 13.3 are particularly interesting. They were obtained with a $Cu_{68}Zn_{16}Al_{16}$ single crystal under strain-driven and stress-driven conditions [18] and a soft machine was especially designed for this experiment to finely monitor the external force due to a dead load hanging from the sample.

The most striking feature of these curves is that almost the entire strain-driven loop is enclosed within the stress-driven one, showing that the dissipated energy is much larger in the second case. Moreover, the strain-driven curve exhibits a yield point upon loading and a reentrant behavior that do not exist with the other device in which there is a macroscopic instability when the martensitic transition starts. Although the microscopic mechanisms at the origin of hysteresis are specific to each particular system, it appears that the same general features are observed in other disordered materials undergoing athermal first-order transition, for instance, in magnets [16].

Analyzing the soft-spin random-field model is again helpful to reach a global (though admittedly crude) interpretation of the experimental observations (note that Ising spins are inappropriate to study a hard-driving situation because the energy

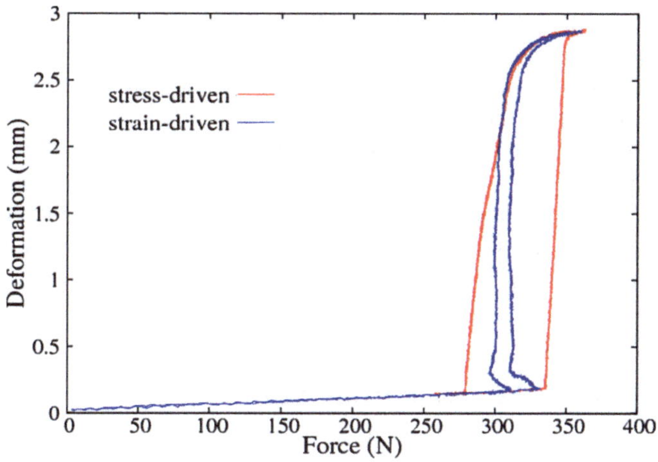

Fig. 13.3 Stress–strain hysteresis loops in a Cu–Zn–Al single crystal obtained under stress-driving or strain-driving conditions

wells have no finite curvature, which induces a degenerate and unphysical behavior [19]). When the magnetization is controlled, the system tries to (partially) minimize its internal energy while satisfying the global constraint $\sum_i s_i = Nm$. It, thus, visits a sequence of single-spin-flip stable states that is different from the one visited in the field-driven case. In the mean-field model, it can be shown that (13.5) is now replaced by

$$s_i - \text{sign}(s_i) = m + \frac{h_i}{k} - \frac{1}{N}\sum_j \text{sign}(s_j). \tag{13.6}$$

The last term in the right-hand side of this equation is an *antiferromagnetic* contribution that plays the role of an infinite-range demagnetizing field. Such a field is often introduced to mimic the effects of boundaries or other long-ranged interactions [20]. It changes the system behavior drastically and leads to self-organized criticality (whereas criticality in the standard field-driven RFIM requires a fine tuning of the disorder). Equation (13.6) shows that this is also a natural ingredient of a hard-driving device. A more sophisticated version of this argument can be found in [21], where a disordered spin model whose critical behavior changes continuously as one moves from soft to hard driving is introduced.

It turns out that the response of the system can be determined exactly when using a very natural relaxation dynamics that states how to go from a metastable state solution of (13.6) to the nearest one when m is changed adiabatically. Remarkably, the response is found to *always* coincide with the contour $\Sigma_Q(m, H) = 0$. In other words, the hard-driving device forces the system to follow the boundary of the domain of existence of the metastable states. In the low disorder regime, where the contour $\Sigma_Q(m, H) = 0$ is reentrant (see Fig. 13.1), the magnetization-driven hysteresis loop is thus also reentrant as observed experimentally. These conclusions are in agreement with numerical calculations performed on the metastable RFIM at finite temperature with a local mean-field theory [22]. We believe that these results reflect the general behavior of a hard-driving device.

Finally, we notice that a very recent work [23] shows the equivalence in the continuum limit of the mean-field RFIM with a demagnetizing factor to the celebrated ABBM model [24] that describes the motion of a single domain wall in a random energy landscape. This unifies two rival mean-field theories and shows that it is not necessary to assume the existence of an interface from the beginning. In fact, in the hard driving, such an interface is spontaneously created and the system self-organizes at the critical depinning threshold [21].

13.3 What Can We Learn from Acoustic Emission Detection?

One of the motivations of the seminal work of [1] was its possible applicability to the description of structural transitions in ferroelastic materials, specifically martensitic transitions. The paper also pointed out the parallelism between the Barkhausen noise in ferromagnets and AE signals. This motivation was later partially forgotten because the model was mainly used to interpret various experimental results in magnetic materials.

Acoustic emission has been used for decades to characterize many different processes [25]. From an engineering point of view, the technique has been quite successful in monitoring and preventing mechanical failure in solids; it is nowadays the base of many nondestructive testing tools. We shall here focus on the applicability of the technique to the study of structural phase transition in solids. In some aspects, this technique plays a role similar to other characterization techniques such as calorimetry or resistivity measurements.

The physics behind the source of AE is still far from being fully understood. When a new domain nucleates or an existing interface moves inside the material, an elastic wave is emitted. It propagates through the material and can be detected at the surface by an appropriate transducer. Typically, the observed pulses are ultrasonic, with frequency components in the range 20 kHz–2 MHz. Within a continuum mechanics description, an AE event can be naively modeled as the sudden creation of a displacement discontinuity [26]. But little is known about the dynamics (acceleration, duration, etc.) of this displacement. A promising recent work [27] may help to clarify this issue.

If the dynamics of this source event were known, the integration of Christoffel equations would allow to predict the AE waves, just like Maxwell Equations are integrated to predict the electric field induced by a sudden magnetization change in the sample. In the magnetic case, the advantage is that one can use detection coils and apply Faraday's law to predict the induced electromagnetic force, thus avoiding an integration that would be otherwise difficult.

Therefore, from the information contained in the detected AE signals, it is very difficult to recover the information about the source. Many of the works that will be mentioned in the following are based on the very simple idea that the maximum amplitude of the detected signals is proportional to the speed of the advancing front [28]. Other fundamental questions remain difficult to answer, as for instance the problem of the spatial localization of the source.

In the study of structural transitions, acoustic emission experiments essentially exploit two basic techniques:

13.3.1 Pulse-Counting Technique

This technique is mostly used to characterize the transition. It consists in counting the number of avalanches (also called events or hits) per unit time ($\mathrm{d}N/\mathrm{d}t$) with an amplitude above a certain threshold. When the transition is driven by changing the temperature or another external parameter, this number of events per unit time (frequency) can be converted into the so-called activity $A(T)$ (number of events per degree, or number of events per force interval, etc.) by simply dividing by the driving rate: $A(T) = \left(\frac{\mathrm{d}N}{\mathrm{d}T}\right) / \left(\frac{\mathrm{d}T}{\mathrm{d}t}\right)$.

With this technique, one gets information about the existence of avalanches and the "density" of metastable states along a particular path and in a certain range of the control parameter. A typical result is shown in Fig. 13.4 in the case of the temperature-driven cubic-tetragonal transition in a Fe–Pd alloy [29]. The curves correspond to a polycrystalline sample (top) and to a single crystal (bottom).

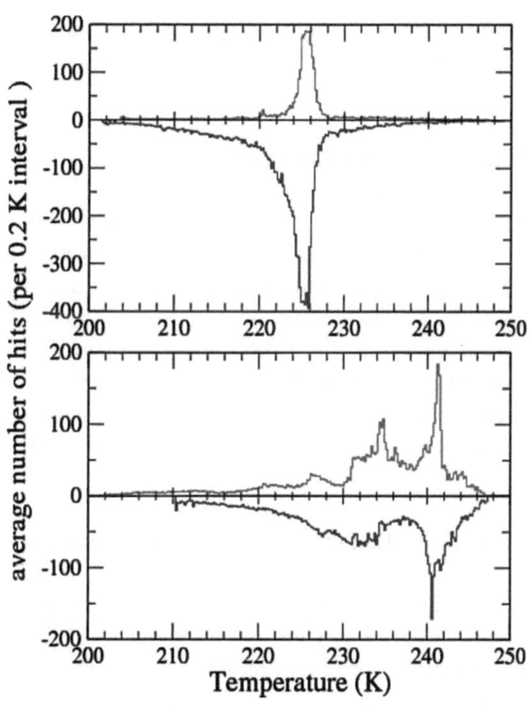

Fig. 13.4 Acoustic emission activity corresponding to a $Fe_{68.8}Pd_{31.2}$ alloy [29]. The figures correspond to a polycrystalline sample (*top*) and to a single crystal (*bottom*). *Red lines* (positive) correspond to data obtained on heating runs at 1 K/min and *blue lines* (negative) to cooling runs at -1 K/min

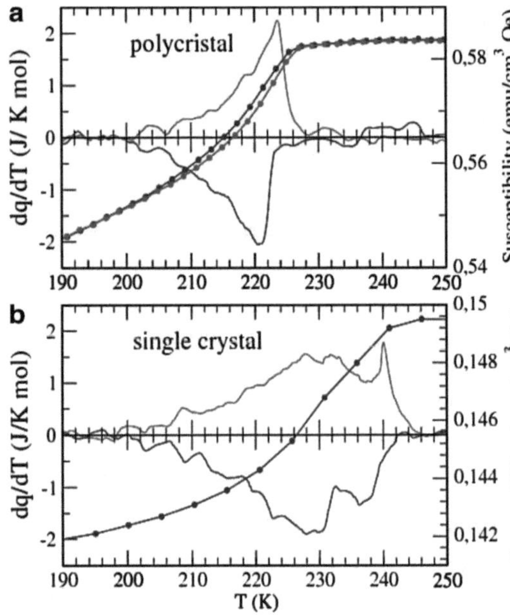

Fig. 13.5 Calorimetric and susceptibility measurements for the same samples as in Fig. 13.4 [29]

For comparison, Fig. 13.5 shows the calorimetric and susceptibility curves for the same samples.

Although the measurement of the activity may seem to provide little extra information, some important conclusions have been obtained by this technique:

13.3.1.1 Transition Temperature

Since AE is much more sensitive than calorimetry, it allows a very accurate measurement of the temperatures at which the transition starts and ends, far beyond the traditional concepts of M_s and A_f temperatures (which correspond to calorimetric estimations of the transformation of 10 and 90% of the sample volume). As an example, Fig. 13.6 shows a magnification of Fig. 13.4 revealing that the starting points of the transition for the polycrystalline sample and the single crystal are the same. It would have been impossible to extract this information from the calorimetric measurements shown in Fig. 13.5.

13.3.1.2 Athermal and Adiabatic Character of the Transition

As was emphasized in Sect. 13.2, in order to exhibit true avalanches a system must behave athermally (thermal fluctuations play no role) and adiabatically (avalanches occur infinitely fast compared to the driving rate). Is it possible to test to what extent these two extreme assumptions are fulfilled in experiments? A first answer comes

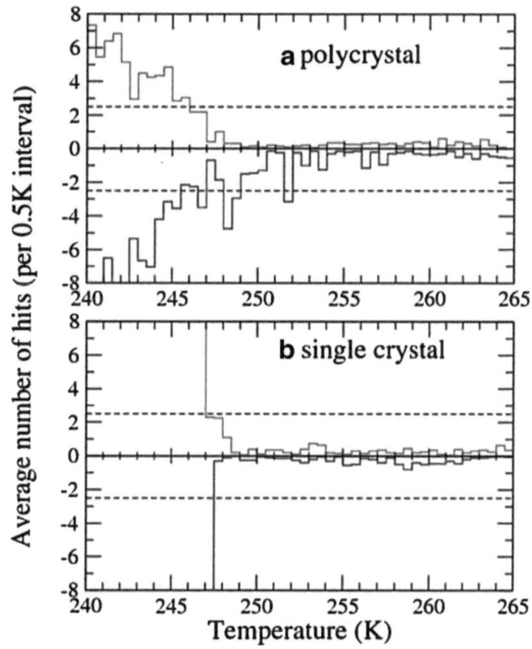

Fig. 13.6 Magnification of Fig. 13.4 showing that the transformation starts at the same temperature in both samples. The *dashed line* indicates the noise level [29]

from the analysis of the dependence of the activity function with the driving rate. As an example, Fig. 13.7 shows the data for a Fe–Pd single crystal recorded at three different rates. Although the overlap is not perfect, it is clear that many features of the curves remain unaffected by a change in the driving rate by two orders of magnitude. This overlap (or scaling) is a clear signature that both assumptions (athermal and adiabatic) are satisfied within the range 0.1–10 K/min.

However, the scaling is expected to only occur in a certain range of the driving rate. It will not be observed at high driving rates due to the overlap of the avalanches that will necessarily reduce their number (nonadiabatic behavior). It will also be lost at very slow driving rates due to the occurrence of thermal relaxations (nonathermal behavior). The activity will then be rate-dependent since the slow driving will increase the probability of thermal relaxation. These upper and lower bounds for the driving rate are often inaccessible experimentally. In some samples, however, it has been possible to observe such changes of behavior and estimate the degree of "athermaliticity" [30, 31].

13.3.1.3 Learning [32]

Two-way shape memory is one of the most interesting properties of some ferroelastic materials exhibiting avalanche dynamics. This property arises from the interplay between the structural transitions and the reorganization of disorder in the system, and only shows up after a convenient training process. What can be

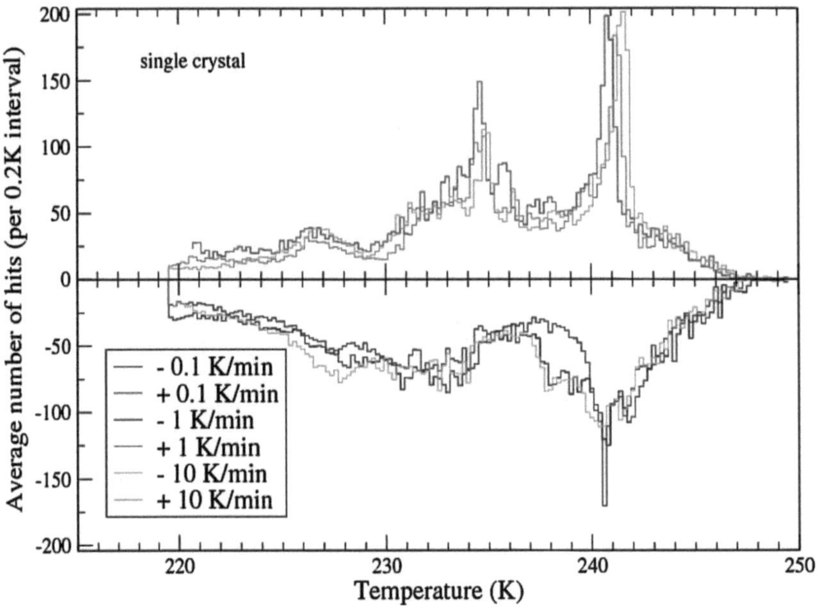

Fig. 13.7 Acoustic emission activity for different driving rates revealing the athermal and adiabatic character of the structural transition in a $Fe_{68.8}Pd_{31.2}$ single crystal [29]

learned about the training process from the AE analysis? Careful measurements in Cu-based shape memory alloys have been performed to investigate the evolution of the acoustic activity. Single crystals are first heat treated in order to "clean" most of the quenched-in disorder (dislocations, vacancies, etc.). The samples are then thermally cycled through the transition by keeping a well-controlled driving rate and fixed minimal and maximal temperatures. As a quantitative measure of the changes occurring from cycle to cycle, the statistical correlation between the activity curves $A(T)$ corresponding to consecutive loops has been calculated. During the initial cycles after the heat treatment, the correlation between consecutive loops is low. But after approximately ten loops, the activity profile tends toward a stable pattern that exhibits a higher correlation between the successive loops. This result shows that the disorder evolves in such a way that the system reaches a final stationary metastable trajectory, which then becomes reproducible.

13.3.1.4 Dependence on the Driving Mechanism

The AE pulse-counting technique is not restricted to thermally induced transitions. For instance, the technique was used some years ago to study the strain-driven martensitic transition in Ni–Mn–Ga alloys [33]. The theoretical studies of the influence of the driving mechanism presented in Sect. 13.2 indicate that significant differences should be observed in the avalanche dynamics when comparing

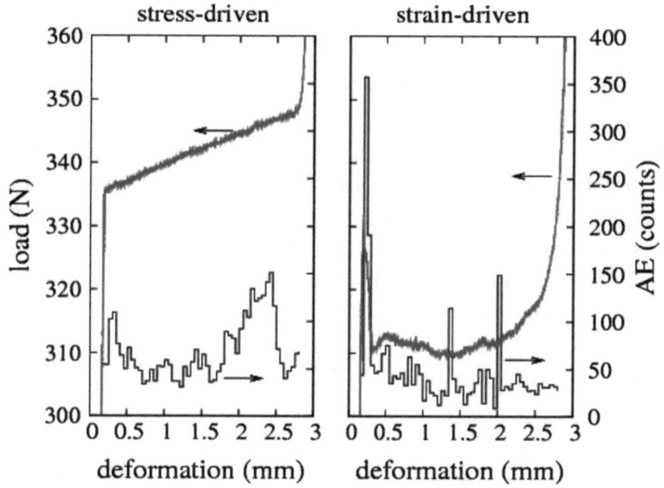

Fig. 13.8 Acoustic emission activity in a $Cu_{68}Zn_{16}Al_{l6}$ sample during stress-driven and strain-driven martensitic transitions [34]

transitions driven by controlling the force/field or the corresponding conjugate displacement. Figure 13.8 shows the activity profile as a function of the deformation for a $Cu_{68}Zn_{16}Al_{16}$ sample under stress-driven and strain-driven conditions. The data correspond to the same experiment as in Fig. 13.3. The stress–strain curves are also shown for comparison. As can be seen, there is an increase of acoustic activity associated with the strong yield point. In contrast, in the stress-driven case, there is no yield point and the activity is more homogeneous during the transition.

13.3.1.5 Correlation with Calorimetry

An interesting issue which still needs some clarification is the fact that the activity curves correlate very well with the calorimetric curves (see, for instance, Figs. 13.4 and 13.5). The main contribution to the calorimetric signal comes from latent heat, and this is naively proportional to the transformed fraction. Therefore, the ratio between the activity $A(T) = \mathrm{d}N/\mathrm{d}T$ and the calorimetric curve $\mathrm{d}Q/\mathrm{d}T$ should be related to the average volume of the individual avalanches. A similar property has been recently found for the case of stress-induced transitions [35]: the simultaneous measurement of the AE frequency ($\mathrm{d}N/\mathrm{d}t$) and the strain changes ($\mathrm{d}\varepsilon/\mathrm{d}t$) reveals a good correlation between both signals. This suggests that it should be possible to define an activity per strain $\mathrm{d}N/\mathrm{d}\varepsilon$ but the low resolution in the stress measurements does not allow to check this point. The rationale behind this interesting hypothesis may be found in some recent results (that will be more commented in the next subsection) that suggest a proportionality relation between the energies of the AE events and the heat released during these individual events [36].

13.3.2 Statistical Analysis of Single Events

The second technique that has been extensively used is based on the detection of a large number of AE signals during the transition and the statistical analysis of their amplitude A, energy E, or duration T. These three magnitudes are easily accessible using data acquisition systems. The motivation of this study is to obtain information about the probability densities $p(A)\mathrm{d}A$, $p(E)\mathrm{d}E$, and $p(T)\mathrm{d}T$. If any of these magnitudes A, E, or T are related to the avalanche size, and avalanches behave critically (as suggested by theoretical models), one expects that the distributions will exhibit a power-law behavior

$$p(A)\mathrm{d}A \propto A^{-\alpha}\mathrm{d}A; \quad p(E)\mathrm{d}E \propto E^{-\varepsilon}\mathrm{d}E; \quad p(T)\mathrm{d}T \propto T^{-\tau}\mathrm{d}T \qquad (13.7)$$

with α, ε, and τ being critical exponents.

The main problem for estimating these probability distributions is usually the lack of statistics. To have a good resolution in the highest decades, an enormous amount of data that are generally not available is required. Accordingly, in most cases, statistical analysis is performed by taking into account the signals recorded during the whole transition, although the process could be nonhomogeneous, as indicated by the activity curves. In fact, some analyses have revealed that there is some change in the histograms when only the initial part of the transition is studied [37, 38]. In some cases, in order to gain statistics, one is even forced to average over different cooling or heating runs. This supposes that the system has reached a stationary trajectory after enough cycles.

Once the data are recorded, the sets of amplitudes, duration, or energies are analyzed assuming a power-law distribution. In most cases, simple histograms show a power-law behavior with exponents typically ranging from 2 to 4. Figures 13.9 and 13.10 show examples of such histograms corresponding to measurements done with a Fe–Pd single crystal.

To obtain a good numerical estimate of the exponents, the maximum-likelihood fitting methods are used. They provide estimates of the critical exponents and error bars that do not depend on the way histograms are represented. Table 13.1 presents a compilation of the exponents found in the literature.

In addition to the exponents, it is also important to study the correlation between the magnitudes measured for each avalanche so as to establish whether they are really independent quantities or not. The usual analysis is done by plotting bivariate cloud maps, like the one in Fig. 13.11 representing the energy E vs. the amplitude A of each individual recorded signal. In many cases, the maps indicate a clear power-law statistical relation between the measured variables. For instance, Fig. 13.11 shows evidence that both magnitudes are related, i.e., $E \propto A^y$. In this case, it is found that $y \approx 2$. These statistical dependences may be contrasted with theoretical results that propose a universal shape function for the temporal profile of the avalanches [23, 41]. The problem with such comparisons is again the uncertainties in relating the pulse recorded by the transducer to the source. In particular, the pulse

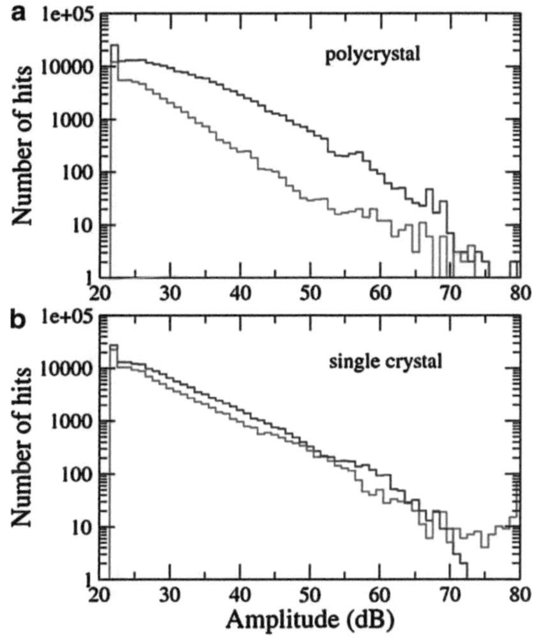

Fig. 13.9 AE amplitude distribution for two $Fe_{68.8}Pd_{31.2}$ samples [29]

duration T (which is determined by an ad hoc threshold) is probably very sensitive to the transducer response.

The main conclusions that have been reached so far from the statistical analysis of individual events can be summarized as follows:

13.3.2.1 Exponent Universality Classes

For the thermally driven transitions, and provided the driving rates are in the correct regime (athermal behavior and no avalanche overlap), the values of the exponent α can be grouped in three "universality classes" that depend on the symmetry of the martensitic phase (but not on sample composition): transitions from cubic to monoclinic structure yield $\alpha \approx 2.8$–3.0, transitions from cubic to orthorhombic are in general less athermal but yield $\alpha \approx 2.4$–2.6, and transitions to a tetragonal structure give an even lower exponent $\alpha \approx 2.3$–2.4. A similar conclusion can be obtained for the exponent ε. When the driving rate is too high avalanches may overlap, which decreases the exponent, and when the driving rate is too slow the exponent for the transitions to orthorhombic structure has been found to also decrease. This is because small avalanches become larger due to thermal fluctuations [31].

Some questions remain to be better understood. First, in some cases, deviations beyond the error bars have been found when comparing the forward and reverse transitions. (One should note that the number of recorded signals may be very

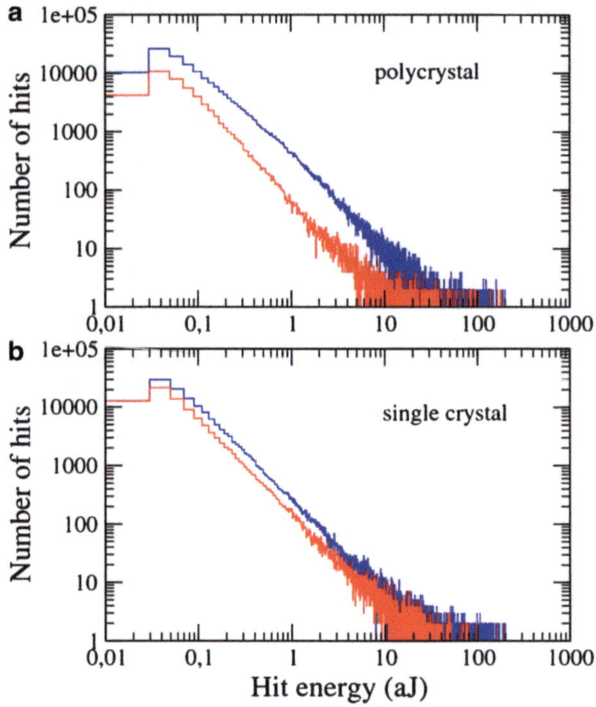

Fig. 13.10 AE energy distribution for two $Fe_{68.8}Pd_{31.2}$ samples [29]

different depending on the direction of the transition since the activity in the two directions is also very different.) For instance for the $Fe_{68.8}Pd_{31.2}$ polycrystalline sample [29], the exponents corresponding to heating ramps are much higher than expected. This increase is not observed in single crystalline samples. The reason for this deviation could be related to internal strains between grains. Secondly, the application of a magnetic field in the case of NiMnGa samples with a strong magnetoelastic coupling has also been shown to alter the exponents associated with the martensitic transition as well as those associated with the premartensitic transition [40, 41].

Finally, it should be mentioned that a comparison between the avalanche exponents obtained from the analysis of the energy E of AE and calorimetric pulses (at extremely low driving rates) has been performed recently [36]. The values coincide within error bars. This reinforces the idea that the energy is proportional to the heat released in each avalanche.

13.3.2.2 Learning Process

The evolution of the exponents with cycling after a heat treatment of the sample has also been studied. The values compiled in Table 13.1 correspond to samples that have been cycled many times so as to reach a stationary AE activity profile. In the

Table 13.1 Critical exponents obtained from AE experiments in metallic alloys exhibiting martensitic transitions, chronologically ordered. Results in the first line correspond to least-squares estimates (exhibiting large error bars), whereas the others correspond to maximum-likelihood estimates

Material	Transition	References	α $p(A) \propto A^{-\alpha}$	τ $P(T) \propto T^{-\tau}$	ε $P(E) \propto E^{-\varepsilon}$	x $A \propto T^{x}$	y $E \propto A^{y}$
CuZnAl	Cubic(bcc)–monoclinic(18R)	[37]	3.6 ± 0.8	3.5 ± 0.8		1.0 ± 0.1	
Four different alloys	Cubic(bcc)–monoclinic(18R)	[39]	3.1 ± 0.2				
Four different alloys	Cubic(bcc)–orthorhombic(2H)	[39]	2.4 ± 0.2				
CuZnAl	Cubic(bcc)–monoclinic(18R)	[31]	2.8 ± 0.9				
CuAlNi	Cubic(bcc)–orthorhombic(2H)	[31]	2.48 ± 0.7				
CuAlMn	Cubic(bcc)–orthorhombic(2H)	[32]	2.27 ± 0.03				
FePd Single crystal	Cubic(fcc)–tetragonal (fct)	[29]	2.26 ± 0.1		1.64 ± 0.1		1.97 ± 0.4
FePd Polycryst. Cooling	Cubic(fcc)–tetragonal(fct)	[29]	2.14 ± 0.1		1.59 ± 0.1		1.92 ± 0.4
FePd Polycryst. heating	Cubic(fcc)–tetragonal(fct)	[29]	2.95 ± 0.1		2.0 ± 0.1		1.95 ± 0.4
NiMnGa	Premartensitic transition	[40]	2.44 ± 0.03	4.3 ± 1.1	1.73 ± 0.02		
NiMnGa	Martensitic transition	[41]	2.6 ± 0.1		1.75 ± 0.1		2
CuZnAl soft driving	Cubic(bcc)–monoclinic(18R)	[34]	2.95 ± 0.02		2.24 ± 0.02		
CuZnAl hard driving	Cubic(bcc)–monoclinic(18R)	[34]	2.67 ± 0.03		1.98 ± 0.03		
CuZnAl heating	Cubic(bcc)–monoclinic(18R)	[36]			2.15 ± 0.05		
CuZnAl cooling	Cubic(bcc)– monoclinic (18R)	[36]			2.05 ± 0.05		

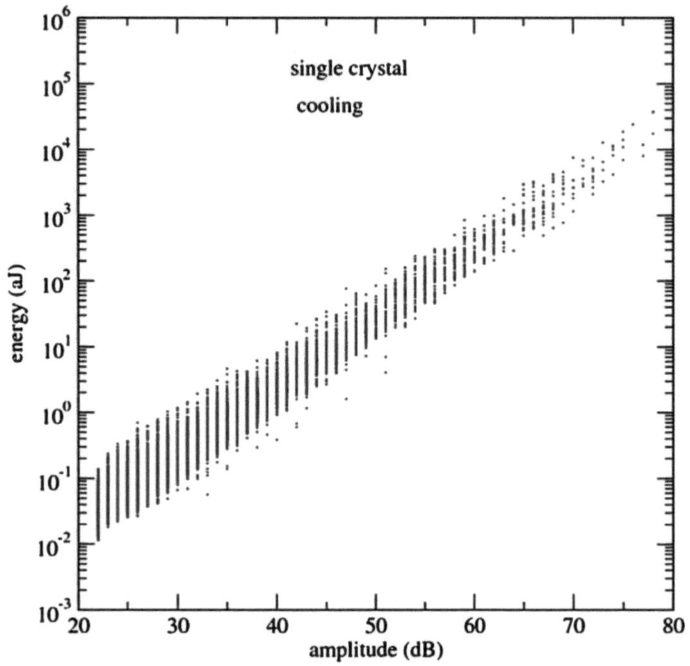

Fig. 13.11 Energy vs. amplitude cloud map of the AE signals recorded in a $Fe_{68.8}Pd_{31.2}$ single crystal [29]

initial cycles, the fitted exponents show an evolution. In some studies, it has even been possible to fit to an exponential correction of the type $p(A)\mathrm{d}A \propto A^{-\alpha}e^{-\lambda A}\mathrm{d}A$. In this case, one finds that λ decreases (in absolute value) toward 0 when increasing the number of cycles [32, 39]. Such an evolution of the avalanche distribution toward a stable power-law distribution with cycling has been recently theoretically understood as arising from the interplay between the reversible phase change and the irreversible development of an optimal amount of plastic deformation [42].

13.3.2.3 Influence of the Driving Mechanism

It has been shown that stress-driven transitions (soft driving) exhibit exponents comparable to thermally driven transitions, whereas strain driving (hard driving) gives a much higher exponent (due to a smaller proportion of large avalanches). Such an increase of the exponent is in qualitative agreement with the theoretical predictions [21]. It should also be mentioned that the recent measurements with an applied magnetic field [40,41] correspond to thermally induced transitions, but in the near future the same experimental setup may enable to measure AE under magnetic driving.

13.3.3 Future Trends for the AE Technique in the Study of Structural Transitions

There is still much to be learned using the techniques that have just been described. For instance, as suggested by theoretical models, one could analyze AE by performing partial hysteresis loops through the (thermally driven, stress-driven, or strain-driven) transition. Comparison with the signal along the main loop could yield some information about the distribution of the metastable states and corroborate the theoretical scenario described in Sects. 13.2.1 and 13.2.2. Besides, it is clear that AE waves contain much more information than the one extracted by the above techniques. Many of the methods that are used at large scales (for nondestructive testing or even geological purposes) could be potentially useful for studying structural transitions. In particular, the simultaneous use of several transducers could provide a precise location of the source. This would be a very powerful technique to analyze the dynamics of bulk structural transitions. Location along one dimension has recently been possible in stress and strain-driven samples with a length of 3 cm. From this, the energy and amplitude at the source were computed [34]. But locating the source in 2D or 3D is still very difficult at such small scales, given the anisotropic properties of the materials and the complex microstructures generated during the transition. Most probably, numerical simulations in conjunction with experiments will be needed to analyze the information extracted from the AE signals and solve the inverse problem. The simultaneous use of several transducers could also be a useful technique to identify the variant inducing each single AE pulse. This possibility was investigated many years ago [26], but has not been further explored. Finally, as again suggested by theoretical models, it would be interesting to study the statistical distribution of the waiting times between consecutive avalanches. It has been shown recently [43] that they may contain a lot of information that could usefully complement the one extracted from the distribution of amplitudes or energies.

13.4 Concluding Remarks

In this review, we have tried to illustrate by some examples the fruitful interaction between theory and experiments over the past few years. Theoretical models, despite their simplicity, have suggested interesting measurements to be made. We also have made some proposals for new experiments. As a final remark, we would like to point out that experiments also suggest that the theoretical description should be improved. In particular, one should study the avalanche properties whose statistical distributions are experimentally accessible and compute the corresponding exponents. So far, focus has been mainly on the size of the avalanches (volume or number of spins), although this information is still inaccessible in structural transitions. For instance, it would be interesting to study the distribution of the speeds of the advancing interfaces and/or the energy released by each avalanche as this would be closer to the amplitude and energy of the pulses that are actually measured. In this respect, the RFIM is too simple because of the exact symmetry between the parent and product phases, which implies the absence of latent heat. Such a symmetry may be valid for some magnetic materials, but not for most of other athermal FOPT. Therefore, a more sophisticated model is needed. Although there have been some attempts recently [44, 45], there is still much to be done in this direction.

Acknowledgments The authors acknowledge fruitful discussions with Ll.Mañosa, A.Planes, F.J. Pérez-Reche, and G. Tarjus. E.V. acknowledges the hospitality of the Physics Department (University of Warwick) during a sabbatical stay supported by the Spanish Ministry of Education (PR2009-0016). We also acknowledge financial support from the Spanish Ministry of Science and Innovation (MAT2010-15114).

References

1. J.P. Sethna, K. Dahmen, S. Kartha, J.A. Krumhansl, B.W. Roberts, J.D. Shore, Hysteresis and hierarchies: dynamics of disorder-driven first-order phase transformations. Phys. Rev. Lett. **70**, 3347 (1993)
2. J.P. Sethna, K. Dahmen, O. Perkovic, in *The Science of Hysteresis*, eds. by G. Berttoti, I. Mayergoyz (Academic, Amsterdam, 2006)
3. E. Kierlik, P.A. Monson, M.L. Rosinberg, L. Sarkisov, G. Tarjus, Capillary condensation in disordered porous materials: hysteresis versus equilibrium behavior. Phys. Rev. Lett. **87**, 055701 (2001)
4. F. Detcheverry, E. Kierlik, M.L. Rosinberg, G. Tarjus, Helium condensation in aerogel: avalanches and disorder-induced phase transition. Phys. Rev. E **72**, 051506 (2005)
5. F. Bonnet, T. Lambert, B. Cross, L. Guyon, F. Despetis, L. Puech, P.E. Wolf, Evidence for a disorder-driven phase transition in the condensation of ^{4}He in aerogels. Eur. Phys. Lett. **82**, 56003 (2008)
6. E. Vives, M.L. Rosinberg, G. Tarjus, Hysteresis and avalanches in the T = 0 random-field Ising model with two-spin-flip dynamics. Phys. Rev. B **71**, 134424 (2005)
7. P. Bortolotti, V. Basso, A. Magni, G. Bertotti, Oriented graph structure of local energy minima in the random-field Ising model. Physica B **403**, 398 (2008)
8. F. Detcheverry, M.L. Rosinberg, G. Tarjus, Metastable states and T = 0 hysteresis in the random-field Ising model on random graphs. Eur. Phys. J. B **44**, 327 (2005)
9. F.J. Pérez-Reche, M.L. Rosinberg, G. Tarjus, Numerical approach to metastable states in the zero-temperature random-field Ising model. Phys. Rev. B **77**, 064422 (2008)
10. V. Basso, A. Magni, Field history analysis of spin configurations in the random-field Ising model. Physica B **343**, 275 (2004)
11. M.L. Rosinberg, G. Tarjus, F.J. Pérez-Reche, The T = 0 random-field Ising model on a Bethe lattice with large coordination number: hysteresis and metastable states. J. Stat. Mech: Theory Exp. 03003 (2009)
12. F. Pazmandi, G. Zarand, G.T. Zimanyi, Self-organized, criticality in the hysteresis of the Sherrington-Kirkpatrick model. Phys. Rev. Lett. **83**, 1034 (1999)
13. M.L. Rosinberg, T. Munakata, Hysteresis and metastability in the mean-field random field Ising model: the soft-spin version. Phys. Rev. B **79**, 174207 (2009)
14. F. Bouchet, T. Dauxois, D. Mukamel, S. Ruffo, Phase space gaps and ergodicity breaking in systems with long-range interactions. Phys. Rev. E **77**, 011125 (2008)
15. Y. Liu, K.A. Dahmen, Unexpected universality in static and dynamic avalanches. Phys. Rev. E **79**, 061124 (2009)

16. G. Bertotti, *Hysteresis in Magnetism* (Academic, New York, 1998)
17. E. Kierlik, J. Puibasset, G. Tarjus, Effect of the reservoir size on gas adsorption in inhomogneous porous media. J. Phys.: Condens. Mat. **21**, 155102 (2009)
18. E. Bonnot, R. Romero, X. Illa, L. Manosa, A. Planes, E. Vives, Hysteresis in a system driven by either generalized force or displacement variables: martensitic phase transition in single-crystalline Cu-Zn-Al. Phys. Rev. B **76**, 064105 (2007)
19. X. Illa, M.L. Rosinberg, P. Shukla, E. Vives, Magnetization driven random-field Ising model at T = 0. Phys. Rev. B **74**, 244404 (2006)
20. S. Zapperi, P. Cizeau, G. Durin, E. Stanley, Dynamics of a ferromagnetic domain-wall: avalanches, depinning transition, and the Barkhausen effect. Phys. Rev. B **58**, 6353 (1998)
21. F. Pérez-Reche, L. Truskinovsky, G. Zanzotto, Driving-induced crossover: from classical criticality to self-organized criticality. Phys. Rev. Lett. **101**, 230601 (2008)
22. X. Illa, M.L. Rosinberg, P. Shukla, E. Vives, Influence of the driving mechanism on the response of systems with athermal dynamics: the example of the random-field Ising model. Phys. Rev. B **74**, 244403 (2006)
23. S. Papanikolaou, F. Bohn, R.L. Sommer, G. Durin, S. Zapperi, J.P. Sethna, Beyond scaling: the average avalanche shape, Nature Phys. **7**, 316 (2011)
24. B. Alessandro, C. Beatrice, G. Bertotti, A. Montorsi, Domain-wall dynamics and Barkhausen effect in metallic ferromagnetic materials: I. Theory. J. Appl. Phys. **68**, 2901 (1990)
25. C.B. Scruby, An introduction to acoustic emission. J. Phys. E: Sci. Instrum. **20**, 946 (1987)
26. Ll. Mañosa, A. Planes, D. Rouby, M. Morin, P. Fleischmann, J.L. Macqueron, Acoustic emission field during thermoelastic martensitic transformations. Appl. Phys. Lett. **54**, 2574 (1989)
27. O.U. Salman, Modeling of spatio-temporal dynamics and patterning mechanisms of martensites by phase-field and Lagrangian methods, PhD dissertation, Université Pierre et Marie Curie (2009)
28. C.B. Scruby, Quantitative acoustic emission techniques ed. by R.S. Sharpe in *Research Techniques in Non-destructive Testing*, vol III (Academic Press, London, 1985)
29. E. Bonnot, Ll. Mañosa, A. Planes, D. Soto-Parra, E. Vives, B. Ludwig, C. Strothkaemper, T. Fukuda, T. Kakeshita, Acoustic emission in the fcc-fct martensitic transition of $Fe_{68.8}Pd_{31.2}$. Phys. Rev. B **78**, 184103 (2008)
30. F.J. Pérez-Reche, E. Vives, Ll. Mañosa, A. Planes, Athermal character of structural phase transitions. Phys. Rev. Lett. **87**, 195701 (2001)
31. F.J. Pérez-Reche, B. Tadi, Ll. Mañosa, A. Planes, E. Vives, Driving rate effects in avalanche-mediated first-order phase transitions. Phys. Rev. Lett. **93**, 195701 (2004)
32. F.J. Pérez-Reche, M. Stipcich, E. Vives, Ll. Mañosa, A. Planes, M. Morin, Kinetics of martensitic transitions in Cu-Al-Mn under thermal cycling: Analysis at multiple length scales. Phys. Rev. B **69**, 064101 (2004)
33. L. Straka, V. Novák, M. Landa, O. Heczko, Acoustic emission of Ni-Mn-Ga magnetic shape memory alloy in different straining modes. Mater Sci Eng A **374**, 263 (2004)
34. E. Vives, D. Soto-Parra, Ll. Mañosa, R. Romero, A. Planes, Driving-induced crossover in the avalanche criticality of martensitic transitions. Phys. Rev. B **80**, 180101 (2009)
35. E. Bonnot, E. Vives, Ll. Mañosa, A. Planes, R. Romero, Acoustic emission and energy dissipation during front propagation in a stress-driven martensitic transition. Phys. Rev. B **78**, 094104 (2008)
36. M.C. Gallardo, J. Manchado, F.J. Romero, J. del Cerro, E.K.H. Salje, A. Planes, E. Vives, Avalanche criticality in the martensitic transition of Cu67.64 Zn16.71Al15.65 shape-memory alloy: A calorimetric and acoustic emission study. Phys. Rev. B **81**, 174102 (2010)
37. E. Vives, I. Ràfols, L. Mañosa, J. Ortín, A. Planes, Statistics of avalanches in martensitic transformations. I. Acoustic emission experiments. Phys. Rev. B **52**, 12644 (1995)
38. I. Ràfols, E. Vives, Statistics of avalanches in martensitic transformations. II. Modeling. Phys. Rev. B **52**, 12651 (1995)
39. Ll. Carrillo, Ll. Mañosa, J. Ortín, A. Planes, E. Vives, Experimental evidence for universality of acoustic emission avalanche distributions during structural transitions. Phys. Rev. Lett. **81**, 1889 (1998)

40. B. Ludwig, C. Strothkaemper, U. Klemradt, X. Moya, Ll. Mañosa, E. Vives, A. Planes, Premartensitic transition in Ni2MnGa Heusler alloys: Acoustic emission study. Phys. Rev. B **80**, 144102 (2009)
41. B. Ludwig, C. Strothkaemper, U. Klemradt, X. Moya, Ll. Mañosa, E. Vives, A. Planes, An acoustic emission study of the effect of a magnetic field on the martensitic transition in Ni2MnGa. Appl. Phys. Lett. **94**, 121901 (2009)
42. F.J. Perez-Reche, L. Truskinovsky, G. Zanzotto, Training-induced criticality in martensites. Phys. Rev. Lett. **99**, 075501 (2007)
43. B. Cerruti, E. Vives, Correlations in avalanche critical points. Phys. Rev. E **80**, 011105 (2009)
44. B. Cerruti, E. Vives, Random-field Potts model with dipolar-like interactions: hysteresis, avalanches, and microstructure. Phys. Rev. B **77**, 064114 (2008)
45. B. Cerruti, E. Vives, Statistics of microstructure formation in structural transitions studied using a random-field Potts model with dipolar-like interactions. J. Stat. Mech. P05009 (2009)

Chapter 14
Entropy–Driven Conformations Controlling DNA Functions

A.R. Bishop, K.Ø. Rasmussen, A. Usheva, and Boian S. Alexandrov

Abstract In memory of Jim Krumhansl we summarize our growing level of understanding of the origins and functional roles of specific nonlinear conformational excitations ("bubbles") in DNA. We present a number of results that point toward the conclusion that DNA is capable of directing major aspects of its own lifecycle, governed by the laws of equilibrium thermodynamics.

First, we discuss a series of experimental and theoretical research results that demonstrate a correlation between DNA bubbles and essential biological processes such as DNA transcription and DNA–protein binding. Specifically, we discuss how, through a synergetic combination of modeling and experiments, we have developed an extended version of the Peyrard–Bishop–Dauxois model, and used it to predict specific properties, such as bubble location, size, and duration, of DNA breathing. Applying this framework, we show a number of examples that demonstrate that specific breathing properties lead to enhancements in transcription activity and DNA–protein binding efficiency.

Second, we show that DNA may be able to apply its complex conformational dynamics to facilitate its own repair. We demonstrate this in the context of specific DNA damage that has been documented to arise from exposure to UV radiation.

Finally, we discuss our ongoing attempts to harness our knowledge of DNA conformation and dynamics and their impact on function to help predict transcription initiation sites in entire genomes. We apply techniques from bioinformatics and statistical learning to incorporate the above features into a more predictive framework.

A.R. Bishop (✉) · K.Ø. Rasmussen · B.S. Alexandrov
Theoretical Division, Los Alamos National Laboratory, Los Alamos, NM, USA
e-mail: arb@lanl.gov; kor@lanl.gov; boian@lanl.gov

A. Usheva
Beth Israel Deaconess Medical Center, Harvard Medical School, Boston, MA, USA
e-mail: ausheva@bidmc.harvard.edu

T. Kakeshita et al. (eds.), *Disorder and Strain-Induced Complexity in Functional Materials*, Springer Series in Materials Science 148,
DOI 10.1007/978-3-642-20943-7_14,

14.1 Introduction

Jim Krumhansl had an early passion for nonlinear science in contexts including signal processing and second sound in helium, as well as the early numerical evidence for "soliton" behavior from Kruskal, Zabusky, and others [1]. This prepared him well for the rapid expansion of the wide and interdisciplinary impacts of *nonlinearity* and *complexity* beginning in the early 1970s with the mathematical formulation of exactly integrable soliton equations, and then generalizing to many classes of nonlinear equations. He was fascinated both by the pseudoparticle behavior of solitons and solitary waves [2], and by the coexistence of these coherent structures with complex, even chaotic, classical and quantum dynamics. This coexistence is the essence of spatiotemporal "complexity," as we have come to appreciate in great detail over the last 40 years. Similarly, it embodies the essential "multiscale" challenges which now set the frontiers of so many fields – the coexistence of scales, and systems of hard and soft regions, emerging self-consistently from the same interactions.

Krumhansl passionately pursued these developing conceptual frameworks, but most notably *applied* them in the contexts of precursor clusters and soft modes at structural phase transitions [3]; topological (kink) excitations in low-dimensional electronic materials [4]; elastic texturing around solid–solid phase transitions [5] (a direct evolution of his prior concerns with structural phase transitions); and conformational excitations in biological macromolecules [6], including DNA [7–9]. All of these and other examples were framed with a "scientific method" motivation; namely, what are the origins of "complexity," how to measure and characterize it, and what are its functional consequences? Although we have made remarkable strides over these last four decades in our modeling, measuring, and simulation tools, these questions continue to dominate the frontiers of all realizations of complexity.

After several years of exploration, *Krumhansl* and colleagues achieved a formulation of elasticity [10–12] faithfully incorporating local bonding constraints in materials (compatibility conditions), which have now evolved into a rather complete framework for understanding textures (twinning, tweed, nucleation, dyadons, etc.) and their multiscale dynamics at solid–solid phase transformations, as well as the effects of structural disorder, external fields, geometrical constraints, etc. This history and its current status are fully covered elsewhere in this book.

Interestingly, Krumhansl's interest in biological macromolecules has a quite analogous "elasticity" history. In particular, we now understand how coupled acoustic and optic lattice dynamics, and how local bonding constraints [13, 14] (much as in the elasticity above) can lead to effective nonlinear dynamics with entropic consequences of conformational nonlinear excitations ("hot spots") and a rich multiscale complexity of self-consistently coexisting hard and soft matter. Indeed, it was this appreciation for elasticity which directly motivated the proposal of a currently topical coarse-grained model of DNA, the *Peyrard–Bishop–Dauxois (PBD) model*, designed to explain the nucleative, first-order properties of denaturation [13, 14].

As in so many other fields of nonlinearity and complexity, recent years have provided true breakthroughs in our ability to synthesize and measure key structural and dynamic scales controllably, and to appreciate their fundamental physical consequences increasingly. In the case of DNA, our ability to control synthesis and characterization at the gene level, and modern rapid throughput gene-sequencing techniques are astonishing capabilities, unavailable to Krumhansl and his contemporaries. Our aim in this report is to honor the inspirational interests and style of Jim Krumhansl by summarizing our growing level of understanding of the origins and functional roles of specific nonlinear conformational excitations in *DNA* – coherent base-pair opening ("bubbles"). It is extremely likely that the same basic concepts extend to many biological macromolecules, but this complete story is very much in its infancy [15, 16].

14.2 Transcription Initiation, Transcriptional Start Sites, and DNA Breathing Dynamics

According to the accepted paradigm, the DNA molecule, and more specifically its sequence consisting of four nucleic acids, viz. adenine, guanine, thymine, and cytosine, is the main carrier of genetic information encoding life, health, and disease. DNA contains and transmits vital information, needed for the very existence of the cell, from the nucleus to the cell body in a process called transcription. In addition, the DNA molecule is able to self-replicate and thus reproduce or clone itself in a process called replication. Finally, the DNA molecule can change its sequence by incorporating parts of helical molecules in a process known as recombination. These three processes are at the heart of biological functions. *Transcription* governs the production of the essential biomolecules necessary to the living cell, and the process is in essence a sophisticated method of decryption of the genetic code stored in the DNA double helix. The tight packaging of the genetic information in the form of DNA duplexes, buried inside the double helix, protects the genetic information from damage, external and internal, such as OH^- and H^+ ions, various carcinogens, and the other dangerous chemical or physical agents, thereby permitting the safe transfer and inheritance of the information from generation to generation.

In the transcription process, a complementary copy of the genetic information stored within the DNA sequence, and more precisely in the specific DNA segments called genes, is translated to a messenger RNA (mRNA) with the aid of RNA polymerases (RNAP).

The RNAP mechanically traverses the DNA template strand in the $3' \rightarrow 5'$ direction, as illustrated in Fig. 14.1. Transcription itself is regulated by various proteins (transcription factors), whose binding to the DNA double helix enhances or suppresses the process. In eukaryotes, the RNAP, and therefore the initiation of the transcription, requires the presence of specific DNA sequences. These regions called

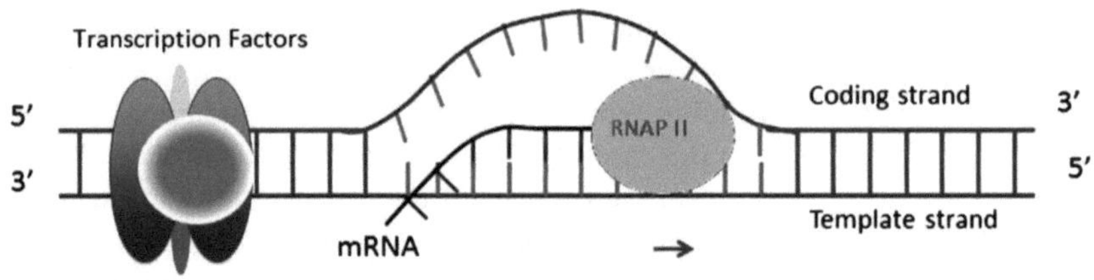

Fig. 14.1 Schematic representation of transcription

core *promoters* are located about 35 base pairs (bp) upstream and/or downstream of the transcriptional start sites (TSSs) and serve as staging areas for the transcription-initiation complexes [17]. In general, there are three types of transcription initiation: (1) dispersed/broad – in which the core promoter contains several weak start sites distributed over a broad region, (2) focused/focal – where the transcription starts from either a single nucleotide or several nucleotides, and (3) mixed – which combines the first two types by having a few closely clustered nucleotides with strong start sites and several weak start sites over a broad region [18, 19]. The majority of mammalian genes have multiple promoters, each of them containing multiple start sites [20].

When the RNAP binds to a promoter site, it must unwind or open part of the DNA double helix in order to expose the template nucleotides. It has been experimentally demonstrated that the binding of the RNAP and/or the transcription-initiation complex with the DNA promoter disrupts the Watson–Crick base pairing in a region of about 10–15 base pairs [21, 22]. On the other hand, in the living cell, the DNA molecule is embedded in water at physiological temperature and because of the thermal motion, there is a finite probability for each DNA base pair to open and reclose spontaneously. This thermal process is referred to as "DNA breathing," as it induces local transient openings of the double helix. DNA *breathing* is well documented by hydrogen-exchange experiments [7, 23]. DNA "heavy" breathing [24] can result in noncanonical structures, such as single-strand hairpins, cruciforms, slipped strands, R-loops, intramolecular triplexes, and others structures, that deviate significantly from the ideal equilibrium Watson and Crick helix [25,26]. Interestingly, it is accepted that the formation of many such non-B-DNA structures that may disrupt normal cellular processes (see, e.g., [25]) originates from transient DNA openings, i.e., local melting, breathing, or bubbles. Importantly, breathing is presented in all DNA molecules and environments, as a result of the thermal energy and associated *fluctuations* of the water in the cell. In other words, specific "soft" segments along the DNA molecule are an unceasing subject of transient destabilizations, or local melting, by the available thermal energy in the system.

Although DNA breathing has been recognized for decades, it was commonly believed that the functional properties of DNA were determined exclusively by its nucleotide sequence. However, the enhanced propensity of DNA core promoters, and specifically of the TSS, for transient local openings, or bubble creation, is independent of any DNA–protein interactions or regulation, and hence it is determined solely by the molecular DNA structure and properties. This suggests that the sequence dependence of DNA openings is not simply a matter of sequence identity, but results from an intricate interplay of intramolecular interactions. At physiological temperatures, some of the thermally induced localized *breathing* in the vicinity of the TSSs at the core promoter can give rise to localized DNA melting extending over 10 bp or more [27, 28], and thus reach the size of transcriptional bubbles. The fact that the exact location and the evolution of these collective openings of consecutive base pairs are determined by the specific sequence then provides motivation to investigate whether *transcription initiation* start sites, and other regulatory sites, possess an enhanced propensity for bubble formation.

The PBD is a one-dimensional nonlinear model that describes the transverse opening motion of the complementary strands of double-stranded DNA [13, 14]. The *Hamiltonian* of this model is

$$H = \sum_{n=1}^{N} \left\{ D_n \left(e^{-a_n y_n} - 1\right)^2 + \frac{k}{2}\left(1 + \rho e^{-\beta(y_n + y_{n-1})}\right)(y_n - y_{n-1})^2 \right\},$$

where the sum is over all N base pairs of the DNA and y_n denotes the relative displacement of the complementary nucleotides of the nth base pair. The first term of the Hamiltonian is the Morse potential, which represents the base pair hydrogen bonds together with the electrostatic repulsion of the backbone phosphates. The parameters D_n and a_n depend on the nature of the base pair (A–T vs. G–C). The second term represents a harmonic potential approximation but with a nonlinear coupling constant, which takes into account the influence of the stacking interactions between consecutive base pairs on the transverse stretching motion. The exponential term effectively decreases the harmonic spring constant K when one of the base pairs is displaced away from its equilibrium position in the double helix: $K_{\max} = k(1 + \rho)$, when $y_n + y_{n-1} = 0$, a condition met, e.g., at equilibrium, and $K_{\min} = k$, when y_n or $y_{n-1} \rightarrow \infty$, i.e., when at least one of the base pairs is out of the double helix stack. This term is essential for simulating nontrivial entropic and long-range cooperative effects important for sharp DNA melting [14]. Although many embellishments of the basic model are possible, the simplest form above is already remarkably successful and similar to simple models of entropy-driven elasticity [29]. The parameters of the model have been previously obtained by fitting simulations to DNA melting curves.

By comparing *experimental* results of S1 nuclease cleavage for the adeno-associated viral P5 promoter and for a P5 mutant promoter, that is known to be transcriptionally inactive, to the predictions of computer simulations performed with the PBD model, it was for the first time shown in [27, 28] that the most active regions strongly correlate with the TSS and other major regulatory sites. To illustrate this, we reproduce the detailed results for the P5 promoter in Fig. 14.2. Figure 14.2 shows that for the P5 (wild-type) promoter, we find a strong correlation between the transcription start site (labeled +1) and the occurrence of large (>10 bp) openings. This correlation is corroborated by S1 nuclease cleavage assay experiments, which,

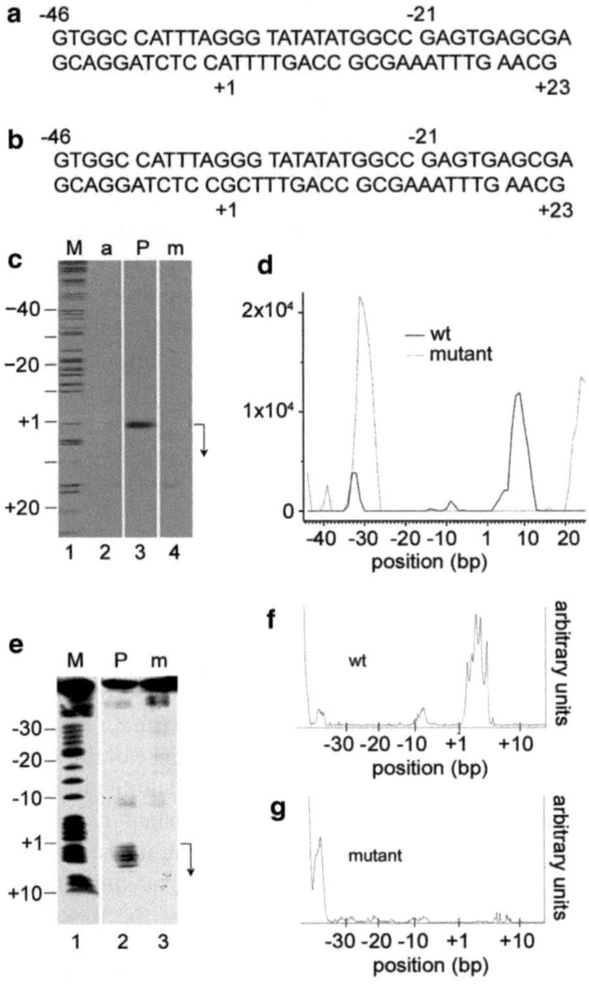

Fig. 14.2 Analysis of P5 promoter and P5 mutant promoter. (**a**) Upper strand, sequence of the 69-bp P5 core promoter. (**b**) Upper strand, sequence of the 69-bp mutant P5 promoter. (**c**) Transcription assay on a 120-bp fragment containing the P5 promoter and P5 mutant fragment in a nuclear extract. The *arrow* on the right indicates the transcription start site and the direction of transcription. The corresponding sequence position is indicated to the left of the marker. Lane 1, GA DNA sequencing reaction was used as a marker (M); lane 2, transcription from the P5 promoter with a-amanitin (a); lane 3, RNA transcription products with P5 promoter (P); lane 4, RNA transcription products from the P5 mutant template (m). (**d**) PBD simulation of the P5 sequence, plotting simulated instances of 2.1 Å-separated openings of 10 bp or more versus the base position in the sequence. The *solid line* represents the results of the wild-type P5 promoter, and the *broken line* represents results with the mutant P5 promoter sequence. (**e**) S1 nuclease cleavage of the P5 promoter and the P5 mutant promoter. The corresponding sequence position is indicated to the left of the panel. Lane 1, lower strand-labeled P5 promoter GA sequencing reaction (M); lane 2, P5 promoter S1 cleavage reaction (P); lane 3, P5 mutant promoter cleavage reaction (m). (**f**) Cleavage density profile of the wild-type (wt) P5 promoter DNA in the S1 nuclease experiment. (**g**) Cleavage density profile of the mutant P5 promoter DNA in the S1 nuclease experiments. From [27]

albeit indirectly, indicate the most active regions, in terms of *thermally–entropically* driven strand separation. For the P5 mutant promoter, the S1 nuclease cleavage assay experiments similarly corroborate the simulation results by the lack of any significant openings around the +1 site. Further Langevin molecular dynamics investigations, based on the PBD Hamiltonian, demonstrated that the maximum in the probability for the large bubble creation is not at the TSSs in the P5 core promoter, but rather at a DNA segment where the so-called TATA box, i.e., the binding site of the TBP transcription factor, is situated [30]. The same work introduced, for the first time, the notion of the average bubble lifetime, and it was found that certain DNA sequences can promote longer lived bubbles, most likely as a result of length scale competition between the nonlinearity and disorder prescribed by the DNA sequence. It was shown that the TSS at the P5 core promoter is one of them. The long-lived bubble situated at the P5 TSS completely vanishes in the P5 mutant. This suggests that different regulation sites in DNA can have different dynamical *patterns* in terms of length, amplitude, and average bubble duration.

Applying a *transfer integral* approach to evaluate the thermodynamic partition function [31, 32] of the PBD model, we investigated the correlations between the major regulatory sites and the DNA segments with enhanced bubble creation in the entire human adenovirus genome [33]. We found a pattern of softness peaks distributed both upstream and downstream from the TSSs, and that early transcriptional regions tended to be softer than late promoter regions. When experimentally reported transcription factor *binding* sites were superimposed on the calculated softness profiles, a close correspondence was observed in many cases, which suggests that DNA duplex breathing dynamics may play a role in protein recognition of specific nucleotide sequences and protein–DNA binding. These results suggest that genetic information is not only stored in explicit codon sequences, but also may be encoded into local dynamic and structural features, and that it may be possible to access this masked information using dynamics calculations.

Using Langevin molecular dynamics simulations, based on the PBD, we derived three dynamic criteria (bubble probability, bubble lifetime, and average amplitude of the DNA strand separation) that are needed to characterize bubble formation fully at the TSSs for eight mammalian gene promoters [34]. We demonstrated the difference in the dynamical pattern of the core promoter, compared with the dynamics of an intron of the same length (Fig. 14.3). Importantly, we observed again that the most stable DNA openings do not necessarily coincide with the most probable openings and the largest average strand displacement.

The dynamic profiles of the considered mammalian promoters differ significantly in overall profile and bubble probability, but the TSS is often distinguished by large (longer than 10 bp) and *long-lived transient openings*. Most importantly, in support of these results are our recent [34] and previous [61] *in vitro* transcription data demonstrating that a DNA template containing an artificial mismatch-based bubble (i.e., a coherent long-lived local DNA opening) is transcribed by human RNAP even in the absence of any other transcription factors, Fig. 14.4. This striking result demonstrates the important role that the specific long-lived bubble formation plays in transcription initiation. Our simulations showed that this is also the case in the P5 promoter in comparison with the P5 mutant promoter, Fig. 14.4, panel B.

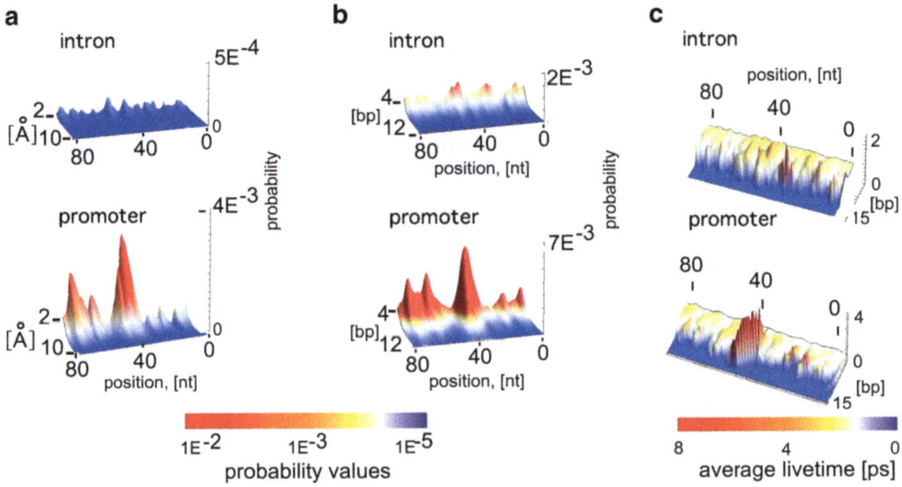

Fig. 14.3 The dynamics patterns of the collagen promoter and intron sequences obtained from the PBD Langevin dynamics simulations. (**a**) Probability for collective opening (vertical axis) of 10 bp within the collagen intron (*top panel*) and collagen promoter (*bottom panel*), as a function of bubble amplitude [Å]. The nucleotide positions in the collagen promoter are labeled relative to the TSS (+1) (horizontal axis). (**b**) Probability for opening (vertical axis) of amplitude threshold > 1Å, as a function of bubble length [bp]. (**c**) Average lifetimes of DNA collective openings of amplitude > 1Å (vertical axis). From [34]

In this way, we assessed the role of DNA breathing dynamics as a determinant of promoter strength and TSS location, and found that the A/T-rich regions, such as TATA boxes, exhibit faster, lower amplitude motions than the TSS regions [34].

The main source of structural and dynamical heterogeneity in G/C-rich sequences presumably originates from a strong difference in the stacking interaction between GG/CC steps on the one hand, and CG/CG and GC/GC on the other [35]. To take into account the important role of the heterogeneous *stacking interactions*, we extended the original PBD Hamiltonian to account for the sequence dependence of the stacking potentials and to reproduce the melting transitions of G/C-rich, homogeneous, and repeat DNA sequences, with high accuracy [36]. We used the fact that large deviations in the melting behavior of repeats and homopolymers, due to the stacking interactions, were experimentally well known already in 1970 [37], and have since been discussed at length in the literature due to the abundance of such sequences in vertebrate genomes [38]. We collected melting data for several DNA oligos and applied Markov chain Monte Carlo simulations [39] to establish stacking force constants for the ten dinucleotide steps (CG, CA, GC, AT, AG, AA, AC, TA, GG, TC). The experiments and numerical simulations confirm that the GG/CC dinucleotide stacking is remarkably unstable, compared to the stacking in GC/GC, CG/CG, and even AA/TT dinucleotide steps [36].

Recently, we designed specific point mutations in the sequence of the well-known constitutive SCP1 promoter [40] using Langevin molecular dynamics simulations

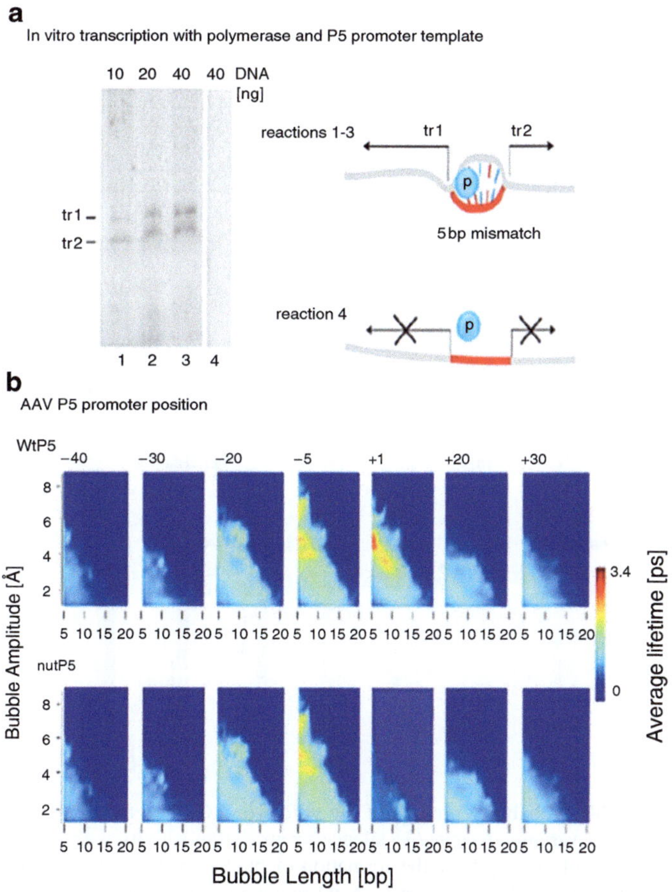

Fig. 14.4 (**a**) Artificial mismatch bubbles enable bidirectional transcription in the absence of transcription factors. The DNA template in reactions 1, 2, and 3 contains 5-bp long mismatches creating a bubble in the region of the transcription start site. The reaction in lane 4 received DNA with no mismatch. The RNA products have been separated by gel electrophoresis based on differences in the size of the transcripts. The positions of the RNA transcripts are shown on the *left*: transcripts are initiated at the bubble and terminated at the 5′ end (tr1) and at the 3′ end (tr2), respectively. A schematic diagram of the experiment is given at the *right*. The promoter region is labeled with *red* and the polymerase (P) with *blue*. (**b**) Bubble lifetime as a function of length and amplitude at physiological temperature, shown for individual base pairs of both the wild-type (wt) P5 and the mutant (mut) P5 variant (transcriptional silent). Each *square* presents the average bubble lifetimes (color scale) at a given amplitude (vertical axis) and length (horizontal axis), for bubbles that contain the base pair given on the *top right*. Transcription starts at base pair +1s. From [34]

based on the extended PBD (EPBD) model [36]. Our experimental and theoretical results demonstrated that DNA's dynamical activity at the TSS can be suppressed by these designed mutations that do not affect basal transcription factor binding DNA, but do suppress the transcription initiation *in vivo*, Fig. 14.5 [41]. Our data

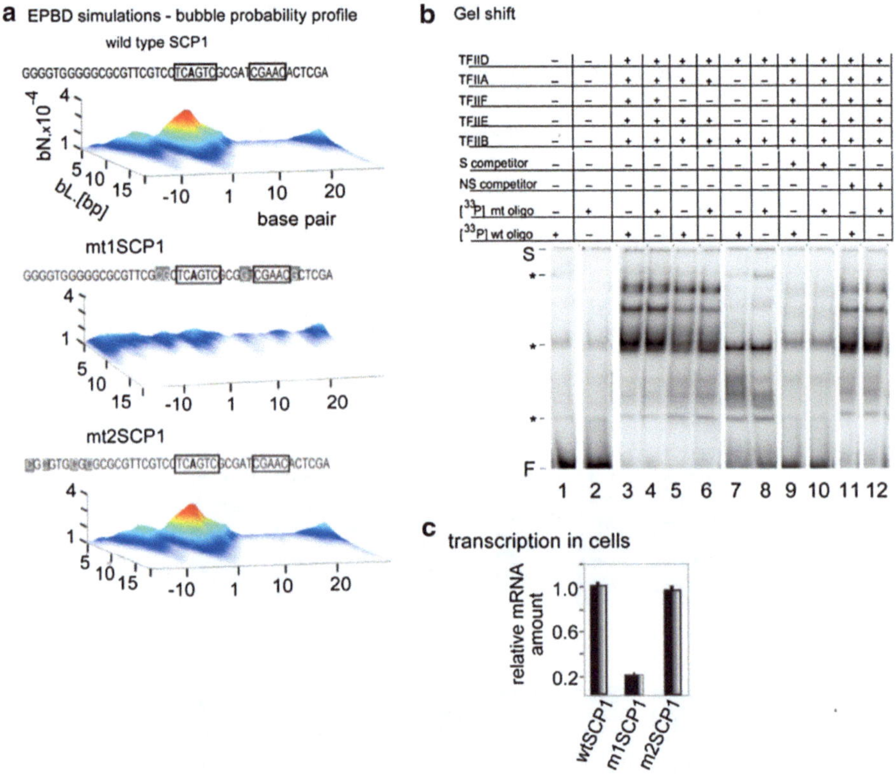

Fig. 14.5 (**a**) Bubble probability profiles of the wild-type SCP1 promoter (wtSCP1), m1SCP1, and m2SCP1 mutant variants *designed* to silence *transcription* activity without affecting *protein binding*. The probability (z-axis) for the formation of bubbles with given length in bp (y-axis) beginning at a given nucleotide position (x-axis) relative to the TSS ("+1"). Mutated residues are indicated with gray boxes (*on top*). Protein-binding sites are indicated with *black frames*. The profile of m2SCP1 is identical to that of wild-type SCP1, as shown at the *bottom*. (**b**) Gel shift reactions. Effect of the m1SCP1 mutations on complex formation between TFIID, TFIIB, TFIIF, TFIIE, and the Inr promoter fragments is shown. Band shift reactions received the wild-type (lanes 1, 3, 5, 7, 9, and 11) and the m1SCP1 Inr box sequences (lanes 2, 4, 6, 8, 10, and 12). Transcription factor samples are as follows: lanes 1 and 2 bovine serum albumin; lanes 3–12 received equal amounts (by weight) of transcription factors. The reactions in lanes 9 and 10 received 3 nM of homologous wild-type cold SCP1 oligonucleotide as a competitor. The reactions in lanes 11 and 12 received 10 nM of unrelated cold oligonucleotide as a competitor. The presence (+) or absence (−) of competitor oligo DNA and basal transcription factors in the reactions is indicated above the lanes. The positions of the gel shift start (S), the free DNA (F), and the nonspecific gel shift products (asterisk) are indicated. (**c**) Transient cell transfection experiments were carried out to measure wtSCP1, m1SCP1, and m2SCP1 promoter activity. The three pUC119-based constructs were transfected by electroporation into HeLa cells. Total RNA was extracted from the cells and subject to Q-PCR-based analysis with pUC119 primers to measure cellular level of promoter-specific RNA transcripts. Data are expressed as fold induction relative to wtSCP1 mRNA level (on the vertical). From [41]

indicate that the dynamics at the TSS are as important as the binding of the basal transcriptional factors for determining the transcription initiation strength and therefore, the DNA breathing may serve as an indicator for the TSS position.

14.3 DNA Repair

During a cell's lifespan, a broad set of processes, external agents, and internal chemical activities may reach and damage the cell's genome by causing substitutions, deletion, modifications, or insertions in the genomic DNA sequence. To survive, the cells have developed a set of repair mechanisms and processes. These processes serve to recognize and correct continuously the abovementioned genomic alterations. In human cells, the normal metabolic activities and environmental factors such as hard radiation, ultraviolet light, and many others can cause as many as 10^6 molecular lesions per cell per day [42]. Part of these lesions can create structural damages in the DNA molecule, which in turn can alter or eliminate the encoded information, silence or alter the gene expression, or modify the cell's replication activities. Other damage and substitutions can provoke dangerous mutations in the cell's genome that can be inherited and endanger its daughter cells, or even kill the cell itself. Hence, in order to defend the cell, the DNA repair processes have to be constantly active and respond to harmful changes in the DNA structure and sequence. It is vital for the cell that repair proteins recognize various defects, substitutions, mutations, etc., rapidly and without errors.

Various experiments have shown that some of the DNA mutation's and base substitutions, such as mismatches, *UV dimers*, methylations, and others [43–45], can change the thermal stability of the DNA molecule, and therefore strongly affect the local breathing dynamics of DNA.

While it is not surprising that such kinds of changes in the DNA structure and local breathing dynamics may have adverse effects on biological function, it is less clear whether the cell's repair process can gain efficiency from the induced change in the dynamics. Therefore, we have numerically simulated the changes in DNA breathing dynamics induced by damage resulting from ultraviolet light [46,47]. UV at a wavelength of 250 nm promotes the formation of DNA dimers, by making a rather strong connection between adjacent thymine bases. If left unrepaired, these dimers may lead to skin cancer.

While the UV light causes adjacent thymine base pairs to form a strong covalent bond, the pairing of the complementary adjacent adenines is weakened. This weakening in turn results in a 13 K decrease in the *melting* temperature of the double-stranded octamer d(GCGTTGCG)d(CGCAACGC) in the presence of a *UV-induced dimer* [44]. Using the PBD model and a constrained Monte Carlo algorithm [48], the parameters of the PBD model were adjusted to reproduce this change in melting temperature for the octamer. An 11% reduction in the dissociation energy of the complementary nucleotides was observed to be necessary to obtain the 13-K reduction (see melting curves in Fig. 14.6).

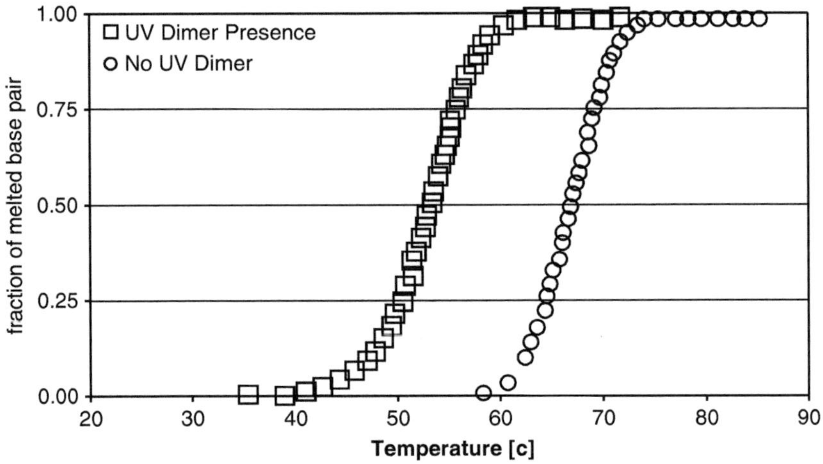

Fig. 14.6 Simulated melting curves of the octamer d(GCGTTGCG)d(CGCAACG). The *circles* show the melting behavior for the unaltered sequence, while *squares* indicate the behavior when the UV dimerization has occurred. From [47]

Besides weakening the interaction between the adjacent adenines, the UV dimer acts as a structural impurity in the DNA sequence and, therefore, effectively changes the stacking interaction between the adjacent thymine base pairs. In addition, a decrease in the spatial and temporal coherence length of the fluctuations also results from the presence of the UV dimer. A random sequence of 64 bp was generated and the octamer, with and without the presence of a UV dimer, was inserted in the middle of this sequence. The Monte Carlo technique was applied to this new sequence and the base pair's average displacements were obtained (see Fig. 14.7). In Fig. 14.7, a three- to fourfold increase in the relative displacements between the base pairs belonging to the two opposite strands at the dimer position is evident in comparison with the case without a UV dimer in the sequence.

The DNA sequence was also investigated with Langevin molecular dynamics, and Fig. 14.8 shows results of these *simulations* in terms of the probability for an opening larger than a given threshold. It can be seen that the probability for the occurrence of large bubbles at the dimer position is approximately 25 times larger than the probability in the absence of a dimer. Finally, the data presented in [46] indicate that there are significant changes in the spatiotemporal characteristics of double-stranded DNA upon *UV dimer* formation between adjacent thymine base pairs. We suggest that this changed dynamics may help enable the *repair proteins* to detect the occurrence of radiation damage rapidly and attach there. The enormously enhanced propensity for opening at the dimer effectively leads to a locally increased temperature, which, by analogy to a simple one-step chemical reaction, increases the binding reaction rate twofold for every 10 K increase in temperature. The enhanced local breathing dynamics not only increases the speed of the binding reactions, but most importantly the appearance of large local openings also exposes the UV dimer

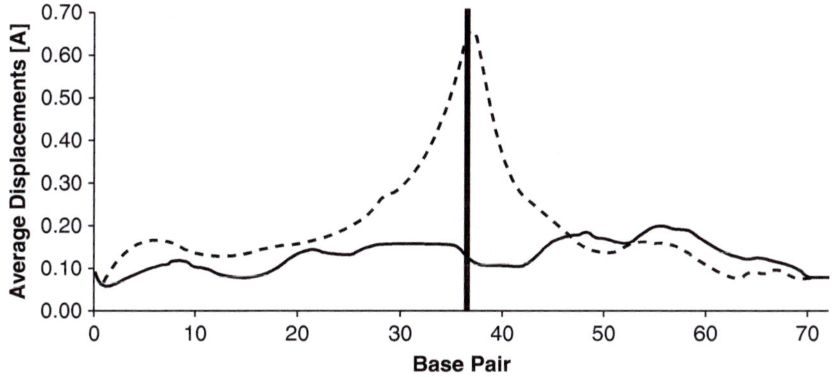

Fig. 14.7 Simulated, average strand displacement of DNA with a dimer (*dashed line*) and without a dimer (*solid line*) shows the enhanced opening in the dimer neighborhood at 37°C. The *solid vertical line* indicates the position of the UV dimer. From [47]

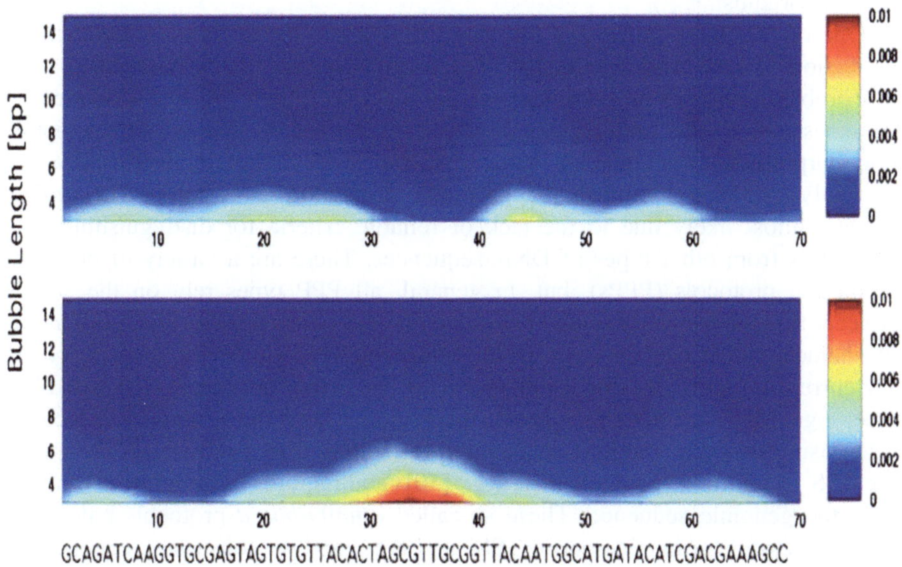

Fig. 14.8 The probability for bubbles with amplitude above 1.5 Å as a function of bubble position (x-axis) and length (y-axis). The *top panel* represents this probability without a UV dimer, and the *bottom panel* represents the probability with a UV dimer present, both at 37°C. From [47]

to repair proteins diffusing in the cell nucleus, and the average time that a given protein senses the exposed nucleotides is in general proportional to the average bubble lifetime. This investigation suggests that DNA may be able to facilitate its own repair through its complex breathing dynamics.

14.4 Bioinformatics and DNA Breathing Dynamics

Bioinformatics has emerged as an application of multivariate statistics, machine learning, and computer science in the growing fields of molecular biology, omics, system biology, and many other sciences concerned with quantifying the processes and mechanisms in living cells. In the late 1990s, bioinformatics' primary use was in genomics and more specifically the whole-genome DNA sequencing, which resulted in the revolutionary sequencing of the human genome. These days, exploding amounts of biological data – from high-throughput sequencing of large numbers of genomes, to enormous quantities of *in vivo* imaging cellular and molecular data, necessitate the creation and advancement of large databases, statistical learning theory algorithms, macro-computational approaches, and mathematical methods to solve the massive amount of formal and practical problems arising from the management and analysis of the data. Nevertheless, the main goal of bioinformatics remains the increase of predictive multiscale understanding of biological processes in living organisms.

One of the main challenges for bioinformatics is the accurate computational prediction of core promoters and their transcription start sites. The need to solve this problem emerges proportionally to the massive amounts of newly sequenced genomes and to the broad genomic diversity defined by the single nucleotide polymorphisms, copy number variations, and methylation. Unfortunately, the presently available bioinformatic promoter prediction tools have very limited success, most likely due to the lack of reliable criteria for distinguishing core promoters from other types of DNA sequences. There are a variety of promoter prediction protocols (PPPs), but, in general, all PPP types rely on the simple concept that the promoter region has some distinct characteristics (or combination of characteristics) that all other DNA sequences lack. Generally speaking, there are currently two major classes of promoter prediction programs. The first class (see, e.g., [49]) attempts to identify the region where the promoter is located, often using only the genomic sequence information. The second class comprises the TSS recognition protocols (see, e.g., [50]) and requires more information than only the genomic sequence. These so-called *affinity-based* protocols have been very successful for single-peaked TSSs and are based on consensus sequences of known transcription factor binding sites and their enrichment regions [50,51]. These affinity-based protocols achieve near-perfect TSS prediction with high accuracy and resolution, for a specific subgroup of mammalian promoters, by relying on motif recognition such as TATA box, initiator, and other transcription factor binding sites. However, it is unclear how to generalize this kind of approaches because the identification of transcription factor binding sites and the notion of the consensus sequences represent a significant challenge themselves [52].

The state-of-the-art PPPs that only require genomic sequence information can be divided into four distinct categories: (1) thermodynamic-based (i.e., relying on entropic, free energy, and similar thermodynamics quantities), (2) motif-based, (3) structure-based (i.e., relying on DNA structural characteristics such as curvature, bending, etc.), and (4) hybrid protocols that combine various of the above characteristics. Thermodynamic protocols are typically used for prokaryotic genomes, while motif-based and structural approaches are more commonly applied for mammalian genomes. Several authors have used thermodynamic quantities (e.g., Gibbs free energy) to predict soft spots in genomic DNA to enable promoter predictions [53–55]. Purely *thermodynamic* approaches typically suffer from limited base-pair resolution because of the required averaging window. This aspect causes insensitivity to single point mutations, which are common in mammalian and especially human genomes. However, in prokaryotes, where static genomic superhelicity plays an important role, some thermodynamic approaches provide accurate results [55]. Furthermore, thermodynamic protocols primarily track the A–T content, since the hydrogen bonds between the complementary nucleotides play a major role in determining the overall DNA softness. This renders thermodynamic considerations less applicable than other bioinformatic methods when considering GC-rich promoters, which dominate the mammalian genomes. *Motif-based* PPPs are the most common approaches for predicting transcription start sites in mammalians. Artificial neural networks, discriminant analysis, genetic algorithms, hidden Markov models, and innovative hybrid machine-learning approaches have been employed to detect specific DNA motifs, leading to some of the currently top-performing PPPs (e.g., [56]). However,*structural* approaches have recently emerged in the realm of promoter prediction with superior performance. These methods make predictions based on structural properties of the DNA molecule. Structural prediction tools such as EP3 [57], ARTS [58], and ProSOM [59] have become some of the best performing protocols. Interestingly, there are almost no hybrid prediction protocols among the top-performing tools [58].

Recently published detailed comparisons [49,59] provide guidelines for comparing PPPs as well as comparing the top-performing promoter prediction tools' ability to locate core promoters within regions of 500–50 bp. As expected, all approaches perform better for larger resolution regions, i.e., in 500 bp. However, even in regions of 500 bp, no approach is able to provide both recall and precision above 0.50. This means that promoter prediction programs can either accurately locate a large number of true promoters (e.g., 51%) at a high rate of false positives (e.g., 89%), or predict very few promoters (e.g., 29%) at a lower (e.g., 19%) rate of false positives. The effectiveness significantly deteriorates as the resolution is decreased to 50 bp. The best promoter prediction with 50 bp accuracy is achieved by ProSOM, with a recall of 0.17 and precision of 0.30. This means that if ProSOM investigates DNA sequences with 100 experimentally verified TSSs, it will be able to locate only 17 of them correctly, while simultaneously making at least 39 false-positive predictions. The poor performance at 50 bp resolution most likely results from recognition criteria that are inadequate at single-base resolution and, for example, do not take into account the importance of DNA local breathing dynamics.

To pursue this possibility, we have leveraged simulated characteristics of the DNA local breathing dynamics at single nucleotide resolution as a criterion in genomic-scale core promoter prediction at 50 bp resolution [60].

We combined characteristics of DNA's conformation dynamics with physical, structural, and sequence-specific characteristics in a novel hybrid core PPP,

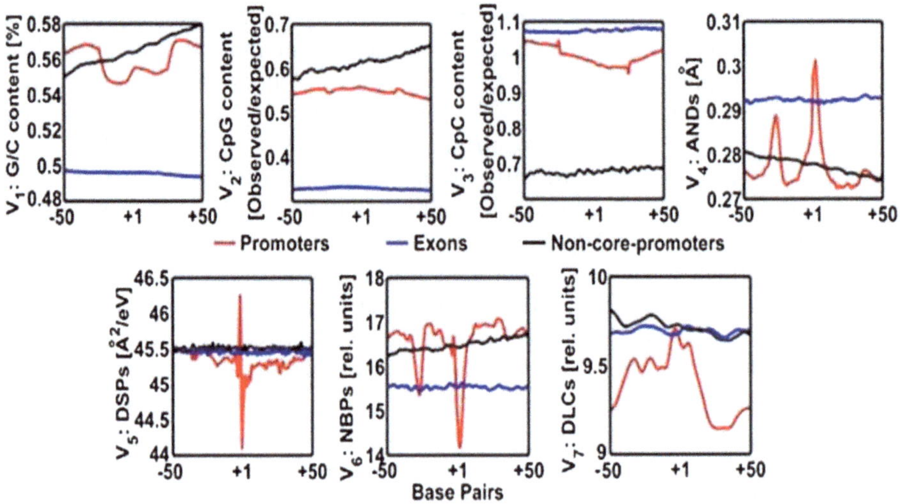

Fig. 14.9 Average profiles of the seven variables used by our new prediction method: (**a**) V1: G/C content, (**b**) V2: normalized CpG content, (**c**) V3: normalized CpC content, (**d**) V4: average nucleotide displacements (AND), (**e**) V5: dinucleotide stacking potential (DSPs), (**f**) V6: nucleosome bending propensities (NBPs), and (**g**) V7: DNA local curvature (DLCs). The +1 site corresponds to the TSS in the promoter sequences and the central nucleotide of the exon and noncore promoter sequences. X-axes represent the basepair position relative to the TSS; Y-axes represent the value for each of the six variables. Promoters are depicted in *red*, exons in *blue*, and noncore promoters in *black*. From [60]

Fig. 14.9. Specifically, our method combines (1) transcription-relevant dynamic DNA characteristics, obtained via Markov chain Monte Carlo simulations, with the EPBD model, and (2) experimentally identified structural characteristics, Fig. 14.9, combined with (3) the motif-based features already used in bioinformatics, and united by (4) a novel use of contemporary unsupervised and supervised learning methods based on multidimensional clustering and statistical learning. As an example in Fig 14.10, we demonstrate that our new method has a superior recognition rate in comparison with existing general PPPs, when applied to human promoters at genomic-scale promoter data set.

14.5 Conclusions

In the visionary spirit of Krumhansl, we have proposed here that DNA is capable of directing major aspects of its own lifecycle, governed by the laws of equilibrium thermodynamics. If validated by experiments, this will be a beautifully efficient use of the extraordinarily robust average (Watson–Crick) DNA structure; sequentially controlling the location, size, and duration of conformational breather excitations (nucleative precursors of the total DNA denaturation) in response to cofactor stimuli (temperature, gene sequence, torsion and bending, pressure, methylation, etc.),

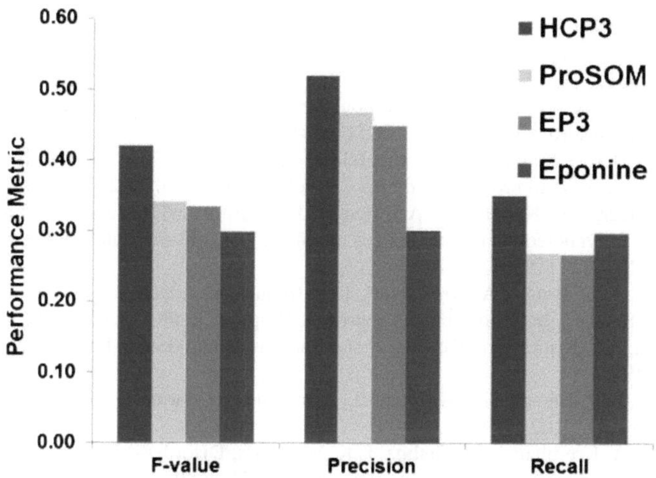

Fig. 14.10 The bar plot shows the *F-value*, *precision*, and *recall* for our new hybrid core promoter prediction method (HCP3) and the three top-performing PPPs on the same data set. The X-axis depicts the performance metric while the Y-axis shows its actual value. The HCP3 is shown as a *blue bar* while all other PPPs are shown by different colors. From [60]

and thereby controlling biological functions (transcription initiation and replication sites, protein attachment sites, drug interaction sites, repair protein attachment, and more) – all without destroying the average DNA template.

Ultimately, there must be feedback mechanisms between all of these functions, DNA's *mesoscale* structure, and DNA's environment – a next frontier! As a step in this direction, we have introduced tools from information theory to characterize the complexity and help direct the search for functional hot spots based on more than gene sequence alone. As important as the newly available genomic sequence knowledge is, it is now evidently incomplete, and the path to a *predictive capability* for biological function requires understanding and control of crucial additional factors. Naturally, achieving this holy grail of understanding and functional prediction would open wonderful vistas of design and control for natural and manufactured systems.

Acknowledgments We gratefully acknowledge all our collaborators with whom we have coauthored the original works summarized here. This work was carried out under the auspices of the National Nuclear Security Administration of the US Department of Energy at Los Alamos National Laboratory under Contract No. DE-AC52–06NA25396.

References

1. D.K. Campbell, A.C. Newell, R.J. Schrieffer, H. Segur (eds.), Solitons and coherent structures. Phys. D **18**, (North-Holland, Amsterdam, 1986)
2. A.R. Bishop, J.A. Krumhansl, S.E. Trullinger, Solitons in condensed matter: a paradigm. Phys. D **1**, 1 (1980)

3. J.A. Krumhansl, J.R. Schrieffer, Dynamics and statistical-mechanics of a one-dimensional model Hamiltonian for structural phase-transitions. Phys. Rev. B, **11**(9), 3535 (1975)
4. B. Horovitz, J.A. Krumhansl, Solitons in the Peierls condensate – phase solitons. Phys. Rev. B **29**, 2109 (1984)
5. G.R. Barsch, J.A. Krumhansl, Twin boundaries in ferroelastic media without interface dislocations. Phys. Rev. Lett. **53**(11) 1069 (1984)
6. A.E. Garcia, J.A. Krumhansl, H. Frauenfelder, Variations on a theme by Debye and Waller: From simple crystals to proteins. Proteins - Struct. Funct. Bioin. **29**(2) 153 (1997)
7. S.W. Englander, N.R. Kallenbach, A.J. Heeger, J.A. Krumhansl, S. Litwin, Nature of the open state in long polynucleotide double helices: possibility of soliton excitations. Proc. Natl. Acad. Sci. U S A **77**, 7222 (1980)
8. A.E. Garcia, C.S. Tung, J.A. Krumhansl, Low-frequency collective motions of DNA double helices – A reduced set of coordinates approach. Biophys. J. **49**(2) A123 (1986)
9. A.E. Garcia, J.A. Krumhansl, Density of states of single and double helical DNA. Biophys. J. **51**, A420 (1987)
10. B. Horovitz, G.R. Barsch, J.A. Krumhansl, Twin bands in Martensites – statics and dynamics. Phys. Rev.B **43**(1), 1021 (1991)
11. S. Kartha, J.A. Krumhansl, J.P. Sethna, L.K. Wickham, Disorder-driven pretransitional tweed pattern in Martensitic transformations. Phys. Rev. B, **52**, 2, 803 (1995)
12. T. Lookman, S.R. Shenoy, K.Ø. Rasmussen, et al., Ferroelastic dynamics and strain compatibility. Phys. Rev. B **67**, 2, 024114 (2003)
13. M. Peyrard, A.R. Bishop, Statistical mechanics of a nonlinear model for DNA denaturation. Phys. Rev. Lett. **62**, 2755 (1989)
14. T. Dauxois, M. Peyrard, A.R. Bishop, Entropy-driven DNA denaturation. Phys. Rev. E **47**, R44 (1993)
15. H. Gohlke, M.F. Thorpe, A natural coarse graining for simulating large biomolecular motion. Biophys. J. **91**, 2115 (2006)
16. A. Katherine, A. Henzler-Wildman, M. Lei, V. Thai, S.J. Kerns, M. Karplus, D. Kern, A hierarchy of timescales in protein dynamics is linked to enzyme catalysis. Nature **450**, 913 (2007)
17. S.T. Smale, J.T. Kadonaga, The RNA polymerase II core promoter. Annu. Rev. Biochem. **72**, 449 (2003)
18. T. Juven-Gershon, J.T. Kadonaga, Regulation of gene expression via the core promoter and the basal transcriptional machinery. Dev. Biol. **339**, 225 (2010)
19. J.A. Stamatoyannopoulos, Illuminating eukaryotic transcription start sites. Nat. Methods **7**, 501 (2010)
20. A. Sandelin, P. Carninci, B. Lenhard, J. Ponjavic, Y. Hayashizaki, D.A. Hume, Mammalian RNA polymerase II core promoters: insights from genome-wide studies. Nat. Rev. Genet. **8**, 424 (2007)
21. A.F. Melnikova, R. Beabealashvilli, A.D. Mirzabekov, A study of unwinding of DNA and shielding of the DNA grooves by RNA polymerase by using methylation with dimethylsulphate. Eur. J. Biochem. **84**, 301 (1978)
22. U. Siebenlist, RNA polymerase unwinds an 11-base pair segment of a phage T7 promoter. Nature **279** 651–652 (1979)
23. M. Guéron, M. Kochoyan, J.L. Leroy, A single mode of DNA base-pair opening drives imino proton exchange. Nature, **328**, 89 (1987)
24. D.M.J. Lilley, DNA opens up – supercoiling and heavy breathing. Trends Genet. **4**(4), 111 (1988)
25. A. Kornberg, T.A. Baker, *DNA Replication*, 2nd edn. (University Science Books, Sausalito, CA, 2005)
26. Y. Lin, S.Y. Dent, J.H. Wilson, R.D. Wells, M. Napierala, R-loops stimulate genetic instability of CTG.CAG repeats. Proc. Natl Acad. Sci. U. S. A. **107**, 692 (2010)
27. C.H. Choi, G. Kalosakas, K.Ø. Rasmussen, M. Hiromura, A.R. Bishop, A. Usheva, DNA dynamically directs its own transcription initiation. Nucl. Acids. Res. **32**, 1584 (2004)

28. G. Kalosakas, K.Ø. Rasmussen, A.R. Bishop, C.H. Choi, A. Usheva, Sequence-specific thermal fluctuations identify start sites for DNA transcription. Europhys. Lett. **68**, 127 (2004)
29. W.C. Kerr, A.M. Hawthorne, R.J. Gooding, A.R. Bishop, J.A. Krumhansl, First-order displacive structural phase transitions studied by computer simulation. Phys. Rev. B **45**, (1992)
30. B.S. Alexandrov, L.T. Wille, K.Ø. Rasmussen, A.R. Bishop, K.B. Blagoev, Bubble statistics and dynamics in double-stranded DNA. Phys. Rev. E **74**, 050901 R (2006)
31. Z. Rapti, A. Smerzi, K.Ø. Rasmussen, A.R. Bishop, C.H. Choi, A. Usheva, Healing length and bubble formation in DNA. Phys. Rev. E Stat. Nonlin. Soft Matter Phys. **73**, 051902 (2006)
32. Z. Rapti, A. Smerzi, K.Ø. Rasmussen, A.R. Bishop, C.H. Choi, A. Usheva, Lengthscales and cooperativity in DNA bubble formation. Europhys. Lett. **74**, 540 (2006)
33. C.H. Chu, Z. Rapti, V. Gelev, M.R. Hacker, B.S. Alexandrov, E.J. Park, J.S. Park, N. Horikoshi, A. Smerzi, K.Ø. Rasmussen, A.R. Bishop, A. Usheva, Profiling the thermodynamic softness of adenoviral promoters. Biophys. J. **95**, 597 (2008)
34. B.S. Alexandrov, V. Gelev, S.W. Yoo, A.R. Bishop, K.Ø. Rasmussen, et al., Toward a detailed description of the thermally induced dynamics of the core promoter. PLoS Comput. Biol. **5**(3), e1000313 (2009)
35. J. Sponer, K.E. Riley, P. Hobza, Nature and magnitude of aromatic stacking of nucleic acid bases. Phys. Chem. Chem. Phys. **10**, 2595 (2008)
36. B.S. Alexandrov, V. Gelev, Y. Monisova, L.B. Alexandrov, A.R. Bishop, K.Ø. Rasmussen, A. Usheva, A nonlinear dynamic model of DNA with a sequence-dependent stacking term. Nucl. Acids. Res. **37**, 2405 (2009)
37. R.D. Wells, J.E. Larson, R.C. Grant, B.E. Shortle, C.R. Cantor, Physicochemical studies on polydeoxyribonucleotides containing defined repeating nucleotide sequences. J. Mol. Biol. **54**, 465 (1970)
38. J.C. Venter, et al., The human genome. Science **291**, 1304 (2001)
39. S. Ares, N.K. Voulgarakis, K.Ø. Rasmussen, A.R. Bishop, Bubble nucleation and cooperativity in DNA melting. Phys. Rev. Lett. **94**, 035504 (2005)
40. T. Juven-Gershon, S. Cheng, J.T. Kadonaga, Rational design of a super core promoter that enhances gene expression. Nat. Methods **3**, 917 (2006)
41. B.S. Alexandrov, V. Gelev, S.W. Yoo, L.B. Alexandrov, Y. Fukuyo, A.R. Bishop, K.Ø. Rasmussen, A. Usheva, DNA dynamics play a role as a basal transcription factor in the positioning and regulation of gene transcription initiation. Nucl. Acids. Res. **38**(6), 1790 (2010)
42. H. Lodish, A. Berk, P. Matsudaira, C.A. Kaiser, M.P. Scott, S.L. Zipursky, J. Darnell, *Molecular Biology of the Cell*, 5th edn. (WH Freeman, New York, NY, 2004) p. 963
43. J. Montgomery, C.T. Wittwer, R. Palais, L. Zhou, Simultaneous mutation scanning and genotyping by high-resolution DNA melting analysis. Nat. Protoc. **2**, 59 (2007)
44. J. Kemmink, R. Boelens, T. Koning, et al., 1H NMR study of the exchangeable protons of the duplex d (GCGTTGCG). d (CGCAACGC) containing a thymine photodimer containing a thymine photodimer. Nucl. Acids Res. **15**, 4645 (1987)
45. J. Ramstein, C. Hélène, M. Leng, A study of chemically methylated deoxyribonucleic acid. Eur. J. Biochem. **21**(1), 125 (1971)
46. K.B. Blagoev, B.S. Alexandrov, E.H. Goodwin, A.R. Bishop, Ultra-violet light induced changes in DNA dynamics may enhance TT-dimer recognition. DNA Repair **7**, 863 (2006)
47. B.S. Alexandrov, N.K. Voulgarakis, K.Ø. Rasmussen, A. Usheva, A.R. Bishop, Pre-melting dynamics of DNA and its relation to specific functions. J. Phys.: Condens. Matter **21**, 034107 (2009)
48. M. Fixman, Classical statistical mechanics of constraints: a theorem and application to polymers. Proc. Natl. Acad. Sci. U. S. A. **71**, 3050 (1974)
49. T. Abeel, Y. Van de Peer, Y. Saeys, Toward a gold standard for promoter prediction evaluation. Bioinformatics **25**, i313 (2009)
50. M. Megraw, F. Pereira, T.J. Jensen, et al., A transcription factor affinity-based code for mammalian transcription initiation. Genome Res. **19**, 644 (2009)
51. C.F. Frith, E. Valen, A. Krogh, et al., A code for transcription initiation in mammalian genomes. Genome Res. **18**, 1 (2008)

52. G. Badis, M.F. Berger, A.A. Philippaki, et al., Diversity and complexity in DNA recognition by transcription factors. Science **26**, 5935 (2009)
53. F. Liu, E. Tostesen, J.K. Sundet, T.K. Jenssen, C. Bock, et al., The human genomic melting map. PLoS Comput. Biol. **3**, e93 (2007)
54. D.G. Dineen, A. Wilm, P. Cunningham, D.G. Higgins, High DNA melting temperature predicts transcription start site location in human and mouse. Nucl. Acids Res **37**, 7360 (2009)
55. C.J. Benham, Duplex destabilization in superhelical DNA is predicted to occur at specific transcriptional regulatory regions. J. Mol. Biol. **255**, 425 (1996)
56. T.A. Down, T.J. Hubbard, Computational detection and location of transcription start sites in mammalian genomic DNA. Genome Res. **12**, 458 (2002)
57. T. Abeel, Y. Saeys, E. Bonnet, P. Rouze, Y. Van de Peer, Generic eukaryotic core promoter prediction using structural features of DNA. Genome Res **18**, 310 (2008)
58. S. Sonnenburg, A. Zien, G. Ratsch, ARTS: accurate recognition of transcription starts in human. Bioinformatics **22**, e472 (2006)
59. T. Abeel, Y. Saeys, P. Rouze, Y. Van de Peer, ProSOM: core promoter prediction based on unsupervised clustering of DNA physical profiles. Bioinformatics **24**, 24 (2008)
60. L.B. Alexandrov, K.Ø. Rasmussen, A.R. Bishop, A. Usheva, B.S. Alexandrov, DNA breathing dynamics for genomic-scale core promoter prediction. Submitted
61. A. Usheva, T. Shenk, YY1 transcriptional initiator: protein interactions and association with a DNA site containing unpaired strands. Proc. Natl. Acad. Sci. U. S. A. **93**, 13571 (1996)

Chapter 15
Conclusion and Outlook

With the Off-Set from the Prototypical Martensitic Materials

Per-Anker Lindgård

Abstract General aspects of functionality are discussed: The ability to change according to variations in the external conditions. The Martensitic materials were among the first metals discovered to show functionality, as a rapid change in shape at a given temperature. This has for thousands of years been of vital technological importance. Understanding this phenomenon forms the basis for understanding related phenomena in the numerous recently discovered materials, discussed in the previous chapters. In these, the changes can be induced and tuned by, for example, magnetic or electric fields. It would extend too far to attempt to survey all this here – and fortunately, since the martensites are sufficiently rich, that the principles – and the possibilities can be gauged from a knowledge of martensites.

Functionality is the ability to respond appropriately to a stimulus. Functionality and complexity are intimately connected. The prime example is found in the biological realm. Complex units of all length scales self-organize and interact – from society, persons, organs, cells, down to nano-scale molecules (proteins). These can do only simple physical things, such as expanding and contracting – or even passively, just be optimally designed (textured) for strength (spider web). Since man picked tools, he used the textured materials provided by biology: wood, bone, and fiber; however, he realized that much stronger materials with distinct properties could be found in the inorganic realm, such as the sharpness of glass (obsidian and flint) and the toughness and heaviness of metal (bronze). A real advance happened with the advent of iron, for three reasons: (1) Iron transforms from a high temperature, easily deformable state to a low temperature hard, self-organized, polyvariant state. (2) Most importantly, the properties of that phase can be tuned by a tiny amount of

P.-A. Lindgård (✉)
Materials Research Division, Risö, DTU, National Laboratory for Sustainable Energy, 4000-Roskilde, Denmark
e-mail: hanne.frederiksen@mail.tele.dk

T. Kakeshita et al. (eds.), *Disorder and Strain-Induced Complexity in Functional Materials*, Springer Series in Materials Science 148,
DOI 10.1007/978-3-642-20943-7_15, © Springer-Verlag Berlin Heidelberg 2012

carbon, readily present in the production phase. This made it possible to design textured materials with desired macroscopic properties, using textures from the visible scale, seen in swords and anchors, self-organizing down to the nano-scale. (3) Finally, and not least, it was cheap and available all over (contrary to the elements in bronze). Tremendous experience and technique with iron were attained over millennia, but it was not before around 1900 that Adolf Martens undertook a scientific study, which formed a basis for an understanding of the phenomenon, probably not advancing the technique – but rather improving the reliability of the production process (a demand by the rail road industry). The well-known structural transformation was hereafter called a Martensitic transformation (MT).

The phase transition in martensites and related shape memory alloys involves lattice strain and in some cases, intra-unit cell shuffle (or phonon) modes. In many materials of technological interest, not only the strain but also other degrees of freedom such as spin and charge couple to strain, leading to a multitude of emergent functionalities, e.g. ferroelectricity, magnetoelasticity, colossal magnetoresistance, and multiferroic behavior. The long-range elastic interaction, in conjunction with dipolar forces, plays a crucial role in determining the nano-scale inhomogeneity that renders these materials with sensitivity to external field as well as desired functionality. I shall mostly focus on martensites and strain to illustrate some key features. Toward the end, I shall comment on the role of spin and charge in various materials.

Coming back to iron, it so happens that the MT in iron is rather atypical for the now more general definition: A temperature (or stimuli such as pressure or magnetic/electric field, chemical environment, etc.)-induced diffusion-less, displacive transition from an open cubic structure (so-called austenite, or β-phase) to a closed packed structure (so-called martensite or α-phase) with many equivalents in different directions elongated structures. These are called variants – or we shall prefer the name domains, in analogy to the magnetic language. Marten's work spurred a large scientific engagement and resulted in the discovery of many materials of various potential uses. The shape memory effect (SME) caused by a forced predominance of some domains gives a macroscopic elongation up to 10% in the low temperature phase, which can then be removed by heating and not regained by a following cooling if a polyvariant phase is formed. This is called a one-way SME. If it reforms with predominant domain distribution and regains its length, it is called a two-way SME. We notice the similarity to the simplest biological response. The effect can be used in any place where one cannot achieve the same by a screw. This is of use, for example, in actuators, sensors, and tightening cords of medical or dental use, and in controlling surfaces. Other applications are the superelasticity and the extraordinary intrinsic damping properties near the transition temperature, T_{M}. We have here emphasized that textured materials, here "Martensitic materials," are primarily of interest for practical use. The ambitious goal of the research is to gain an understanding, which enables one to compose and design materials with some of the functionalities found in the biological realm, but with the larger strength and larger parameter space (stimulus: change in temperature, pressure, magnetic/electric field, etc.) found in metals. Some model materials may be useful for gaining a basic understanding, but we do not expect a new state of matter. Yet, this may not be true. "No texture" as in glass is also a texture. It seems possible in some Martensitic materials to promote a domain structure on nano-scale, where the difference between domains and domain walls becomes blurred, such that the structure appears disordered and glass-like, but in a cubic framework. This is indeed a new state of condensed matter. This possibility has been given a lot of attention in this book.

Let us go back to basics and see what the simplest possible model system reveals. Such a system is pure Zr, which has a Martensitic transformation from the open *bcc* to the closed-packed *hcp* structure at 1,135 K, well below the melting point at 1,993 K. The first question is, why do most simple materials crystallize into the *bcc* structure below the melting point. This was answered quite generally by Alexander and McTague [1], who argued that condensed matter is best understood in terms of density waves (or strain waves), i.e., in reciprocal space, with wave vectors $\mathbf{q}$. By considering a Landau expansion, it was demonstrated that the third-order term would favor a structure where three equal amplitude, interacting density waves form equilateral triangles with $\mathbf{q}_1 + \mathbf{q}_2 + \mathbf{q}_3 = 0$. The dimension is given by the inter-atomic interactions (assuming not strongly directional bonds) entering in the second-order term. The structure with most triangles is the *fcc*. Hence it is *bcc* in real space. For Zr, the MT is weakly first order and shows no complete softening of any mode, and precursor phenomena in the dynamical neutron scattering. The most striking effect is macroscopic deformation (shape memory) due to a uniform strain in the *hcp* phase. However, attempts to explain the transition by Landau expansion in terms of uniform strains were not successful. Following a similar path as above, we argued [2] that to understand the MT in Zr, it was not sufficient to consider uniform strains. One needed to consider the internal strain corresponding to the shuffle mode (phonon) with the wave vector $\mathbf{q}_\mathrm{N} = 1/2[110]$ at the N-point zone boundary in the Brillouin zone. Again, the third-order term in the Landau expansion plays a crucial role. The symmetry allowed term couples the two shuffle modes ($\varepsilon_2, \mathbf{q} = \mathbf{q}_\mathrm{N}$) with a uniform mode ($\varepsilon_1$, $\mathbf{q} = \mathbf{0}$) such that $\mathbf{q}_\mathrm{N} + (-\mathbf{q}_\mathrm{N}) + \mathbf{0} = 0$. The free energy can be written as follows:

$$F = F_0 + {}^1\!/_2 c\varepsilon_1^2 + {}^1\!/_2 \rho a^2 \omega_\mathrm{N}^2 \varepsilon_2^2 + V_3 \varepsilon_1 \varepsilon_2^2 + V_4 \varepsilon_2^4 + V_4' \varepsilon_1^4 + V_6 \varepsilon_2^6, \tag{15.1}$$

where c, ρ, a, and ω_N are temperature-dependent constants in the *bcc* phase: elastic constant, density, lattice constant, and the frequency for phonons at the N-point. The anharmonic terms are not expected to be strongly temperature dependent. F_0 is the vibrational and electronic free energy, assumed to be the same in both phases. The transition path follows $\varepsilon_1 = -V_3\varepsilon_2^2/c$ and this gives (assuming ε_1 small) the equilibrium free energy:

$$F^\mathrm{e} = {}^1\!/_2 \rho a^2 \omega_\mathrm{N}^2 \varepsilon_2^2 + (V_4 - {}^1\!/_2 V_3^2/c)\varepsilon_2^4 + V_6 \varepsilon_2^6. \tag{15.2}$$

This free energy could explain the observations in general terms. By measuring the Youngs and the rigidity moduli for Zr between 300–1,300 K, Ashida et al. [3]

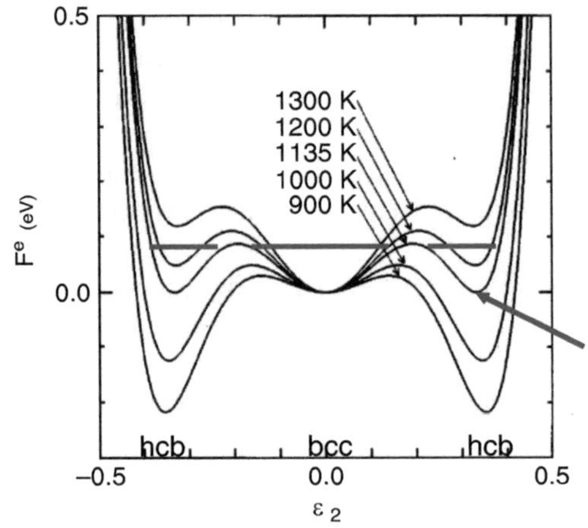

Fig. 15.1 The free energy [2,3] for Zr with no adjustable parameters (after [3]). At exactly $T_M = 1{,}135$ K, the minima for the *bcc* and *hcp* phase change relative depth, as required for the first-order transition. However, it is remarkable that the energy barrier is not much higher than $k_B T_M$, indicated by the *horizontal line*. Therefore, a mixed phase or average phase is expected just above T_M, and essentially a "continuous" first-order transition

succeeded in measuring all parameters in the proposed Landau expansion. Hence, the free energy could be calculated as a function of temperature with no adjustable parameters. It demonstrated that the transition is driven not only by a slightly softening ω_N mode, but also by the smallness and softening of the uniform mode elastic constant c. This will make the effective fourth-order term become negative and drive the transition without any complete softening of modes, as argued in [2]. The resulting free energy, shown in Fig. 15.1, is highly instructive. The condition for a first-order transition is that the depths of the minima for the *bcc* and *hcp* phases are the same. This occurs exactly at the observed transition temperature, $T_M = 1{,}135$ K. Even more interesting is that the energy barrier is almost the same as $k_B T_M$ (k_B is the Boltzmann constant). Hence, large fluctuations are possible – and essentially a mixed or average phase is to be expected just above T_M, where pronounced metastable *hcp* minima occur. To minimize the effect of the uniform strain in a cubic matrix, the *hcp* fluctuations preferentially should form twins, i.e., a tweed structure. The scale of this depends partly on the accommodation (strain) energy and the domain wall energies. These effects are not included in (15.2), and could disfavor the presence of the minority phase slightly. Below T_M, the *hcp* minimum quickly deepens, so fluctuations of the *bcc* phase are expected to be small, as also observed. When a multi-domain structure is found, it means the system is not in an equilibrium state. A single domain obviously has the minimum thermodynamic energy, but it cannot be found due to the kinetics of the transformation. Other domains are formed independently – and the system cannot resolve the competition. It is interesting that in pure Zr, dynamic, i.e., overdamped phonons, but no static precursors, i.e., Bragg-like peaks, are observed by neutron scattering [4,5]. This indicates that the "tweed

Fig. 15.2 Molecular dynamics simulation of an ordered AB alloy (after [9]). From *left* to *right* is shown the time evolution (pico-seconds) after a quench below $T_M = 150\,K$ of the four martensititic variants (*blue*, *purple*, *green*, and *yellow*) in the pre-martensitic (dynamical) twinned *bcc* phase, *dark* and *light red*. The different *red* regions show an average of the *bcc* phase and two alternative *hcp* variants. Notice the overall shape change of the simulated region, and also, that the *hcp* phase nucleates at the surface. The simulated region is a $40 \times 40\,nm$ nano-particle with open boundary conditions

structure" has a very small dimension and fluctuates dynamically at the rather high temperatures involved. By high static pressures, one can reduce T_M by $\sim$20% and thereby slow down the kinetics. By this method, Zr was studied by X-rays [6] and one found a state which appeared to be a glassy or an amorphous phase. It was by more measurements found not to be the case [7, 8]. And an interpretation was preferred in terms of rapidly growing domains near the temperature for the *bcc* to ω-phase transition (i.e., to a different hexagonal structure, related to the ω-point phonon, $\mathbf{q}_\omega = 2/3[111]$). However, it is mentioned here, because it would be most interesting to find a metallic glass phase in a pure (or almost pure) material.

A tweed structure above T_M is found in a pure material by a molecular dynamics (MD) simulation in 2D of a perfect AB binary alloy [9]. Figure 15.2 shows the structure, as it develops in time at a temperature near T_M. The red part to the left is *bcc* phase, which is strained in a pre-martensitic tweed pattern. It consists of domains with an average between the *bcc* and *hcp* phases, in two complementary variants (dark and light red). Notice the rapid change in the texture of the pre-martensitic phase. It is a dynamic texture. To the lower right is the approaching front of a fully developed martensite. The color indicates the four different variants. The authors emphasize that the martensitic fine structure is intimately related to the kinetics of the transition (i.e., depending on the quench rate). They also emphasize that nucleation occurs at the surface. Notice further, the deformation of the sample, which is initially a $40 \times 40\,nm^2$ nano-particle. The time scale is in pico-seconds in this model study. By using pair potentials, it is possible to study rather large systems. With the chosen interactions, $T_M = 150\,K$ is rather low. For Zr, with $T_M = 1{,}135\,K$, one would expect much more thermal noise – and would need a certain time average to obtain a picture like Fig. 15.2, as done by Morris and Ho [10]. Using pair potentials, they MD simulated a 3D sample of Zr, and found several interesting effects. In particular, they found a pre-martensitic phase, which also has "average *bcc*/*hcp*" characteristics. One can also mention a MD simulation of phonons in Zr under high pressure [11]. Such studies are indeed instructive, but might be too simple a model for metals, where the electronic properties and inter-atomic interactions depend on the actual lattice structure. A first-principles calculation at $T = 0$ is of limited use, because generally the *bcc* structure is not stable. It shows imaginary phonon frequencies, in particular, near the N-point (and the ω-point). It is stabilized by the anharmonic phonon interactions, renormalizing the phonon frequencies. One may, therefore, use the $T = 0$ calculations to determine these nonlinear coupling terms of the third and fourth order. This was done successfully for Zr [12]. In a sense, this is a refined Landau expansion, where one is coupling a distribution of strain waves (all phonons) instead of just concentrating on the two order parameters. The effect of temperature is then taken care of by anharmonic phonon theory – and is simply a consequence of the high phonon population at elevated temperatures (i.e., depending on a self-consistent, not just random, disorder). For Zr, T_M is more than ten times higher than the Debye temperature, so all phonons – especially the low frequency ones – are very highly excited. This, of course, sets a limit for the reliability of the anharmonic theory. To get a realistic first-principles description, valid at high temperatures, the proper disordered structure must be invoked. A successful attempt has been made [13] yielding phonon frequencies in good agreement with those measured in *bcc* Zr. Further work in this direction is clearly a task for the future.

The experimental experience [4, 5] from Zr is that one needs a tiny amount of impurities to provide a static tweed structure above T_M. In [3], the influence of hydrogen impurities was investigated as a concentration-dependent change of the parameters in (15.2). This may be sufficient for the light element H, but for, say, oxygen, the important role as nucleation site must further be considered. This means, one must leave the reciprocal space and consider the actual behavior in real space. This holds true in particular for disordered or non-stoichiometric alloys. Here, there will be a statistical distribution of the constituents. Since T_M and other properties depend strongly on small tuning parameters (such as the C impurities in Fe), a disordered or non-stoichiometric alloy will always be a heterogeneous system. Hence, with respect to the martensitic transformation, there will be a distribution of local T_M values. Such a phase was suggested [14] to be analogous to a spin glass, for which there is a distribution of interactions. But the characteristic signatures in the response function (a frequency-dispersed dip in the elastic modulus) corresponding to a glass transition have not been found.

An important task for the future is to investigate and understand the strong tuning capacity of certain impurities (or alloy constituents). An ideal example to start with could be the Zr_{1-x}–Ti_x alloy system. Because of the chemical similarity of the constituents, it forms for any x. Pure Zr and Ti have almost the same T_M, but at $x = 1/2$, T_M is reduced by a factor of two. An attempt [15] to explain this was made, in terms of the influence of the mass difference ($m_{Zr}/m_{Ti} = 1.9$) on the nonlinear phonon coupling. It cannot be an effect of entropy of mixing, which should influence the melting curve more directly, but the melting temperatures interpolate essentially linearly. More work needs to be done to understand the influence of the constituents in alloys. As an example of the dependence, we mention the almost stoichiometric

$Ni_{0.5+x}Ti_{0.5-x}$ system, with $0 < x < 0.02$. Here, Sarkar et al. [16] found a metallic phase near the martensitic transition with glass-like properties in the response functions. The real space structure is interpreted as nano-size ω-phase domains distributed randomly and uncorrelated in the *bcc*, which freezes below T_M into a similar structure with domains up to 20–25 nm. The transition is not related to the *bcc* → *hcp* transition, which has strain deformations up to 10%, but to the weaker transition to the ω-phase with only 1% deformation. It is interesting that it is rather similar to what appears to happen in the abovementioned study of pure Zr under pressure. This could motivate further study of that model system and, in particular, to search for the characteristic glass signatures in the response function. One could possibly hamper crystallization by introducing a trace of suitable impurities. A martensitic glass is not a normal (metallic) glass, but a "glass" confined in crystal symmetry.

15.1 Outlook

The functional materials are by definition primarily interesting for their usefulness in applications. For this purpose, it is not enough with high performance. If we look at the cousin, the permanent magnet, it is not $SmCo_5$, but the inferior NdFeB which has made it to all sorts of applications surrounding us in every day life – and it has taken 40 years after the discovery of the compound. The applications of martensitic materials are still rather specialized. There is, therefore, considerable room for a search for materials, which perform well also on parameters such as price, abundance, stability, bio-compatibility, environment, etc. A further study of model systems (such as Zr mentioned above, and also others) may provide useful information for a deeper understanding. However, with the ever-increasing performance of computers, it is becoming possible to include the realistic state of a complex disorder, especially on the nano-scale. Computer simulations and full-scale, first-principle calculations are candidates for future developments. It is important that such works are completed by calculating the wave vector and frequency-dependent response functions. Disorder happens in real space; therefore, reciprocal space (scattering) experiments are useful, but not ideal. However, they give reliable average information about the sample that can be related to theory. Real space methods tend to pick out interesting, but not necessarily typical phenomena. On the experimental side, it is important to further develop methods to characterize the complex materials on the nano-scale. It would be highly interesting to image a possible martensitic glass experimentally and to seek to define this state relative to a normal, locally disordered glass and a micro (nano)-crystalline state. Materials purposely composed of martensitic nano-particles (in a suitable matrix, possibly a polymer matrix) might also be interesting, since nucleation appears at the surfaces and also because properties of nano-particles may differ significantly from the bulk (discrete bands for phonons and electrons – due to quantum effects). Several interesting phenomena are found in magnetic nano-particles [17], which have been investigated for decades. With respect to applications, it is important to analyze if functionality is only required locally. The sword needs a sharp, hard edge, but the rest can be of another quality (inferior in that respect). A revolution in making durable, sharp razor blades was made just 40 years ago. Breaking with the quest for finding better and better steels, one focussed on improving just the cutting edge, namely, by sputtering materials on to it, at first Cr and then Cr/Pt, but now using a zoo of materials, including ceramics. This increased the corrosion resistance and hence the durability of the sharpness dramatically; just the functionality needed. It is now a science in itself (or rather business, since all is patented – of course, based on knowledge, materials, and techniques derived from basic science) with constantly made new developments. For the martensites, we have emphasized that the properties can be tuned by small amounts of tracer elements. Hence, it is possible to – so to say – "spray paint" or even print a designed pattern of different properties to a surface or a sample, down to a very small scale. Thereby, one may be able to design samples, at least surfaces, that respond to a stimulus in a desired way. There seems to be endless possibilities for the future with such designed heterogeneity for the martensites, now not made by magical blacksmiths but by knowledge-based nano-scale technology.

The complex, functional materials discussed in this book are inorganic materials, mainly metals or metallic alloys. The emphasis is on the fundamental properties and in the possibility of having a new glassy state. The coupling of spin, charge, and lattice degrees of freedom in the presence of dipolar and long-range elastic interactions often results in competing phases and nano-scale heterogeneity. This causes very high sensitivity to external fields, resulting in large response and cross-response, as found in multiferroics, ferroelectrics, etc. This opens up the possibility of tuning the phenomena and making the phenomena susceptible to a large range of external stimuli, as is discussed in detail in several of the preceding chapters. Finally, note that the martensitic transition is also found in other materials than metals and alloys, and hence, also in polymers and ceramics. For applications, competition may well come from such materials. But the general understanding of the phenomena should be useful for the entire class. The preceding chapters all contribute to this.

References

1. S. Alexander, J. McTague, Should all crystals be *bcc*? Landau theory of solidification and crystal nucleation. Phys. Rev. Lett. **41**, 702 (1978)
2. P.-A. Lindgård, O.G. Mouritsen, Theory and model for Martensitic transformations. Phys. Rev. Lett. **57**, 2454 (1986)
3. Y. Ashida, M. Yamamoto, S. Naito, M Mabuchi, T. Hashino, Calculation of the Lindgård and Mouritsen's free energy using recently measured moduli of elasticity for hydrogen in zirconium. J. Appl. Phys. **80**, 3259 (1996)
4. W. Petry, Dynamical precursors in Martensitic transitions. J. de Phys. III Suppl., Colloque C2–15, **5** (1995)

5. W. Petry et al., Phonon-dispersion of the bcc phase of group-iv metals.2. Bcc zirconium, a model case of dynamic precursors of martensitic transitions. Phys. Rev B **43**, 10948 (1991)
6. J. Zhang, Y. Zhan, Formation of zirconium metallic glass. Nature **430**, 332 (2004)
7. J. Zhang, Y Zhan, Formation of zirconium metallic glass. Nature **437**, 1957 (2005)
8. T. Hattori, H. Saitoh, H. Kaneko, Y. Okajima, K. Aoki, W. Utsumi, Does bulk metallic glass of elemental Zr and Ti exist? Phys. Rev. Lett. **96**, 255504 (2006)
9. O. Kastner, J. Ackland, Mesoscale kinetics produces Martensitic microstructure. J. Mech. Phys. Solids **57**, 109 (2009)
10. J.R. Morris, K.M. Ho, Molecular dynamic simulation of a homogeneous *bcc->hcp* transition. Phys. Rev. B **63**, 224116 (2001)
11. V.Yu. Trubitsin, E.B. Dolgusheva, Anharmonic effects and vibrational spectrum of *bcc* Zr under pressure studied by molecular dynamics simulations. Phys. Rev. B **76**, 024308 (2007)
12. Y. Ye, Y. Chen, K.-M. Ho, B.N. Harmon, P.-A Lindgård, Phonon-phonon coupling and the stability of the high-temperature *bcc* phase of Zr. Phys. Rev. Lett. **58**, 1769 (1987)
13. P. Souvatzis, O. Eriksson, M.I. Katnelson, S.P. Rudin, Entropy driven stabilization of energetically unstable crystal structures explained from first principles theory. Phys. Rev. Lett. **100**, 095901 (2008)
14. S. Kartha, T. Castan, J.A. Krumhansl, J.P. Sethna, Spin-Glass nature of tweed precursors in Martensitic transformations. Phys. Rev. Lett. **67**, 3630 (1991)
15. P.-A Lindgård, What determines the Martensitic transition temperature in alloys? J. de Phys. III Suppl. Colloque, C2–29, **5** (1995)
16. S. Sarkar, X. Ren, K. Otsuka, Evidence for strain glass in the ferroelastic-martensitic system Ti_{50-x} Ni_{50+x}. Phys. Rev. Lett. **95**, 205702 (2005)
17. P.V. Hendriksen, S. Linderoth, P.-A. Lindgård, Phys. Rev. **48**, 7259 (1994)

Index

AC mechanical susceptibility, 234
Acoustic, 259, 262
Acoustic emission, 10, 258
Adiabatic, 260
Adiabatic shear band, 147
AF correlations, 72
Affinity-based, 286
Aliovalent doping, 114
Amorphous layer, 122
Analytical, 193
Anharmonicity, 39
Anharmonic phonon theory, 298
Anisotropic stress, 215
Anisotropy, 239
Anisotropy factor, 232
Antiferromagnetic, 160
Antiferromagnetic correlations, 27
Antiphase boundaries, 164
Antiphase domain, 169
Applications, 299
Athermal, 250, 260
Avalanches, 250, 251, 256, 259, 262, 264, 269

Band Jahn–Teller effect, 23, 26
Barium zirconate titanate, 119
$BaTiO_3$, 115
B2-disordered state, 164
Binding, 279
Bioinformatics, 286
Bloch wall, 168
Breathing, 276, 277

Calorimetric, 260
Calorimetry, 263
$CaTiO_3$, 4
$CeCoIn_5$, 96, 108
Charged defects, 126
Charge-density waves, 106
Charge-ordered (CO) insulating phase, 152
Chemical disorder, 168
Chemical order parameter, 168
Chirality, 165
Cluster-spin glass transition, 202
Co_2FeSi, 20
Coherent, 143
Coherent potential approximation (CPA), 21
Colossal magnetoresistance, 152
Combinatorial study, 24
Co_2MnGe, 21
$Co_2MnSi(001)/MgO$, 22
Competitive interactions, 177
Complex, 177, 196
Complexity, 253, 274
Compressibilities, 184
Compressive stress, 32
Computer simulations, 193, 299
Conical dark-field, 146
Conventional $L2_1$ ($Fm\underline{3}m$) Heusler structure, 25
Conventional and inverse magnetocaloric effects, 68
Core–shell structure, 129
Corrected microscopes, 144
Correlation length, 236
Crossover, 205
CrSb, 21
Cubic materials, 232
Curie temperature, 114
Curie–Weiss, 116

T. Kakeshita et al. (eds.), *Disorder and Strain-Induced Complexity in Functional Materials*, Springer Series in Materials Science 148,
DOI 10.1007/978-3-642-20943-7,

Damping, 213
Dark-field image, 161
Degree of $L2_1$ long-range order, 52
Degree of long-range order, 51, 54, 55, 58, 62
Delocalization, 138
Demagnetization energy, 162
Depletion, 139
Depletion layer, 123
Designed, 282
DFT, 76
Dielectric, 113
Diffraction contrast, 165
Diffuse neutron scattering, 71
Diffuse scattering, 171
Dipolar interactions, 228
Disorder, 178, 240, 250, 253, 254
Disorder–order, 201
Displacement field, 237
DNA, 275
Domain, 203, 294
  boundary engineering, 2
  engineering, 115
  pattern symmetry, 128
  structures, 125
  wall, 115, 125
    energy, 162
    movement, 126
    width, 162
Doping, 234
Double domain, 159
Double exchange, 101
DTA, 10
Dynamical matrix, 33, 38

Edwards and Anderson, 180
EELS, 139
EELSMODEL, 139
EFTEM, 140
Eigenvector, 140
Elastic anisotropy, 244
Elastic constants, 37, 232
Elastic modulus, 205, 229
Electron diffraction, 136, 170
Electron holography, 152
Electron tomography, 144
Embryos, 137
Energy barriers, 123, 236
Entropy, 221
Entropy-change, 70
Ergodic, 184, 208
Ergodicity-breaking, 208
Euler–Lagrange minimization, 237
Exchange constants, 74
Exchange interactions, 58, 61, 62
Exchange stiffness constant, 158
Experimental, 277
Exponents, 264, 268, 269
Extrinsic contribution, 126

FC, *see* Field cooling (FC)
Fe_2CoGa, 25
Fe_2CoZn, 43
$Fe_2Co_{1-x}Fe_xGa$, 43
Fe–Pd–Cu, 24
Fermi energy, 91
Fermi level, 124
Fermi surface sheets, 34
Ferroelastic materials, 244
Ferroelastics, 201
Ferroelastic transition, 241
Ferroelectrics, 188, 201
Ferroic glasses, 209
Ferroic materials, 228
Ferroic switching, 9
Ferromagnetic shape memory alloy (FSMA), 49, 151
Ferromagnetic transition, 202
Ferromagnets, 258
FFLO, *see* Fulde–Ferrell–Larkin–Ovchinnikov (FFLO)
Field cooling (FC), 160, 188
  and ZFC, 184
First-order phase transformation, 159
First-order transformation, 137
First-order transitions, 181, 183
First-principles calculations, 20, 299
Fluctuations, 265, 276
FM correlations, 72
Focussed-ion-beam (FIB), 143
Foucault mode, 153
Free energies of Ni_2MnGa, 40
Free energy, 295
Fresnel mode, 153
Frustration, 178
FSMA, 49
Fulde–Ferrell–Larkin–Ovchinnikov (FFLO), 108, 109
Full-Heusler alloys, 20
Full-potential code FPLO, 41
Functionality, 293
Functional materials, 227
Functional properties, 131
F-value, precision, 289

Geometric phase analysis, 138

Ginzburg–Landau, 105, 235
Glass, 187, 201, 295
Glass transition, 201
Glassy behaviour, 177, 244
Glassy state, 235
Goldschmidt tolerance factor, 118
Goodness of fit, 145
Grain boundary, 122
Growth rate, 160

HAADF STEM, 146
Habit plane, 140
Half-Heusler alloy, 20
Hamiltonian, 277
Harmonic Hamiltonian, 40
Heat capacity, 239
Heisenberg model, 27
Heterogeneity, 300
High-resolution, 136
Hologram, 154
HRSTEM, 144
Hubbard model, 99, 100
Hysteresis, 141, 250–254, 257, 269

Incommensurate, 106–108
Interchange energy, 51, 52
Interface, 140
Intermediate phase, 185
Internal stress, 130
Interstitial defects, 121
Inverse ($F\underline{4}3$ m) Heusler structure, 25
Inverse freezing, 185
Inverse Heusler structure, 35
Ising, 183, 256

Jahn–Teller effect, 92, 100, 101

Kohn anomaly, 39
Krumhansl, 274

$LaAlO_3$, 12
Lagrangian tensor, 236
$LaMnO_3$, 96, 105
Landau expansion, 295
Landau–Ginzburg, 3, 181
La_2NiO_4, 99
$La_2NiO_{4+\delta}$, 96
La_2NiO_{4+}, 96
Latent heat, 221
Lattice anomaly, 169
Lattice modulations, 137, 169
Layered manganite, 161
Learning, 261, 266
Lindhard function, 39
Long-lived transient openings, 279
Long-range interactions, 227
$L2_1$-ordered state, 164
Lorentz microscopy, 152

Magnetic, 182
Magnetic analogy, 190
Magnetic domain walls, 153
Magnetic domains, 153
Magnetic exchange interactions, 21, 28
Magnetic exchange length, 162
Magnetic-field-induced reverse martensitic transformation (MFIRT), 50
Magnetic-field-induced strain (MFIS), 49, 80
Magnetic flux, 154
Magnetic phase transformations, 156
Magnetic shape-memory, 68
Magnetic shear stress, 86
Magnetic vortices, 165
Magnetization process, 156
Magnetocaloric effects, 20
Magnetocrystalline anisotropy, 42, 81, 86
Magnetocrystalline anisotropy constant, 158
Magnetoelastic interaction, 171
Magnetoelastic tweed, 172
Magnetoelectric coupling, 228
Magnetoelectrics, 228
Magnetoresistance, 103
Magnetostriction, 164
Magnetostructural transition, 30
Manganites, 96, 100, 101, 103, 107, 110, 228
Martens, 294
Martensite, 141, 181, 182, 294
Martensite variants, 164
Martensitic, 256, 258, 262, 267
  alloys, 177
  transformation, 23, 34, 203, 294
  transitions, 68, 243
Materials science, 196
Mechanical loss, 205
Melting, 283
Mesoscale, 289
Mesoscopic symmetrics, 127
Metal–insulator transition, 97
Metastability, 179, 236
Metastable, 178, 250, 254
MFI effects, 41
MFIS, 50, 51

Microstructure, 231
Micro-wires, 142
Minimalist models, 193
Modelling, 189
Modulated 5M martensite, 22
Molecular dynamics (MD) simulation, 297
Monte Carlo simulations, 29
Morphotropic phase boundary, 127
Motif-based, 287
Mott insulator, 96, 98
Mott transition, 95, 98
MSMA, 20
Multiferroic materials, 227
Multiferroics, 2, 110

Nano-beams, 142
Nanoceramic, 129
Nanocrystalline, 142
Nanodomains, 203, 230
Nano-particles, 299
Nanoscale textures, 243
Nanostructure, 234
$NdGaO_3$, 3
Negative temperature dependence, 221
Ni–Al, 230
Ni_3Al, 31
Ni-excess phase diagram, 32
Ni-free MSMA, 43
NiMn, 25
Ni–Mn-based Heusler alloys, 68
Ni_2MnGa, 34
NiMnSb, 21
Ni–Mn–Z, 29
$Ni_{2+x}Mn_{1-x}Z$($Z=$ Ga, In, Sn, Sb), 30
Ni_3Sb, 31
Ni_4Ti_3, 139
Ni–Ti, 142
Nonequilibrium, 251
Non-ergodic, 184, 191
Nonergodicity, 208
Nonlinearity, 274
Nucleation and growth, 159

Operation temperatures, 41
Optimization, 146
Order–disorder, 62
Order–disorder phase transformation, 51, 54–56
Oxygen vacancies, 4

Paraelectric, 122
Parameter estimation, 145
Pattern formation, 169
Patterns, 279
Peierls, 96, 102
Perovskite, 188
Peyrard–Bishop–Dauxois model, 274, 277
Phase diagrams, 31, 240
Phase separation, 152
Phase shift, 154
Phase transition, 249
Phenomenologically, 181
Phenomenological theories, 130
Phonon dispersion curves of $L2_1$ Ni_2MnGa, 38
Phonon softening, 34, 41
Piezoelectric, 113
Pinning force, 160
PMN, 188
PMN-PT, 127, 189
Point defects, 201, 229
Polaron, 103
Potts, 185
Power-law, 264
Precipitates, 137
Precursor effect, 169
Precursors, 136, 205, 233, 296
Prediction, 286
Predictive capability, 289
Premartensite phase, 37
Premartensitic tweed, 203
Promoters, 276
Protein binding, 282
Pseudoelasticity, 24
Pseudogap, 38
Pseudo-spin, 182
Pseudo-spin glass, 184
PTCR, 116
PT-PMN, 189, 191
Pyroelectric, 113
PZN–PT, 127

Quantification, 144
Quenched disorder, 177

Random exchange, 191
Random-field, 191
Rare-earth dopants, 118
Re-entrance, 185
Recall, 289
Reconstructed phase image, 155
Relaxor ferroelectric materials, 119
Relaxor ferroelectrics, 177, 234
Relaxors, 188, 201

Repair proteins, 284
Response function, 96
RFIM, 191, 251, 254, 257, 258, 270
Rubber-like behavior, 24

Saint-Vénant compatibility condition, 237
SCAILD, 40
Schottky defect, 116
SDW, *see* Spin-density waves (SDW)
Self-accommodating structures, 141
Shape memory, 187, 295
  alloys, 20, 232
  effect, 211, 228
  materials, 135
Shear modulus, 171
Sherrington–Kirkpatrick, 181
Shuffle mode phonon, 295
Signal-to-noise, 146
Simulations, 284
Single domain, 159
Size effects, 129
Slater–Pauling behavior, 22
Softening of elastic constants, 37
Softening of modes, 296
Spatial resolution, 156
Spectroscopy, 139
Spin-density waves (SDW), 97, 98, 106
Spin glass, 177, 203, 298
Spin model, 27
Spinodals, 187
SPR-KKR code, 27
Stacking interactions, 280
Statistical physics, 178, 195
St Venant, 182
Stimulus, 294
Strain, 242
  crystal, 205
  field, 138, 171
  fluctuations, 236
  glass, 184, 201, 203, 205, 207, 209, 211, 213, 215, 217, 219, 221, 223, 225, 242
  glass transition, 201, 203, 205, 207, 209, 211, 213, 215, 217, 219, 221, 223, 225
  liquid, 205
Stress, 243
Stress-induced, 142
Stripes, 104
Structural, 287
Structural phase transitions, 181
Substitutional alloy, 188
Superconductivity, 108
Supercooling, 160
Superelasticity, 24, 142, 211, 228
Superexchange, 104
Surface effects, 129

Temperature, 238
Temperature-change, 70
Tetragonal distortion, 28
Tetragonal $L1_0$ phase, 33
Tetragonal shear modes, 38
Textured materials, 293
Textures, 239
The SPR-KKR code, 21
Thermal activation, 215
Thermally–entropically, 279
Thermodynamic, 287
Thin film, 143
Third-order anharmonic contribution, 40
$Ti_{49}Ni_{51}$, 232
Ti–Ni-based alloys, 232
$Ti_{50-x}Ni_{50+x}$, 203, 234
$Ti_{50}Ni_{48}Fe_2$, 232
$Ti_{50}Ni_{50-x}Fe_x$, 204
$Ti_{50}(Pd_{50-x}Cr_x)$, 205
Tracer elements, 300
Transcription, 275, 282
Transcription initiation, 277
Transfer integral, 279
Transitions, 260, 263, 265
Transition temperature, 239
Transmission electron microscopy (TEM), 135, 152
Tuning capacity, 298
Tweed, 136, 231
Tweed structure, 297
Twin walls, 4
Twinning, 140, 164
Twins, 125, 140, 183, 240

Uniform strain, 295
Universality, 265
UV dimers, 283, 284
UV-induced dimer, 283

Variants, 294
Vegard's law, 119
Vibrational entropy, 230
Vienna ab initio simulation package, 20
Vogel–Fulcher, 205
Vogel–Fulcher relation, 234

180° wall, 163
WO_3, 7

XYZ polarization analysis, 71

Z-contrast, 146
Zero-field cooling (ZFC), 82, 159, 188
Zero-field-cooling/field-cooling (ZFC/FC), 208, 241
ZFC, *see* Zero-field cooling (ZFC)
ZFC/FC, *see* Zero-field-cooling/field-cooling (ZFC/FC)
Zinc-blende, 21

MIX
Papier aus verantwortungsvollen Quellen
Paper from responsible sources
FSC® C105338

If you have any concerns about our products,
you can contact us on
ProductSafety@springernature.com

In case Publisher is established outside the EU,
the EU authorized representative is:
Springer Nature Customer Service Center GmbH
Europaplatz 3, 69115 Heidelberg, Germany

Printed by Libri Plureos GmbH
in Hamburg, Germany